Advanced Land Warfare

Advanced Land Warfare

Tactics and operations

Edited by

Mikael Weissmann

Swedish Defence University

Niklas Nilsson

Swedish Defence University

OXFORD
UNIVERSITY PRESS

OXFORD
UNIVERSITY PRESS

Great Clarendon Street, Oxford, OX2 6DP,
United Kingdom

Oxford University Press is a department of the University of Oxford.
It furthers the University's objective of excellence in research, scholarship,
and education by publishing worldwide. Oxford is a registered trade mark of
Oxford University Press in the UK and in certain other countries

Published in the United States of America by Oxford University Press
198 Madison Avenue, New York, NY 10016, United States of America

British Library Cataloguing in Publication Data
Data available

Library of Congress Control Number: 2022950956

ISBN 978-0-19-285742-2

DOI: 10.1093/oso/9780192857422.001.0001

Printed and bound by
CPI Group (UK) Ltd, Croydon, CR0 4YY

Foreword

Dr Meir Finkel,
Brigadier General (reserve) of the Israel Defense Forces (IDF), Head of Research and former Director of the Dado Center for Interdisciplinary Military Studies/IDF-J3, as well as former Director of the Concept Development and Doctrine Department of the IDF Ground Forces Command, Israel.

Whilst human fighting was first conducted on land, and land forces were the first to fight, the sea, and later air, space, and information domains were rapidly inhabited by military operations. In the last decades, tremendous improvements in intelligence-based standoff fires seem to dominate operations conducted by military organizations, from the USA in Afghanistan to Hamas and Israel in recent engagements. In many cases, the initial phases of war were mainly conducted from the air, or through the air. Later phases were short land manoeuvres followed by long periods in which land forces sought unsuccessfully to pacify conflicts or enhance local governance efforts.

In many security establishments around the world, this created the notion that land forces and land manoeuvres were either not relevant for future conflicts, or incapable of providing answers to salient security challenges. The perception of a limited need for land warfare elevated the role of special forces in order to accompany and assist the air war. This notion was enhanced by the accepted but hidden assumption that the national fight for land as a resource or asset for political negotiation was an obsolete idea in the age described at the beginning of the 1990s as the "end of history". The debate on the roles of land forces in future conflicts is still at its height.

I would argue that many of the assumptions underpinning this debate are misleading. Overemphasizing the counterinsurgency and peacekeeping roles of land forces neglects their main role in beating determined adversaries, unimpressed by standoff fires. Moreover, the struggle to conquer or protect land (depending on perspective) is arising again. Examples are the Russian operations in Ukraine, Chinese operations in the South China Sea, and the efforts of both nations to gain strongholds in Syria and Djibouti. Whilst these examples relate to great power competition, climate change will probably force more local struggles over water sources, or even inhabitable land, mainly in Africa and Asia.

More than 100 years ago, the military theorist JFC Fuller suggested that the relations between offensive and defensive capabilities could be described by a sinusoid curve, demonstrating the never-ending struggle of opposing sides in military history. Surprisingly to some more recent theorists, state and non-state military actors have in various ways adapted to the dominance of standoff fires. Adaptation includes going underground as in the case of Hamas in Gaza or Iran's nuclear facilities, jamming and disrupting standoff fires, or developing their own capabilities in this regard, in a manner that balances the previously dominant side (e.g. Hezbollah vs Israel, China vs the USA). Most western militaries have recognized this deadlock during the last decade; either formally or through reactions to recent incidents, which testify that they are deterred. It may be said that cyberwar, with its kinetic effects, is the successor of standoff fires. However, this seems to be correct only as long as the conflict has not crossed the threshold of war.

As hostile activities related to land will continue, whether as part of global competition or local conflicts, and as one sided supremacy in standoff fires is already denied in many regions of the world, what are the roles of land forces in national security—in their current form and in the future?

My argument is twofold. First, as history has revealed time after time, our ability to envision future conflicts, and what roles arms and capabilities will play in them, is quite disappointing. Recent examples are Israel's false perception prior to the Second Lebanon War that counter terrorism was the dominant feature of warfighting and western false perceptions concerning the Russian use of forces in most domains before the Ukraine War. The employment of land forces may be the final step in a military confrontation, after cyberwar and standoff war (and again, military history teaches us that 'bombing to win' has never worked against a determined opponent). However, when it comes, land warfare should be decisive and conclusive. Thus, building land forces is like an expansive earthquake insurance policy, in a region where substantial earthquakes occur every fifty years on average, but have not occurred in the last sixty years. When land forces can be a game changer, you better have them ready. It is not easy for political leaders to invest money in capabilities with delayed gratification, yet otherwise, someday, history will judge them for disregarding received knowledge and experience.

Second, land forces should be built to rapidly manoeuvre, destroy enemy forces, and conquer ground. However, this requires many adaptations, most crucially regarding the ability to find the enemy, which is already adapted to aerial supremacy; and the ability to survive threats posed by an enemy with advanced intelligence assets and fire capabilities, without the advantage of pre-battle effective standoff strikes. The ability to do both is essential in order to gain political support for the design and employment of land forces. Otherwise, nations will have to make serious compromises regarding national security interests.

In order to accomplish these two basic abilities, land forces must not only become better integrated (with other services, domains, and within themselves) and more lethal, survivable, and logistically efficient; they must also assimilate emerging available assets in fields like AI and Cyber at a faster pace than previously. This requires not only substantial investment in resources, but also an open minded and exploratory approach, in contrast to the common but sometimes overexaggerated perception of military organizations as conservative entities.

This volume offers a fresh view on the present and future of land forces. First, it presents various aspects of land warfare including command, combat logistics, and interoperability. These aspects are approached mainly through the lens of existing challenges and ways to overcome them. The second part presents case studies that illuminate how different national land forces, each with their distinct cultures, organizations, and roles in their nation's security apparatus, are struggling to cope with changing geopolitical threats, adversary tactics, technological potential, and internal social and resource constraints. Altogether, much can be learned and absorbed from this multifaceted and rich overview on one of the most challenging, debated, and fundamental issues in current security affairs.

Contents

Abbreviations   xv
Contributors   xix

1. **Approaching Land Warfare in the Twenty-first Century**   1
   Niklas Nilsson and Mikael Weissmann
   Approaching Land Warfare   1
   Development of Land Warfare   3
   Current and Future Challenges   7
   Character of War and Transformation of the Battlefield   9
   Structure of the Volume   11
   Chapter-by-chapter Synopsis   12

I LAND WARFARE

2. **The Future of Manoeuvre Warfare**   25
   Christopher Tuck
   Introduction   25
   Manoeuvre Warfare   26
   The Future Importance of Manoeuvre Warfare   29
   Alternative Perspectives   35
   Conclusion   42

3. **Commanding Contemporary and Future Land
   Operations: What Role for Mission Command?**   43
   Niklas Nilsson
   Introduction   43
   Mission Command: Culture or Method?   44
   Commanding Land Operations in the Contemporary and
     Future Operational Environment   48
   The Informational Challenge   52
   Command, Communication, and Technology   56
   Conclusion   62

4. **Combat Logistics in the Twenty-first Century: Enabling
   the Mobility, Endurance, and Sustainment of NATO
   Land Forces in a Future Major Conflict**   63
   Christopher Kinsey and Ronald Ti
   Introduction   63
   The Character and Scope of Military Logistics   65

The Applications of Combat Logistics   68
AM, Robotics, and AI in Future Force Structures   80
Conclusion   84

5. **The Command of Land Forces**   **87**
   Jim Storr
   Introduction   87
   Purpose   89
   Products   91
   Processes   93
   Structures   96
   Systems   97
   People   99
   Summary and Conclusions   102

6. **Tactical Tenets: Checklists or Toolboxes**   **105**
   B. A. Friedman and Henrik Paulsson
   Introduction   105
   What Is 'Tactics'?   106
   The History of Tactical Theory   106
   Purpose of Theory   113
   From Principles to Tenets   114
   Relationship with Strategy   118
   Tactical Theory in Practice   120
   Conclusion   122

7. **Urban Warfare: Challenges of Military Operations on
   Tomorrow's Battlefield**   **125**
   Mikael Weissmann
   Introduction   125
   Approaching Urban Warfare   128
   Future Challenges for Urban Warfare   131
   Conclusion: Eleven Takeaways about Urban Warfare   148

8. **Emerging Technologies: From Concept to Capability**   **153**
   Jack Watling
   Introduction   153
   Autonomous Systems   154
   Layered Precision Fires   161
   High Fidelity Layered Sensors   163
   Artificial Intelligence   166
   Combining Emerging Capabilities   169

9. **Interoperability Challenges in an Era of Systemic Competition**   173
   Andrew Curtis
   Introduction   173
   Understanding Interoperability   176
   Interoperability in an Era of Systemic Competition   179
   The UK's Approach to Interoperability   184
   Conclusion   190

10. **The Moral Component of Fighting: Bringing Society
    Back In**   193
    Tua Sandman
    Introduction   193
    Why Soldiers Fight   195
    How Society Slips from View   201
    Considering the Role of Society   207
    Conclusion   211

11. **Military Health Services Supporting the Land
    Component in the Twenty-first Century**   215
    Martin C. M. Bricknell
    Introduction   215
    What is a Military Health System?   218
    The Twenty-first-century Context   222
    COVID-19—A Game Changer?   225
    The Future of Health Services Support to the Land Component
       in Military Operations   226
    Conclusions   230

II CASE STUDIES

12. **The Operational Cultures of American Ground Forces**   233
    Bruce I. Gudmundsson

13. **People's Liberation Army Operations and Tactics in the
    Land Domain: Informationized to Intelligentized Warfare**   257
    Brad Marvel
    Introduction   257
    Part I: Building the Modern People's Liberation Army   258
    Part II: From Informationized to Intelligentized Warfare   263
    Bringing it all Together: The *Stratagem*   273
    Ongoing Challenges and Lessons   274
    Conclusions   278

14. A Strategy of Limited Actions: Russia's Ground-Based
    Forces in Syria   279
    Markus Balázs Göransson
    Introduction   279
    Previous Research on Russia's Ground-based Forces in Syria   281
    The Russian Ground-based Contingent   283
    The Strategic Functions of Russia's Ground-based Contingent   293
    Conclusion   298

15. The Role of Israel's Ground Forces in Israel's Wars   301
    Eado Hecht and Eitan Shamir
    Introduction   301
    Background   301
    Perception of the Threat   303
    Israel's National Security Strategy   304
    Evolution of the Ground Forces   305
    Conclusion   318

16. Tactics and Trade-Offs: The Evolution of Manoeuvre in
    the British Army   321
    Alex Neads and David J. Galbreath
    Introduction   321
    Importing Manoeuvre into British Military Thought   322
    Between Warfighting Capability and Counter-insurgency   327
    Reinventing Manoeuvre after Afghanistan: Army 2020 and
      Organizational Change   334
    From Army 2020 to Integrated Manoeuvre   341
    Conclusion   349

17. Caught between a Rock and a Hard Place: The French
    Army, Expeditionary Warfare, and the Return of
    Strategic Competition   353
    Olivier Schmitt and Elie Tenenbaum
    Introduction   353
    French Military Interventions: An Important Part of the Army Culture 355
    A Wide-ranging Operational Experience   356
    The Force Structure and Main Capabilities   361
    The French Army Doctrine   364
    From Network-centric Warfare to High Intensity   368
    Providing Soldiers for Future Wars   369
    Conclusion   370

18. **Trends in the Land Warfare Capability of Poland and the Visegrád States, 1991–2021**   373
    Scott Boston
    Introduction   373
    Background: From Warsaw Pact to NATO   374
    How Land Force Capabilities Changed   376
    Looking to the Future   386
    Conclusion   389

III CONCLUSIONS

19. **Towards a Versatile Edge: Developing Land Forces for Future Conflict**   393
    Mikael Weissmann and Niklas Nilsson
    Introduction   393
    Dynamics of Twenty-first-century Land Warfare   394
    The Continuum of Land Operations   398
    Locating Land Operations   403
    Towards a Versatile Edge: Securing Land Warfare Capability   407
    Ways Forward: Practical Implications and Agenda for Future Research   410

Index   413

Abbreviations

2D	two-dimensional
2DB	Second French Armoured Division (*Deuxième Division Blindée*)
3D	three-dimensional
A2/AD	anti-access/area denial
ABCA	Australia, UK, Canada, USA, and New Zealand
ACR	Army of the Czech Republic
AESA	active electronically scanned array
AI	artificial intelligence
ALMRS	Autonomous Last Mile Resupply System
AM	additive manufacturing
APC	armoured personnel carrier
AR	augmented reality
ARRC	Allied Rapid Reaction Corps
BESA Center	Begin Sadat Center for Strategic Studies
C2	command and control
CA-BDE	combined-arms brigade (PLA)
CA-BN	combined-arms battalion (PLA)
CBRN	chemical, biological, nuclear, or radiological
CGI	computer generated imagery
CIS	communication and information systems
COIN	counterinsurgency
COMINT	communications intelligence
COPD	Comprehensive Operational Planning Directive (NATO)
CPC	Communist Party of China
CPX	command post exercise
CSP	capability sustainment programme
CT	computed tomography
CTC	combat training centre (PLAA)
DHC	deployed hospital care
DLODs	defence lines of development
DoD	Department of Defense (USA)
EBAO	effects-based approach to operations
EBO	effects-based operations
eFP	enhanced forward presence
EMC	European Medical Command
EMP	electromagnetic pulse
EMS	electromagnetic spectrum
EU	European Union

FHP	force health protection
FM	Field Manual
FRES	future rapid effects system
GA	Group Armies (PLA)
GEOINT	geospatial intelligence
GPS	global positioning system
GTIA	Groupements Tactiques Interarmes
HIMARS	High Mobility Artillery Rocket System
HNS	host nation support
HQ	headquarters
HSS	health services support
HUMINT	human intelligence
IDF	Israeli Defence Force
IED	improvised explosive device
IFRI	Institut français des relations internationales
IFV	infantry fighting vehicle
IMINT	imagery intelligence
IopC	Integrated Operating Concept
ISAF	International Security Assistance Force (NATO)
ISIS	Islamic State of Iraq and the Levant
ISR	intelligence, surveillance, and reconnaissance
ISTAR	intelligence, surveillance, target acquisition, and reconnaissance
IT	information technology
JADO	Joint All Domain Operations (NATO)
JFC-NF	Joint Forces Command, Norfolk (NATO)
JLSG	Joint Logistic Support Group (NATO)
JOA	Joint Operational Area (NATO)
JSEC	Joint Support and Enabling Command (NATO)
KFOR	NATO-led Kosovo Force
LEP	Life Extension Programme
LE TacCIS	Land Environment Tactical Communication and Information Systems
LRPF	long range precision fires
LSCO	large-scale contingent operation
LWRG	Land Warfare Research Group
MASCAL	mass casualty events
MASINT	measurement and signatures intelligence
MCDP	Marine Corps Doctrine Publication
MDB	multi-domain battle
MDI	multi-domain integration
MDO	multi-domain operations
MEDEVAC	medical evacuation
MEL	main events list
MHS	military health service
MIV	mechanised infantry vehicle

ML	machine learning
MMCC	Multinational Medical Coordination Centre
MoD	Ministry of Defence (UK)
MRLS	multiple rocket launcher systems
MOUT	military operations in urban terrain
MTF	medical treatment facility
NATO	North Atlantic Treaty Organization
NCO	noncommissioned officer
NCW	network-centric warfare
NEC	network enabled capability
NORTHAG	NATO Northern Army Group
NRF	NATO Response Force
NSCR	National Security Capability Review
NWCC	NATO Warfighting Capstone Concept
OEM	original equipment manufacturer
OPFOR	opposing force
OSINT	open-source intelligence
PHC	primary health care
PHEC	pre-hospital emergency care
PLA	People's Liberation Army (China)
PLAA	PLA Army
PMC	private military company
PME	professional military education
POI	point of injury
PSO	peace support operation
QDR	Quadrennial Defense Review (US DoD)
RAF	Royal Air Force (UK)
RAS	robotic and autonomous systems
RGPP	'General Review of Public Policies' (*Révision Générale des Politiques Publiques*)
RMA	Revolution in Military Affairs
RSOM	reception, staging, and onward movement
RUSI	Royal United Services Institute for Defence and Security Studies
SACEUR	Supreme Allied Commander Europe (NATO)
SDR	Strategic Defence Review
SDSR	Strategic Defence and Security Review
SIGINT	signals intelligence
SOF	Special Operations Forces
STANAVFORLANT	NATO's Standing Naval Force Atlantic
STRATEVAC	strategic MEDEVAC
TACEVAC	tactical MEDEVAC
TAR	tactical augmented reality
TC	Theater Command (PLA)
TRADOC	US Army Training and Doctrine Command
TTP	tactics, techniques, and procedures

UAS	unmanned aerial systems
UAV	unmanned aerial vehicle
UGV	unmanned ground vehicles
UKADGE	UK Air Defence Ground Environment
UNIFIL	United Nations Interim Force in Lebanon
UORs	urgent operational requirements
URSA	urban reconnaissance through supervised autonomy
US, USA	United States (a,n)
USMC	United States Marine Corps

Contributors

Dr Mikael Weissmann is Academic Head / Deputy Head, Land Operations Division, Swedish Defence University.

Dr Niklas Nilsson is Associate Professor in War Studies, Land Operations Division, Swedish Defence University.

Scott Boston is a senior defense analyst at the RAND Corporation, based in Arlington, Virginia.

Lt Gen (Rtd) **Professor Martin C. M. Bricknell** is Professor of Conflict, Health and Military Medicine at King's College London.

Air Commodore (Rtd) **Andrew Curtis**, OBE, is Associate Fellow at RUSI, UK, and an independent defence researcher.

B. A. Friedman is based at the Marine Corps Warfighting Laboratory, USA.

Professor **David J. Galbreath** is Professor of International Security, University of Bath, UK.

Dr Markus Göransson is Senior Lecturer, Swedish Defence University.

Dr Bruce I. Gudmundsson is advisor to the Marine Corps Tactics and Operations Group, Twentynine Palms, California.

Dr Eado Hecht is Senior Research Fellow at the Begin Sadat Center, Bar Ilan University.

Dr Christopher Kinsey is Reader in Business & International Security, Kings College, London, UK.

Brad Marvel is Senior Research Analyst at the US Army Training and Doctrine Command (TRADOC), USA.

Dr Alex Neads is Assistant Professor of International Security, Durham University, UK.

Henrik Paulsson is based at the Swedish Defence University.

Dr Tua Sandman is Assistant Professor in War Studies, Land Warfare Division, Swedish Defence University.

Professor **Olivier Schmitt** is a professor (wsr) at the Center for War Studies, University of Southern Denmark.

Dr Eitan Shamir is Director of the Begin Sadat Center and an Associate Professor at Bar Ilan University.

Dr Eado Hecht is Senior Research Associate with the Begin Sadat Center for Strategic Studies (BESA Center).

Jim Storr, former British Army officer, is now an independent defence consultant, UK.

Dr Elie Tenenbaum is the Director of the Security Studies Center at the French Institute of International Relations (Ifri) Colonel **Ronald Ti,** visiting lecturer Baltic Defence College and PhD candidate, Defence Studies Department, King's College London, United Kingdom.

Dr Christopher Tuck is Reader in Strategic Studies, the Department of Defence Studies, King's College, London.

Dr Jack Watling is Senior Research Fellow for Land Warfare at the Department of Military Sciences of the Royal United Services Institute in London.

1

Approaching Land Warfare
in the Twenty-first Century

Niklas Nilsson and Mikael Weissmann

Approaching Land Warfare

International politics has become ever more volatile in the last decade, increasing the risk of large-scale military violence. Yet the precise character of future wars will depend on a range of factors that relate to adversaries, allies, technology, geographical scope, and multiple domains of warfighting. Few would question that land forces will also be important in the foreseeable future. Recent wars in Ukraine, Syria, Mali, Yemen, and Nagorno-Karabakh have shown that land forces remain a crucial feature of warfare. However, as the battlefield transforms, so do the mission, purpose, and utilization of land forces. Indeed, the future conduct of land warfare is subject to serious and important questions in the face of large and complex challenges and security threats.

Indeed, the last two decades have seen far-reaching changes in land force employment. In particular, the counterinsurgency missions in Iraq and Afghanistan implied a wholly different operational reality for armies, in terms of adversaries, equipment availability, and tactics, compared to the type of large-scale land war anticipated during the Cold War. In Europe following the 2014 annexation of Crimea, armies have begun to adapt to the task of defending against a peer-adversary, and this change undoubtedly has far-reaching consequences, not only for the required size of land forces, but also for battle-planning methods. Although the reinvention of Cold War tactical concepts may seem obvious, these must be adapted to the current and future realities of, for example, technological complexity, a fragmented and potentially geographically dispersed battlefield, and increasingly lethal, precise, and long-distance weapons systems.

Taking aim at the evolving role of land forces, this volume pays particular attention to the changes that have taken place in the art of commanding

Niklas Nilsson and Mikael Weissmann, *Approaching Land Warfare in the Twenty-first Century*. In: *Advanced Land Warfare*.
Edited by Mikael Weissmann and Niklas Nilsson, Oxford University Press. © Niklas Nilsson and Mikael Weissmann (2023).
DOI: 10.1093/oso/9780192857422.003.0001

and executing combat and the role of rapid technological innovation and information dissemination in shaping warfare. Whilst looking forward, the volume also considers it pertinent to revisit established military theory and thinking (some of it neglected in recent years) with lessons learned from contemporary land warfare.

When analysing the state of the field and current trends in land warfare, a number of central themes emerge that will undoubtedly be crucial in the thinking, concepts, and practice of land warfare in the years to come. The role of manoeuvre warfare, command, and military theory are among these. Will manoeuvre warfare maintain its status as the supreme method of land warfare, or will it fade into the background in favour of other, emerging methods of force employment? To what extent is classic and contemporary military theory pertinent for interpreting and describing the realities of current and expected future combat? What method of command will be most suited to future Western tactics and operations? In particular, how is mission command, a key component of manoeuvre warfare, likely to evolve in the future? What should twenty-first-century combat logistics look like?

Emerging technologies are transforming warfare. The technological innovations expected to play increasingly important roles on future battlefields include artificial intelligence, sensors, unmanned air and ground systems, and cyber capabilities. These technologies are currently evolving at a rapid pace and will need to be integrated with evolving land forces' tactical practices, whilst they may also prompt the development of countermeasures by peer adversaries. Further, the environment in which armies fight may see a considerable change in the future. In particular, urban environments have been predicted to play an enlarged role in future wars, not least due to advances in target location and long-distance fire capabilities, which may diminish the chances of survival of land forces in open terrain. Nevertheless, armies will need to prepare for a range of different operational environments. If armies, particularly in the European context, are presently undergoing a decisive re-transformation into territorial defence forces after decades of primarily solving expeditionary tasks overseas, deployment in expeditionary operations will remain a distinct possibility. There is also a need to extend multi-domain capabilities and interoperability.

Discussions of future wars often focus on technological developments. However, we should not lose sight of the fact that war is a fundamentally human endeavour, and that its character will be shaped by the actions of the people fighting it. Psychological, cultural, and social issues need to remain at the centre of any discussion of land warfare. Among other things, the cohesion of military units is a critical factor in the ability of units to function under the extreme pressures of combat, as is the need for an efficient medical support system.

The present volume explores the issues described above from a thematic and an empirical perspective. It provides various perspectives on key contemporary developments in land warfare, but also presents case studies on land tactics and operations in different national contexts. In the latter case, several actors of military importance for the foreseeable future—the USA, the United Kingdom, France, Israel, China, and Russia—are at focus. Thus, a consideration of their respective approaches to land tactics will be instructive. This volume also includes a chapter covering trends in the land warfare capability of Poland and the Visegrád Group since the end of the Cold War. But first, let us briefly consider the evolution of land warfare.

This chapter is structured as follows. First, the development of land warfare is briefly outlined, before key current and future challenges in the operational environment are examined. In the following section, the future character of war and the transformation of the battlefield is addressed. Thereafter, the structure of the volume and its chapters are outlined.

Development of Land Warfare

Some authors have described the evolution of warfare as a generational development in five steps. These generational leaps begin with a first generation of ancient warfare between massed land formations. Second-generation warfare denotes the emergence of modern tactics due to the early development of firearms and later indirect fire. Third-generation warfare was enabled by technological innovations facilitating speed and manoeuvrability, permitting the utilization of indirect methods and tactics aiming to surprise, shock, and collapse—rather than annihilate—opposing forces. Fourth-generation warfare denotes a change in the character of war after the end of Cold War superpower competition, including a de-monopolization of state-controlled military force and a blurring of the boundaries between combatants and civilians. Fifth-generation warfare, finally, shifts the focus from kinetic force to the informational environment, where narratives and perceptions take centre stage, enabled by emerging technologies such as artificial intelligence, automation, and robotics.[1]

---

[1] William S. Lind and Gregory A. Thiele, 4th *Generation Warfare Handbook* (Kouvola: Castalia House, 2015); Daniel H. Abbott, *The Handbook of Fifth Generation Warfare* (Ann Arbor: Nimble Books, 2010).

The twentieth century saw a rapid evolution of warfare, fuelled by tactics and concepts developed during, between, and after the two world wars. The First World War induced the development of modern tactics, including defence in depth and infiltration techniques, necessary to avoid the massive destructive capacity of industrial-era artillery, the stagnation of direct tactical approaches into fortified trench lines, and devastating attrition warfare. These new conditions sought flexibility in offensive and defensive warfare, whilst dispersion and mobility would limit exposure to indirect enemy fire. In modern warfare, a premium was put on both offensive and defensive combat, based on cover, dispersion, small-unit independent manoeuvre, suppressive fire, and presenting the opponent with insoluble dilemmas through combined weapons integration.[2]

Whereas these innovations granted tactical successes, it rarely proved possible to exploit the advances made into strategic victories. Thus, in the interwar period, and particularly in the Soviet Union, the development of operational art and the operational level of war formed as a means for the large-scale coordination of tactics in pursuit of strategic aims. During the Second World War, Germany exploited the potential of mobile armoured units with concepts for operational-level mobile warfare and operational defence in depth.[3]

These concepts developed further during the Cold War, as the rival superpowers prepared to fight a massive war on the European continent. It was particularly the Soviet numerical advantage in terms of land forces, which grew over time, that prompted the US army to introduce the AirLand Battle doctrine in 1982. The doctrine later developed into the manoeuvre warfare concept that constituted an operational solution to the strategic problem presented by the large numerical superiority of the opponent, the Soviet Union, in the operational theatre. Manoeuvre warfare aimed to offset this disadvantage by fighting across the depth of the operational area, relying on speed, movement, and combined weapons to create unexpected and perilous dilemmas for the opponent by means of warfare across the opponent's whole formation and attacks against weak points. The concept thus rewards tactical prowess and speed rather than material resources, mass, and tolerance for attrition.[4]

---

[2] Stephen D. Biddle, *Military Power: Explaining Victory and Defeat in Modern Battle* (Princeton and Oxford: Princeton University Press, 2004).

[3] Christopher Tuck, 'Modern Land Warfare', in *Understanding Modern Warfare*, edited by David Jordan, James D. Kiras, David J. Lonsdale, Ian Speller, Christopher Tuck, and C. Dale Walton 2nd edn (Cambridge: Cambridge University Press, 2016).

[4] Richard Lock-Pullan, 'How to Rethink War: Conceptual Innovation and AirLand Battle Doctrine', *Journal of Strategic Studies* 28, 4 (2005): 679–702.

The concept of Revolution in Military Affairs (RMA) originated in Soviet military theorist Nikolai Ogarkov's work on military–technical revolutions, in the 1970s and 1980s, and made its way into Western military thinking above all through the work of Andrew Marshall, head of the Office of Net Assessment.[5] The concept nevertheless became highly influential among Western military powers following the overwhelming US victory in the 1991 Gulf War. Although technological innovations have always been an important aspect in defence planning, the Gulf War pioneered an understanding that technology enabled a completely new type of warfare. This idea made a major breakthrough in the USA and among other Western military powers during the 1990s. The main argument of RMA claims that progress, not least in computer technology and sensor systems, enables an unprecedented degree of coordination of military strikes through, for example, network-centric warfare, target identification, and precision bombing. The different parts of the military force may be integrated through a 'system of systems', where digitized command systems, coupled with supreme reconnaissance and situational awareness and long distance precision strike capabilities, would allow the achievement of war objectives with attacks against critical vulnerabilities and minimal losses to the attacking side. In the 1990s, several thinkers presumed these developments would eliminate the Clausewitzian 'fog of war', the unpredictability of battle and frictions that counteract effective planning and command, leading RMA advocates to question many of the 'eternal truths' which had formed the basis of operational thinking since the Second World War.[6]

The RMA concept was in large part discredited following conflicts during the 1990s. The succession wars in former Yugoslavia and the post-Soviet countries, as well as in sub-Saharan Africa, suggested that warfare in the post-Cold War world order had reverted to pre-modern features of tribal competition for territory and resources, where violence targeted civilians more often than enemy combatants. Moreover, the major US and NATO engagements at the turn of the century, in Kosovo, Afghanistan, and Iraq, vividly demonstrated the limitations of technological advantage and surgical precision strikes as means for achieving conclusive victory.[7]

[5] Stephen P. Rosen, 'The Impact of the Office of Net Assessment on the American Military in the Matter of the Revolution in Military Affairs', *Journal of Strategic Studies* 33, 4 (2010): 469–482.

[6] Dima Adamsky and Kjell I. Bjerga, 'Introduction to the Information-Technology Revolution in Military Affairs', *Journal of Strategic Studies* 33, 4 (2010): 463–468; Eliot A. Cohen, 'Change and Transformation in Military Affairs' *Journal of Strategic Studies* 27, 3 (2004): 395–407; Benjamin M. Jensen, 'The Role of Ideas in Defense Planning: Revisiting the Revolution in Military Affairs', *Defence Studies* 18, 3 (2018): 302–317.

[7] Mary Kaldor, *New and Old Wars*, 3rd edn (Cambridge: Polity, 2012); Patrick A. Mello, 'Review Article: In Search of New Wars: The Debate about a Transformation of War', *European Journal of International*

These developments have underscored the enduring significance of land operations across the conflict spectrum. At the same time, the conduct of land operations has become increasingly complex. The NATO Allied Joint Doctrine for Land Operations highlights the multiple functions that land forces serve aside from combat; they operate among civilian populations and infrastructure, in an increasingly intense and mediatized information environment and are often key to enabling the activities of other agencies in the framework of a comprehensive approach. Aside from combat, they have a strong symbolic importance, since deployment signals long-term political–strategic commitment.[8] Moreover, the increasing emphasis on integration and synergies across warfighting domains has acquired new heights of ambition, particularly with the US Army's Multi-Domain Operations concept, which envisions the ability to coordinate effects beyond joint land–air–sea operations to also include space and cyberspace as warfighting domains, and emphasizes the electromagnetic spectrum and information environment as key dimensions of modern warfare.[9]

Taken together, the future battlefield envisioned is one where land forces are simultaneously expected to maintain the capability to perform a wide variety of tasks, ranging from peacetime activities to high-intensity warfare, placing a premium on proficiency in manoeuvre warfare and the exercise of mission command.[10] They must simultaneously positively manage relations with civilian populations in complex conflict environments, adopt and utilize high-technological systems for communication, reconnaissance, and kinetic effect, retain the capacity to operate without these systems if needed, and contribute to extensive joint operations with other services, agencies, allies, and partners. It is no exaggeration that the future of land warfare, and the demands placed on land forces, will become ever more daunting as we approach the mid-twenty-first century. So, what are the key challenges in the current and future operational environment?

Relations 16, 2 (2010): 297–309; Edward Newman, 'The "New Wars" Debate: A Historical Perspective Is Needed', *Security Dialogue* 35, 2 (2004): 173–189; Rupert Smith, *The Utility of Force: The Art of War in the Modern World* (London: Penguin Books, 2019).

[8] *Allied Joint Doctrine for Land Operations (AJP-3.2)*, NATO Standardization Office, Edition A (Brussels: NATO, 2016).

[9] James C. McConville, *Army Multi-Domain Transformation: Ready to Win in Competition and Conflict*, US Army (Washington, DC: Headquarters, Department of the Army, 2021).

[10] Niklas Nilsson, 'Land Operations and Competing Perspectives on Warfare', *Comparative Strategy* 40, 4 (2021): 372–386.

Current and Future Challenges

A new operational environment is developing, posing new challenges for future warfare and combat. The changing character of war, with a compression of time ('the death of distance') and the information domain as the centre of gravity, has become widely recognized. Cyber and space have become domains in their own right, and Artificial Intelligence (AI), Machine Learning (ML), and other types of technologies have come to the forefront of military discussions and thinking.

It is also clear that future combat will take place in urban terrain, including in megacities, posing new challenges for land forces.[11] Furthermore, the new operational environment brings challenges in both cross-domain and cross-conflict-spectrum fighting, as the grey zone between peace and war has grown.[12] The former calls for multi-domain operations and a need for interoperability, whilst at the same time handling warfare in an operating environment that is often situated in the grey zone between peace and war.

Whilst breakthroughs in technology are at the centre stage when evaluating the future operational environment and battlefield, it is also important to recognize that we live in a time of a trembling world order. There is an ongoing shift of economic, political, and military power from the West to the East, from the USA and Japan to China, and from the North to the South, which changes the global balance of power and, in the long run, risks undermining the existing world order.[13] Opinions may differ regarding the end result of this power struggle, but it is a fact that the world will change. The resulting new reality, whether one likes it or not, will be where tomorrow's wars and battles take place.

The military will here have to deal with the new requirements and challenges that come from myriad actors seeking new roles. This applies not only to smaller countries such as Iran, North Korea, and Belarus, and major

[11] Anthony King, *Urban Warfare in the Twenty-First Century* (Cambridge and Medford: Polity Press, 2021).

[12] Mikael Weissmann, 'Conceptualizing and Countering Hybrid Threats and Hybrid Warfare: The Role of the Military in the Grey Zone', in *Hybrid Warfare: Security and Asymmetric Conflict in International Relations*, edited by Mikael Weissmann, Niklas Nilsson, Björn Palmertz, and Per Thunholm (London: I.B. Tauris, 2021); Niklas Nilsson, Mikael Weissmann, Björn Palmertz, Per Thunholm, and Henrik Häggström, 'Security Challenges in the Grey Zone: Hybrid Threats and Hybrid Warfare', in *Hybrid Warfare: Security and Asymmetric Conflict in International Relations*, edited by Mikael Weissmann, Niklas Nilsson, Björn Palmertz, and Per Thunholm (London: I.B. Tauris, 2021).

[13] Mikael Weissmann, 'Capturing Power Shift in East Asia: Toward an Analytical Framework for Understanding "Soft Power"', *Asian Perspective* 44, 3 (2020): 353–382; Astrid H. M. Nordin and Mikael Weissmann, 'Will Trump Make China Great Again? The Belt and Road Initiative and International Order', *International Affairs* 94, 2 (2018): 231–249.

powers including Russia and China, but also countries such as India, Turkey, Brazil, Indonesia, Qatar, and Dubai. Moreover, existing and emerging patterns of alliances and alignments imply that local developments can easily attain global effects.

It is also of great importance to monitor and develop strategies for dealing with the growth of non-state actors. How these develop, and what they do, has a very large direct and indirect impact on the development of the operational environment and the battlefield. This of course concerns the need to deal with direct antagonistic actors such as ISIL/ISIS and Al-Qaeda and various forms of proxy-based intelligence, crime, sabotage, subversion, and terrorism. The proliferation of private military companies and the participation of private actors in warfare and conflicts should also be mentioned here, as their role and the size of this sector have grown and there is no indication that change is underway.[14] Private actors have become an integral part of states' military operations and warfare. At the same time, they risk changing the way military operations and warfare take place and, in the long run, challenging state monopolies and roles, by increasingly enabling companies, individuals, and other non-state actors with monetary assets to acquire their own military capabilities.

Technology breakthroughs, both emerging and disruptive, have transformed and will continue to transform the operation environment. These breakthroughs, especially regarding sensor technology, artificial intelligence, and machine learning, have a direct impact on land operations and land warfare. It is already clear that future operations will be more digitized and connected, with the cyber and space domain of foremost strategic importance. At the same time, there is an inherent problem with technology development in relation to warfare; distinguishing revolutionary technology from one-day wonders. A broad perspective is necessary when the future is uncertain. Land forces need to be attentive and adaptable, both utilizing technology to their advantage, understanding how to defend against opponents' technologies, and, not least, identifying which technologies are important, maybe even revolutionary, and which are irrelevant.

It is also clear that the informational environment will be an important centre of gravity in the future operation environment. It is often said that future wars will be decided in the information environment, that 80–90 per cent of future wars will be about strategic communication, and that the struggle for narrative is central and ongoing. Without debating the finer points, it is

---

[14] Peter W. Singer, *Corporate Warriors: The Rise of the Privatized Military Industry* (Ithaca, NY, London: Cornell University Press, 2008); Sean McFate, *The Modern Mercenary: Private Armies and What They Mean for World Order* (New York: Oxford university Press, 2014); Christopher Kinsey, *Corporate Soldiers and International Security: The Rise of Private Military Companies* (London: Routledge, 2006); Joakim Berndtsson and Christopher Kinsey, eds, *Routledge Research Companion to Security Outsourcing* (London: Routledge, 2016).

clear that the information environment will be important for land forces to understand and manage.

To understand the future challenges for future land operations, one must also consider the direct impact of rapid urbanization, with the global trend of migration to cities, not least megacities, and the opportunities and challenges this entails. This fact, together with other global megatrends, such as climate change and limited natural resources, and subsequent demographical and societal changes, will alter who fights, how, and why, as well as the fundamental fighting conditions. These are all global megatrends that reshape our world, being development processes with major consequences for all actors, including land forces. These megatrends, together with technological breakthroughs and an ongoing power shift, will create circumstances to which actors in future land operations must adapt, respond, and contribute to shaping.

Character of War and Transformation of the Battlefield

One of the most important revolutions on the battlefield is the proliferation of high-quality sensors, which, in combination with the digitalization of the battlefield and AI and ML developments, increase battlefield transparency, as both can and will help manage information flows for a viable command and control system. Sensors, encompassing a wide range of technologies and devices, including radars, acoustic, thermal, optics, seismic, magnetic, active sensors, smart sensors, nano sensors, and wearable sensors, may potentially disperse the 'fog of war', making real-time information about the enemy and one's own forces available to commanders (and sometimes even individual soldiers).

The use of unattended ground sensors has permitted high-tech forces, like the USA and NATO, to enhance intelligence, surveillance, and reconnaissance abilities to a degree making adversaries' cover and concealment limited at best. This is also why extensive R&D investments are now made to develop new forms of concealments. Cheap and manoeuvrable micro- and nano-drones are also being developed for use in reconnaissance and surveillance, as is wearable sensor technology, to provide location and navigation data and uninterrupted communication between troops and UAVs in areas where GPS signals are weak or absent.[15] The possibility for uninterrupted communication should not be underestimated, as without communication the information from sensors will be non-existent or of limited practical use.

[15] Margarita Konaev, *The Future of Urban Warfare in the Age of Megacities*, Focus stratégique 88 (Paris: Ifri, March 2019).

Tomorrow's wars will often be fragmented and dispersed, taking place on a multi-territorial battlefield across borders, and often far-flung. Nor is there a clear distinction between the battlefield and elsewhere. This is true in terms of geography, since there are seldom clear borders for battlefields, and in relation to what domain the battle takes place in. This relates not only to the traditional domains of air, sea, and land, but also to the cyber and possibly the space domains. Besides domains, the information dimension is crucial, since here the narrative battle of war, combat, battle, and victory plays out. The battlefield often also includes many types of fighters, ranging from armed groups to regular forces, as well as an assortment of allies, supporters, friendly forces, non-supporters, neutrals, inactive hostiles, and unknowns, in addition to the clear enemy, further complicating future operations.[16]

There is also heterogeneity of actors on the new battlefield, including not only regular and irregular, but also a range of private and hybrid actors with unknown masters, as well many civilians who may, or may not, be friends or foes, or whose loyalty shifts over time.

Tomorrow's battlefield will also be complex in the sense that one must prepare to fight high- as well as low-tech opponents, and prepare to meet not only non-peer opponents, but also peers or near-peers. Similarly, as noted, one must also prepare for cross-domain hybridization, where fighting occurs in all five domains as well as in the information environment simultaneously, not because one wishes, but because one must.

Challenges related to hybrid threats and hybrid warfare must also be managed. It has become clear that the battlefield of the future exists in the grey zone between war and peace. In this grey zone, non-kinetic effects are found to replace, or combine with, kinetic effects. A synergistic assortment of military and non-military activities exists, ranging from different forms of strategic communication, through measures like intrusions, special operations, sanctions, and subversions, and to the use of masked soldiers, like the so-called green men in Crimea, cyberattacks, sabotage, and terror or proxy warfare, before passing the threshold of war.[17]

It is also clear that future combat will take place in dense urban areas, including in megacities. To prepare for urban warfare has become an accepted necessity, driven by several mutually supporting trends. Urbanization and

---

[16] *Allied Joint Doctrine for Land Operations*, 1–5.
[17] Weissmann, 'Conceptualizing and countering hybrid threats and hybrid warfare'. See also 'The U.S. Army in Multi-Domain Operations 2028', US Army, accessed 13 September 2021, https://api.army.mil/e2/c/downloads/2021/02/26/b45372c1/20181206-tp525-3-1-the-us-army-in-mdo-2028-final.pdf; 'Joint Concept Note 1/20, Multi-Domain Integration', Ministry of Defence, accessed 13 September 2021, https://assets.publishing.service.gov.uk/government/uploads/system/uploads/attachment_data/file/950789/20201112-JCN_1_20_MDI.PDF.

technology are driving forces, the former making cities the clear centre of gravity and the latter creating an irregular turn and urbanization of insurgency as urban areas create the defensive advantage needed for irregular forces to survive. To this can be added the changing character of war, outlined above. In short, asymmetrical warfare, in which the weaker force seeks defensive advantage in urban areas, will become a necessity, in particular in the global South, as megacities and feral cities alike grow larger, sometimes even with cross-border megaregions creating further complexity.[18] Urban operations will also need to meet the challenges from cross-domain and cross-conflict-spectrum fighting, as the grey zone between peace and war has grown. Cities, as the interconnected hubs of population and power, are the nexus of this grey zone. This includes dealing with threats and attacks below the threshold of war.

One further parameter increasingly apparent on today's battlefields is the exponential increase in information flows. Thus, access to potentially important information has increased drastically, whilst prioritization, processing, and analysis of almost unlimited amounts of information has become increasingly resource intensive. Operational assessment requires tools for managing the dynamics between information flows, continuous assessment, information dissemination, and forward-looking operational advice in an environment with basically unlimited information. Here, information flows from a range of information sources must be managed.

Structure of the Volume

This volume aims to synthesize the best of theory, practice, and professional experience. To this end, each chapter will be written by a leading international scholar or practitioner. In relating to the realities of the modern battlefield, the volume will address several critical questions about land tactics and operations, combining a conceptual basis with empirical examples of tactical thinking and practice. It emphasizes the importance of understanding the perspectives of various national armies.

By drawing on the knowledge and insights of leading war scholars, many with military experience, the volume aims to provide a current understanding of the central issues of land warfare. The project will be led by members of the

---

[18] Jeremiah Rozman, *Urbanization and Megacities: Implications for the U.S. Army*, ILW SPOTLIGHT 19–3 (Arlington: The Institute of Land Warfare, the Association of the United States Army, 2019); Konaev, 'The Future of Urban Warfare in the Age of Megacities'; Joel Lawton and Lori Shields, *Mad Scientist: Megacities and Dense Urban Areas in 2025 and Beyond* (Fort Eustis: United States Army, Training and Doctrine Command (TRADOC) G-2, 2016).

Land Warfare Research Group (LWRG) at the Swedish Defence University, and brings together contributions by distinguished scholars and practitioners in Europe, the USA, and beyond.

Part I of the volume comprises nineteen chapters divided into two parts. After this Introduction, the first part contains an introduction and ten conceptual chapters, followed by Part II with seven country-based case-studies and a concluding chapter tracing the patterns, practices, and implications going forward.

The first two conceptual chapters address, respectively, the future of manoeuvre warfare and mission command in the emerging operational environment. Chapters 4 and 5 focus on combat logistics in the twenty-first century and the present state of command and challenges in contemporary armies. Chapter 6 explores several tactical tenets and the utility of military theory. Thereafter follow three chapters exploring several dimensions likely to be central on future battlefields: urban warfare, emerging technologies, and interoperability. Chapter 11 addresses the moral component of land warfare from a perspective that transcends the issue of unit cohesion, exploring the link between soldiers' motivation to fight and the society of which they are part. Finally, the focus moves to the military health service's role in the twenty-first century.

The second part of the volume consists of eight country-based case studies of land tactics and operations. They address the divergent cultures of land forces in the USA; the constitution and tactics of China's People's Liberation army; lessons learned by Russia from land operations in Syria; the successes, failures, and adaptive capability of Israel's Defence Forces; and the United Kingdom's balancing act between strategic ambition and financial and material constraints in the development of the British Army. The penultimate chapter focuses on the French army, expeditionary warfare, and the return of strategic competition, whilst the final chapter looks at post-Cold War trends in the land warfare capability of Poland and the Visegrád States.

Finally conclusions are drawn, outlining *the integrated versatility model* as a way to capture the needs to secure the versatile edge of land warfare capabilities ready for tomorrow's battlefields.

Chapter-by-chapter Synopsis

Commencing Part I, Chapter 2, 'The Future of Manoeuvre Warfare', is written by Dr Christopher Tuck, Reader in Strategic Studies, the Department of Defence Studies, King's College, London. Dr Tuck assesses the future relevance of manoeuvre warfare, a key philosophical and doctrinal concept in the debate on the effective conduct of land operations. Tuck argues that the relevance of manoeuvre warfare is likely assured, although its relevance cannot be assumed to be coterminous with effectiveness. Despite its prominence, manoeuvre warfare is a contested idea. This chapter explores contending views on its future: manoeuvre warfare might be of continued relevance, because it is context-agnostic; it might be of greatly increased relevance, because of developments in the character of conflict and the emergence of concepts such as Multi-Domain Operations; or manoeuvre warfare might be largely irrelevant to the reality of future operations, its survival saying more about military norms, values, and perceptions.

Chapter 3, 'Commanding Contemporary and Future Land Operations: What Role for Mission Command?', is written by Dr Niklas Nilsson, Associate Professor in War Studies and Co-Convenor of the LWRG at the Swedish Defence University. Nilsson engages the adaptation of Western land forces in the face of an evolving operational environment that places varying and frequently contradictory demands on command systems. The chapter examines the concept of mission command, a decentralized command philosophy with adjacent methods and practices that is formally embraced by land forces across the West, in light of ongoing trends in the evolution of warfare and military operations. The chapter starts with a discussion of mission command respectively in terms of a culture or command philosophy, and as a set of methods and practices of command. Nilsson then considers the role and future utility of mission command in light of developments in three broad areas that are of central importance to the evolution of military command. These are, first, general trends in the current and future operational environment with implications for the command of land operations, with a focus on the US Army's concept of Multi-Domain Operations. The second area concerns the ever-increasing demands for information management, and the daunting challenge it poses for any military command system. Third, developments in information technology over the last decades and the more recent but very rapid shift toward artificial intelligence and automation have opened new horizons, as well as vulnerabilities, to military command.

Chapter 4, 'Combat Logistics in the Twenty-first Century: Enabling the Mobility, Endurance, and Sustainment of NATO Land Forces in a Future Major Conflict', shifts the focus to combat logistics. Here, Dr Christopher Kinsey, Reader in Business & International Security, Kings College, London, UK, and Colonel Ronald Ti, visiting lecturer Baltic Defence College and PhD candidate, Defence Studies Department, King's College London, UK, reinforce the ongoing importance of combat logistics in NATO, discuss new and old challenges as the Alliance prepares for large-scale combat operations, and comment on the potential effects on combat logistics of emerging, disruptive technologies. The chapter first sets the scene by outlining the character and scope of combat logistics and placing it within the context of conflict between NATO and a peer–near-peer adversary. Critical theatre-wide challenges facing NATO, particularly in sustainment and mobility, are then highlighted, before the chapter focuses on the so-called 'last tactical mile', which is a metaphor for the operational area in closest proximity to the encountered threat. Finally, the chapter concludes with a summary and brief notes regarding how deficiencies may be addressed through technology.

In Chapter 5, 'The Command of Land Forces', Jim Storr, former British Army officer, is now an independent defence consultant, observes that commanders agreed unanimously that land force headquarters are too big and take too long to produce overly long orders. But how, and why? The chapter considers the purpose of land force command systems, the products they generate, the processes they use, their structures, the systems they use, and the people within them. It is argued that command systems are not primarily technical but that they are socio-technical entities. They, and the land forces they direct, would be more effective if they were much smaller and operated much faster. This would require abandoning much explicit process and changing how information systems are used. It would also require higher levels of individual training for fewer, more carefully selected staff officers, and removing most senior staff officers in headquarters. Critically, it would require command post exercises to be genuinely free-play, two-sided, and to take place in real time. Looking more closely at who is promoted to senior ranks would expose some unpleasant realities.

Chapter 6, 'Tactical Tenets: Checklists or Toolboxes' is written by B. A. Friedman based at the Marine Corps Warfighting Laboratory, USA, and Henrik Paulsson at the Swedish Defence University. The chapter focuses on defining tactics as a practice, tactical theory as a field of study, and its relationship to strategy. A brief history of tactical theory, focused on classical tactics prior to modern times, is presented to set the stage for the most common tactical theory thereafter, the principles of war. The chapter then proposes a recapitulation of the principles of war as tactical tenets as an analytical tool. Although almost every military organization has adopted the principles of war, no version is identical and few conceptions of the principles of war use them as an analytical tool, instead just listing them. The tactical tenets can be seen as a common toolset to foster analysis and comparison of military organizations. Finally, tactical tenets is applied on two case studies as a proof of concept as an analytical tool: the United States Marine Corps and the Swedish Army.

Chapter 7, 'Urban Warfare: Challenges of Military Operations on Tomorrow's Battlefield' takes on the challenges of military operations in urban terrain (MOUT). The chapter is written by Dr Mikael Weissmann, Academic Head & Deputy Head, Land Operations Division, Swedish Defence University and Co-Convener of the LWRG. This chapter addresses the daunting challenge of urban warfare on tomorrow's battlefield. In the first section, it provides a brief background of the urban warfare phenomenon. It approaches urban warfare by asking why the field has now emerged after a long period of relative neglect. Thereafter, the chapter outlines the different challenges to and expectations for urban operations on the battlefields of today and tomorrow. A number of key challenges are addressed: the impact of rapid urbanization, multi-domain operations, grey zone problems, the impact of technology on urban operations, and the urbanization of insurgency. Observing that urban areas will be an increasingly important arena for future land warfare, the chapter argues that urban operations and warfare should acquire a greater significance in our understanding of the operational environment. With large cities being the centre of gravity for political and economic interaction and although urban warfare is a nightmare that one reasonably hopes to avoid, it is not always possible to choose the battlefield and it is therefore better to prepare thoroughly for this eventuality. Finally, to help with the preparation, the chapter presents eleven lessons about urban warfare.

In Chapter 8, 'Emerging Technologies: From Concept to Capability', Jack Watling, Senior Research Fellow for Land Warfare at the Department of Military Sciences of the Royal United Services Institute in London, examines several emerging technologies, widely anticipated to transform land warfare, unpacks the practicalities of their employment, and considers how this is likely to shape their eventual use. Critically it outlines why the frictions involved in employing them make some of the visions of military futurists unrealistic. The four technologies to be considered in sequence are autonomous systems, layered precision fires, high fidelity sensors, and artificial intelligence. The chapter concludes by considering these capabilities in combination, and their collective impact on established principles in land warfare.

Chapter 9, 'Interoperability Challenges in an Era of Systemic Competition' is written by Air Commodore (Rtd) Andrew Curtis, OBE, is Associate Fellow at RUSI, UK, and an independent defence researcher. Curtis explores the future challenges for interoperability in an era of systemic competition, beginning with an assessment of what interoperability is, its characteristics, and its benefits. This analysis is centred on NATO's approach to interoperability and how that has influenced the actions and activities of its member states. Curtis then examines the issues surrounding the pursuit of interoperability in an emerging era of systemic competition. Recognizing the impact that the latest evolution of the American way of war—Multi-Domain Operations (MDO)—will have on the development of Western military capability in the coming decade, Curtis considers what the future may hold for the various characteristics of interoperability. Finally, the chapter outlines the UK's approach to interoperability, driven as it has been by the demands of the Cold War, expeditionary operations, and now the outcome of its recent Integrated Review.

In Chapter 10, 'The Moral Component of Fighting: Bringing Society Back In', Dr Tua Sandman, Assistant Professor of War Studies, Swedish Defence University, approaches theories of victory in battle and combat tactics focusing on the oft-included moral dimension. It is argued that the question of how to win a war or battle cannot merely centre on the physical means to fight, or conceptual problems of how to fight. To understand and shape the outcome of land operations, one must also consider the moral component of fighting, essentially the will to fight. Morale, combat motivation, and cohesion are thus typically regarded as integral and critical aspects of how to achieve advantage. The chapter aims to unpack the literature on combat motivation and moral cohesion, seeking to advance our conceptual understanding of willingness to fight.

In Chapter 11, 'Military Health Services Supporting the Land Component in the Twenty-first Century', former Surgeon-General of the British Armed Forces, Lt Gen (Rtd) Professor Martin C. M. Bricknell is Professor of Conflict, Health and Military Medicine at King's College, London, examines the dual tasks of a military health service (MHS): to enable military personnel to be a 'medically ready force', and to provide a 'ready medical force' that supports armed forces during combat and other operations. Armies have the largest number of personnel exposed to risk, suffer the highest number of casualties, and have the largest medical services. Military medicine was transformed during the combat operations in Iraq and Afghanistan, resulting in the highest probability of survival for military casualties in history. MHSs have also been involved in humanitarian missions, the response to the Ebola outbreak in 2014, United Nations peacekeeping missions, and as part of national responses to the COVID pandemic. The future land battlefield may cause high casualty rates and unfamiliar threats to field medical services. The concepts of prolonged field care and prolonged hospital care describe the new approaches that will be necessary if medical planning guidelines cannot be met. Advances in medical information technology, additive printing and autonomous vehicles may also enhance medical care on the future battlefield.

The second part, Case Studies, starts with Chapter 12, 'The Operational Cultures of American Ground Forces' by Dr Bruce I. Gudmundsson, advisor to the Marine Corps Tactics and Operations Group, Twentynine Palms, California. The chapter explores the common origin and subsequent interplay of the two very different ways of thinking, teaching, and fighting at work in the US Army and Marine Corps of the twentieth century. In particular, it looks at the introduction, from Germany, of the 'applicatory method' and the subsequent evolution of its various components, some of which became rigid formats and others which inspired an approach to the art of war that was rich in creativity, innovation, and self-directed action. The chapter also describes the two very different views of 'doctrine' at work in American ground forces, as well as the effect of the 'futuristic fad' phenomenon on American military culture, as well as the experience of four very different wars during the second half of the twentieth century. The chapter will be of interest to students of US Armed Force and American military history, as well as those studying the role of military manuals, the manoeuvre warfare movement, and the relationship between teaching methods and operational styles.

Chapter 13, 'People's Liberation Army Operations and Tactics in the Land Domain: Informationized to Intelligentized Warfare' is written by Brad Marvel, Senior Research Analyst at the US Army Training and Doctrine Command (TRADOC), USA. The chapter argues that China's People's Liberation Army (PLA) is perhaps the most carefully observed and studied military in the world, with forty years of near-constant reform that radically altered the composition and capabilities of the PLA, transforming it from a poorly equipped and trained revolutionary mob to a modernized and professionalized military. The modern PLA presents a true multi-domain capability set, an emerging joint backbone, and a unique operational structure built upon decades of relentless study and experimentation. Indeed, the PLA's modernization efforts are not yet complete: new operational concepts and new systems are under development and are being integrated on a seemingly daily basis. The chapter outlines the historical background and the impetus for change that shaped Chinese military thinking, along with the strategic and political dynamics that influenced the PLA's era of modernization. It then moves into a detailed discussion of the PLA's current and future operational concepts, describing the modern Chinese way of war.

In Chapter 14, 'A Strategy of Limited Actions: Russia's Ground-based Forces in Syria' Dr Markus Göransson, Senior Lecturer, Swedish Defence University, considers the role of Russia's ground-based contingent within the overall Russian military operation in Syria. It identifies six key strategic functions of the contingent, which was small in size but highly diverse in its composition. The functions reach beyond those of base security and support to the aerial forces that spearheaded Russia's operation, and also include the ability to carry out high-value tasks, providing capacity-building to allied forces, facilitating ally coordination and supporting escalation management. Importantly, Russia's ability to operate forces with different degrees of deniability/officiality gives it greater flexibility in managing allies, adversaries, and third-party actors alike.

Chapter 15, 'The Role of Israel's Ground Forces in Israel's Wars' is written by Eado Hecht, Senior Research Associate at the Begin Sadat Center, Bar Ilan University together with Eitan Shamir, Director of the Begin Sadat Center and an Associate Professor at Bar Ilan University. The authors follow the transformation of the Israeli Defence Force (IDF) from an underground militia into an infantry-based state army during the 1948 war, and its subsequent evolution into an army based on armoured units able to conduct combined arms, high-tempo mobile operations in the 1956 and 1967 wars. Following lessons learned in the 1973 war, the IDF increased combined arms training. During the 1990s, new technologies enabled more precise targeting from afar, leading to a new concept that emphasized precision attacks, mostly by the air force. Gradually, the new concept evolved into a belief that campaigns can be won with standoff fire systems alone. However, significant results proved elusive with this means. The enemy improved its ability to disappear in underground shelters, often dug under civilian habitations. The ground forces' setbacks in the 2006 Second Lebanon War set in motion a debate within the IDF that continues to this day. On one side, advocates of improving standoff fire technologies as a substitute for manoeuvre argue that manoeuvre should be limited and sensor-saturated, in order to rapidly discover enemy locations and pass them on to fire-forces. Their opponents argue that, although new technologies improve fire capabilities, they do not enable fire to fully replace aggressive large-force manoeuvres to find and defeat the enemy whilst conquering territory. The IDF's latest multiyear force build-up plan, *Tenufa* (Momentum), seems to be an attempt to find a middle ground between these two approaches.

Chapter 16, 'Tactics and Trade-Offs: The Evolution of Manoevre in the British Army' is written by Professor David J. Galbreath, Professor of International Security, University of Bath, UK, and Alex Neads, Assistant Professor of International Security, Durham University, UK. The chapter argues that the future trajectory of land warfare in the United Kingdom stands at a crossroads. For decades, the British Army has been a reliable and enthusiastic proponent of US-led digital transformation, adapting expensive US concepts to British budgets and organizational preferences. Indeed, the desire to maintain operational currency with the US military lies at the heart of British defence doctrine, even as the UK has increasingly struggled to afford the full spectrum of capabilities such a policy implies. Now, with the character of warfare evolving once again, this old paradox presents new challenges for the British Army as it attempts to rejuvenate its warfighting capabilities in a fashion fit for the future. On the one hand, the UK Ministry of Defence's new *Integrated Operating Concept* mirrors the essential contours of the US's *Multi-Domain Operations*, presaging a further step-change in manoeuvrist doctrine. On the other, the British Army's ageing fleet of conventional platforms—from main battle tanks and infantry fighting vehicles to artillery systems and communication suites—are verging on obsolete, raising profound questions about where the technological crux of future tactical capability should lie. This chapter reveals the complex trade-offs and path dependencies inherent in the construction of British military manoeuvre. Charting the evolution of UK doctrine through professional debates over concepts and capabilities, it illuminates the uncomfortable interaction between martial thinking and material reality, strategic ambition and financial constraint, at the heart of the British Army's emergent approach to land warfare.

In Chapter 17, 'Caught between a Rock and a Hard Place: The French Army, Expeditionary Warfare, and the Return of Strategic Competition', Professor Olivier Schmitt, Center for War Studies, University of Southern Denmark, and Elie Tenenbaum, Director of the Security Studies Center at the French Institute of International Relations (Ifri), explore the transformations of the French army, and its impact on army tactics, broadly understood. The first section discusses the importance of foreign interventions for the army, and details some lessons learned of three decades of expeditionary warfare. The second section details the institutional, doctrinal, and capability changes in the French army. Assessing future challenges for the French Army, Schmitt and Tenenbaum conclude that the advent of a new era of strategic competition and the foreseeable reflux of Western interventionism is a key challenge for the identity of the French army. It has been designed, since the end of the Cold War, as a combat-ready expeditionary force best fitted to low or medium intensity stability and contingency operations. The new strategic environment is being taken into account and already translates in evolving tactics, doctrine, and capability development. This transformation, however, will take time as it challenges both the operational experience and the cultural heritage of a French army that finds itself, more than ever, at a crossroads for defining its future role in the strategic landscape.

Chapter 18, 'Trends in the Land Warfare Capability of Poland and the Visegrád States, 1991–2021', is written by Scott Boston, senior defense analyst at the RAND Corporation. Boston provides an overview of the transition of Poland, the Czech Republic, Hungary, and Slovakia from Warsaw Pact member states to NATO membership and their contributions to NATO and other multi-national missions in the years since 1999. The chapter then compares some important selected aspects of Warsaw Pact and NATO forces, focusing on the nature of the changes needed to fully adopt the system of land warfare typical of modern Western states, in the context of the rapid change in the security environment in Europe. Finally, Boston considers some of the implications of the continuing evolution of combined arms tactics and operations, with a focus on the mission to deter or defeat an adversary possessing a modern combined arms land force. Boston concludes that from the end of the Cold War to the beginning of the 2020s, the military forces of the Visegrád States have followed a winding and occasionally abrupt path from mass conscript forces subject to the control of a foreign power to smaller but more modern and flexible land forces capable of contributing to international missions and collective defence. As this work continues, it will be instructive to see how these armies make their own way toward developing the forces and capabilities they need to meet their nations' aims in the future.

Finally, the concluding chapter outlines the findings of the chapters and the volume. The authors outline a framework for a versatile approach to land warfare. First, they establish a structure of the myriad elements and factors influencing land forces, presenting a continuum of land operations modelling the use of conventional capacity and kinetic effects at different levels of conflict intensity and the role of land forces visualizing the heterogeneity of possible conflict environments where land forces may be deployed.

Thereafter, the chapter presents two schematic models; the first locates land forces in the broader operating environment by outlining how the strategic environment, conflict intensity, interoperability, and multi-domain operations are constitutive enablers and/or constraints to activities in the land domain. The second outlines how the capabilities of forces in the land domain need to be understood as a function of the interaction between own capabilities, the adversary, the human- and physical terrain, and the information environment. The multidimensional demands placed on land forces in contemporary and future operational environments necessitate a conscious multi-pronged approach to the development of land warfare capabilities, aimed at gaining a versatile edge on tomorrow's battlefields. In turn, this concerns both the build-up and construction of capabilities, and the means by which they are deployed and utilized in future conflict. The chapter argues that the achievement of *versatility* should be a crucial aim of contemporary land forces. As outlined in *the integrated versatility model*, versatility builds on two interrelated and mutually reinforcing qualities in a military organization, *adaptability* and *flexibility*. Together, they compose the underlying preconditions for truly versatile land forces.

PART I
LAND WARFARE

2

The Future of Manoeuvre Warfare

Christopher Tuck

Introduction

Manoeuvre warfare has, since the 1980s, been one of the central conceptual and doctrinal lenses through which Western armies have viewed the
effective conduct of military operations. However, the changing character of
war, driven in turn by the accelerating rate of global change,[1] raises questions regarding how relevant manoeuvre warfare might be in the future and
what forms it might take. There appears to be no consensus. For some, '[t]he
continued relevance of maneuver warfare in current and future conflicts is
indisputable'.[2] For others, it is irrelevant: 'Maneuver warfare is bunk. No competent soldier … should embrace it'.[3] Others see the concept as of some utility,
but within narrow limits: 'a select tool for a specific problem, rather than a
general method of war'.[4] Which of these views, then, is correct?

This chapter argues that manoeuvre warfare will remain relevant in the
future, although that 'relevance' may not translate necessarily into 'coherence' or 'effectiveness'. In making this argument, the chapter is divided into
three parts. First, the chapter discusses the component elements of manoeuvre warfare; second, the discussion identifies and assesses the key agents of
change that have a bearing on the nature and utility of manoeuvre warfare,
establishing the parameters of a potential increase in its relevance and efficacy; finally, through engaging with contemporary debates, the discussion
explores contending views on the likely impact of this change on the theory
and practice of manoeuvre warfare in the future.

---

[1] *Global Strategic Trends: The Future Starts Today*, 6th ed. (UK Ministry of Defence, 2018), 13.
[2] William J. Harkin, 'Maneuver Warfare in the 21st Century', Marine Corps Association Blog, 16 August
2019, https://mca-marines.org/blog/gazette/maneuver-warfare-in-the-21st-century/.
[3] Daniel P. Bolger, 'Maneuver Warfare Reconsidered', in *Maneuver Warfare: An Anthology*, edited by
Richard D. Hooker, Jr (Novato, CA: Presidio Press, 1993), 21.
[4] Carter Malkesian, 'Airland Battle and Modern Warfare', *International Forum on War History:
Proceedings, 120.*

Christopher Tuck, *The Future of Manoeuvre Warfare*. In: *Advanced Land Warfare*. Edited by Mikael Weissmann and Niklas
Nilsson, Oxford University Press. © Christopher Tuck (2023). DOI: 10.1093/oso/9780192857422.003.0002

Manoeuvre Warfare

Manoeuvre warfare emerged as a distinct theory of warfare for a variety of reasons.[5] These conditions become important when we consider circumstances today and in the future. The first reason was the NATO debate in the 1970s on the implications of the quantitative superiority of Warsaw Pact forces. Critics argued that existing approaches, as exemplified by the then current doctrine of Active Defence, were too static and attritional in focus. Against an enemy with superiority in numbers and plenty of firepower, attritional approaches were seen as a recipe for defeat. This problem provided an incentive for doctrinal reform. The second reason was the US military's post-Vietnam shift away from irregular warfare and back to conventional operations, and the desire of the Army and Marine Corps to reinvent themselves intellectually.[6] This provided the opportunity for reform by creating an institutional openness to change. The third was the 1973 Yom-Kippur War, and the lessons that could be learnt about the sources of the eventual Israeli decisive military success. This suggested potentially profitable avenues for reform. Finally, there emerged a body of intellectual thought, exemplified by authors such as William Lind and Colonel John Boyd, that explored the implications of these previous factors and suggested concepts for a new approach.[7] This intellectual effort resulted in the creation of the concept of manoeuvre warfare.

Manoeuvre warfare derives much of its meaning from its position as the proposed opposite to attrition warfare. Attrition, as a style of war, focuses on battle, mass, firepower, systematic and sequential activity, cumulative action, and the physical wearing down of an adversary. Attrition is a direct approach. Success is measured in terms of relative casualties and territory taken.[8] Manoeuvre warfare is positioned by its proponents as the antithesis of this. Manoeuvre warfare is indirect; it seeks to avoid enemy strengths and focus on identifying and attacking enemy weaknesses. It emphasizes dislocation, disruption, and the undermining of enemy will and cohesion rather than the physical destruction of the adversary. Explicitly, manoeuvre is presented as a superior approach than attrition, the latter being characterized as

---

[5] See Shimon Naveh, *In Pursuit of Military Excellence: The Evolution of Operational Theory* (London: Frank Cass, 1997), chs 6 and 7; Walter E. Kretchik, *U.S. Army Doctrine: From the American Revolution to the War on Terror* (Lawrence, KS: University Press of Kansas, 2011), 197–211.

[6] See Richard Lock-Pullan, '"An Inward Looking Time": The United States Army, 1973–1976', *Journal of Military History*, 67, 2 (April 2003): 483–511.

[7] Frans Osinga, *Science, Strategy and War: The Strategic Theory of John Boyd* (London: Routledge, 2007), 3.

[8] Daniel Moran, 'Geography and Strategy', in *Strategy in the Contemporary World*, edited by John Baylis, James Wirtz, Colin S. Gray, and Eliot Cohen (Oxford: Oxford University Press, 2007), 126.

incremental, costly, and time-consuming.[9] Three themes, in particular, lie at the heart of manoeuvre warfare approaches: system-based thinking; tempo; and non-linearity.

System-based thinking conceptualizes the enemy as a structure of integrated sub-parts reliant for their effective functioning on critical nodes and such intangibles as cohesion, will, and decision-making. Enemies can be defeated, therefore, by collapsing their system, long before they are physically destroyed. This mind-set emphasizes the importance of the targeting of the enemy's critical vulnerabilities: the discovery, and then leveraging, of enemy weaknesses.[10]

Tempo can be defined as the speed of friendly forces relative to the enemy. Manoeuvre warfare approaches conceptualize warfare as an iterative, time-competitive phenomenon based on the continuous adversarial interplay between action and reaction.[11] Manoeuvre warfare sees success in war as a function of superior tempo. Superior tempo comes from being able to identify opportunities and exploit them more quickly than the adversary, a situation that creates the basis for undermining the adversary's moral, physical, and conceptual cohesion and bringing about their systemic collapse.

Achieving superior tempo and the systemic collapse of the enemy requires a non-linear approach to warfare. Non-linearity embraces uncertainty, friction, and disorder. Commanders must accept that they cannot wholly understand and control events.[12] Consequently, in manoeuvre warfare, the emphasis is on agility, flexibility, surprise, individual initiative, and moral courage in order to exploit emerging circumstances without waiting for orders from above: 'All patterns, recipes and formulas are to be avoided'.[13] In that vein, manoeuvre warfare puts an emphasis on de-centralized decision-making, 'mission command', as the best way of coping with uncertainty and disorder, and the fluidity of combat.

The 'manoeuvre' element in manoeuvre warfare may involve physical manoeuvre, although even here, relative speed of manoeuvre, and not just position, is important.[14] However, manoeuvre also has much wider connotations. The 'manoeuvre' in manoeuvre warfare is focused on attaining positions of advantage: but these positions of advantage may be temporal,

---

[9] Marine Corps Doctrine Publication (MCDP) 1, *Warfighting* (U.S. Marine Corps, 1997), 4–5.

[10] Lt Col H. T. Hayden (ed.), *Warfighting: Maneuver Warfare in the U.S. Marine Corps* (London: Greenhill, 1995), 50.

[11] Harkin, 'Maneuver Warfare in the 21st Century'.

[12] William S. Lind, *Maneuver Warfare Handbook* (London: Routledge, 2018), 4–8.

[13] Ibid., 7.

[14] William S. Lind, 'The Theory and Practice of Maneuver Warfare', in *Maneuver Warfare: An Anthology*, edited by Richard D. Hooker, Jr (Novato, CA: Presidio Press, 1993), 4.

psychological, and/or cognitive rather than physical.[15,16] For the US Marine Corps, for example, manoeuvre is conducted 'in the physical and cognitive dimensions of conflict to generate and exploit psychological, technological, temporal, and spatial advantages over the adversary.'[17] In that sense, manoeuvre warfare can also be conceptualized as a philosophy of war, 'manoeuvrism' or 'a manoeuvrist approach', of general applicability across all levels of conflict and in non-physical domains. This philosophy focuses on applying to operations at all levels principles such as surprise, seizing the initiative, preemption, momentum, simultaneity, exploitation, and a focus on the psychological impact of actions.[18] Indeed, successful manoeuvrism may involve pre-empting the need at all for battle.[19]

Manoeuvre warfare was codified by the US Army in 1982 in its doctrine of *AirLand Battle*, and in the US Marine Corps in its 1989 *Fleet Marine Force Manual 1*.[20] Manoeuvre warfare became soaked into the fabric of Western military doctrine. Manoeuvre warfare, as embodied in US doctrine in the 1980s and 1990s, involved 'non-linear maneuver battles',[21] that focused on avoiding force-on-force attrition, and attacking instead enemy will and cohesion through 'powerful initial blows from unexpected directions and then following up rapidly to prevent his recovery'; it embodied 'rapid, unpredictable, violent, and disorienting' actions.[22] The doctrine advocated flowing around enemy strength, attacking in depth, seizing and maintaining the initiative, isolating and fragmenting the enemy, and destroying their cohesion. Success would come from surprise, tempo, audacity, concentration, agility, synchronization, and aggression.[23,24] Even the doctrinal revision of 2008, shaped by the experience of stability operations in Iraq and Afghanistan, continued to focus on the centrality of manoeuvre warfare themes, including the importance of tempo, surprise, speed, and relentless pressure to shock the enemy and break their will; the embracing of uncertainty and the friction

---

[15] Jerry Gay, 'Modernizing ISR C2 Part I: Multi-Domain Maneuver as the Foundation', Over the Horizon: Multi-Domain Operations and Strategy (21 November 2018), https://othjournal.com/2018/11/21/modernizing-isr-c2-part-i-multi-domain-maneuver-as-the-foundation/

[16] Robert Leonhard, *The Art of Maneuver: Maneuver-Warfare Theory and AirLand Battle* (New York, NY: Ballantine, 1991), 18.

[17] 'How We Will Fight', *Marines*, 16 May 2022, https://www.mccdc.marines.mil/MOC/Operation-Concept-pg/.

[18] Army Doctrine Publication, *Land Operations* (Land Warfare Development Centre, March 2017), ch. 5.

[19] Leonhard, *The Art of Maneuver*, 20.

[20] John L. Romjue, *From Active Defense to AirLand Battle: The Development of Army Doctrine, 1973–1982*, TRADOC Historical Monograph Series (U.S. Army, 1984), https://www.tradoc.army.mil/wp-content/uploads/2020/10/From-Active-Defense-to-AirLand-Battle.pdf.

[21] Field Manual (FM) 100–05, *Operations* (Headquarters: Department of the Army, 1982), 1–1.

[22] Ibid., 2–1.

[23] FM 100–05, *Operations* (Headquarters: Department of the Army, 1993), 6–19.

[24] Ibid., 7–1 to 7–3.

of war; and the necessity of decentralized approaches to command and control.[25] Key elements of this manoeuvre warfare approach have survived in doctrines through to the present day. The philosophy, spirit, and key concepts of manoeuvre warfare have permeated Western doctrine, even those of the non-land domains.[26]

Manoeuvre warfare, therefore, is 'a state of mind bent on shattering the enemy morally and physically by paralyzing and confounding him, by avoiding his strength, by quickly and aggressively exploiting his vulnerabilities, and by striking him in a way that will hurt him most'.[27] For advocates of this style of war, the attractions are obvious. Contemporary and historical theory and practice appear to demonstrate its superiority over attritional approaches. Military doctrine argues that manoeuvre warfare, therefore, is 'a philosophy for generating the greatest decisive effect against the enemy for the least possible cost to ourselves—a philosophy for 'fighting smart'.[28]

The Future Importance of Manoeuvre Warfare

For advocates of the theory of manoeuvre warfare, its future relevance is self-evident. This is because manoeuvre warfare is simply a codification of a successful approach to war as old as warfare itself.[29]

Manoeuvre warfare 'is the modern term for an ancient concept' and as such its applicability demonstrably transcends changes to the character of war.[30] Whilst this theory might have emerged from the study of a specific military problem, the challenge posed by the quantitative superiority of Warsaw pact forces, proponents of the concept of manoeuvre warfare argue that the concept is a codification of actual historical best practice. With references to such luminaries as Sun Tzu, Carl von Clausewitz, J. F. C. Fuller, and Basil Liddell Hart, manoeuvre warfare enthusiasts argued that the concept's principles are validated by the existing corpus of classical military and strategic thinking, including Sun Tzu's views on the importance of deception, and Liddell Hart's advocacy of the effectiveness of indirect approaches in war and the salience of psychological over material factors. At the same time, enthusiasts drew on a range of historical examples to demonstrate

[25] FM 3–0, *Operations* (Headquarters: Department of the Army, 2008), 3–3 to 3–4.
[26] Gay, 'Modernizing ISR C2 Part I: Multi-Domain Maneuver as the Foundation'.
[27] MCDP 1, 95.
[28] Ibid., 96.
[29] Lind, *Maneuver Warfare Handbook*, 4.
[30] Paul Barnes, 'Maneuver Warfare: "Reports of my Death Have Been Greatly Exaggerated"', *Modern War Institute*, 9 March 2021. https://mwi.usma.edu/maneuver-warfare-reports-of-my-death-have-been-greatly-exaggerated/

that manoeuvre warfare approaches lie at the root of most major military victories. These historical examples include the campaigns of Alexander the Great and Napoleon Bonaparte, German blitzkrieg, the Inchon landing in the Korean War, and Israeli success in the Arab-Israeli wars.[31] Manoeuvre warfare doctrines also seemed to receive practical validation through US successes in the Gulf War of 1990–1991, and the early conventional phase of the invasion of Iraq in 2003.[32]

However, it may well be that the importance of manoeuvre warfare will increase in the future, as contemporary trends create conditions in which an attritional style of war is even less attractive to Western armies. Something of a consensus has emerged in Western thinking on the main trends of significance for the development of warfare in the future, their implications, and solutions.[33] These trends include: acceleration; equalization; informationalization; hybridization; and the expansion and blurring of the domains of warfare.

The idea of acceleration encompasses the observation that the world seems to be changing at an unprecedented rate. Many of the obvious changes are geopolitical in character, such as the progressive shift in power from the West to the East. However, a significant feature of this debate is the acceleration in the rate of technological change. This process is delivering significant changes in the realms of such things as firepower, sensor technologies, power systems, human augmentation, robotics, computing, and artificial intelligence.[34] For many, we are in the midst of a military revolution: 'an historical inflection point'[35] in which the 'pervasiveness of information and rapid technological development have changed the character of war'.[36] 'Equalization' describes the erosion over past decades of Western technological, maritime, air, space, and electro-magnetic superiorities. These advantages increasingly have been eroded as a result of developments in Russian and Chinese conventional military capabilities, especially the threat posed by their A2/AD (Anti-Access/Area Denial) systems: multi-domain, multi-level defensive systems designed to deter or defeat Western forces at the longest possible ranges.

---

[31] Harkin, 'Maneuver Warfare in the 21st Century'.

[32] Barnes, 'Maneuver Warfare'.

[33] See, for example, *The Operational Environment and the Changing Character of Warfare*, TRADOC Pam 525-92 (U.S. Army Training and Doctrine Command, 2019), 5–8; *Global Strategic Trends*, 125–145; AFC Pam 525-2 *Future Operational Environment: Forging The Future in an Uncertain World, 2035–2050* (Army Futures Command, 2020), 2–6.

[34] TRADOC Pam 525–92, 16.

[35] *Army Futures Command Concept for Maneuver in Multi-Domain Battle 2028*, AFC Pam 71-20-1 (Army Futures Command, 2020), 10, https://api.army.mil/e2/c/downloads/2021/01/20/2fbeccee/20200707-afc-71-20-1-maneuver-in-mdo-final-v16-dec-20.pdf

[36] Speech, Chief of the Defence Staff, 30 September 2020, https://www.gov.uk/government/speeches/chief-of-the-defence-staff-general-sir-nick-carter-launches-the-integrated-operating-concept.

The challenge is exacerbated by the proliferation of advanced technologies, including drones and cyber, to other state and non-state actors.[37] This technological change has also led to the increasing 'informationalization' of warfare reflected in the growing significance of themes like networking, big data analytics, automated decision-support, surveillance, electronic warfare, information manoeuvre, and, especially, the significance of information dominance.[38]

The previous themes have developed momentum in the context also of 'hybridization', a loose term describing the ways in which conventional warfare is likely to be preceded, accompanied, or replaced, by the orchestrated application of a whole range of other potent activities across the spectrum of conflict, including information warfare, cyber-attacks, deniable operations, and the use of proxies. This form of conflict seeks to target political as well as military objectives, including public will and alliance cohesion where applicable.[39] Linked to this, there has been a widening of the domains relevant to warfare. Non-traditional domains, in the form of the space and cyber realms are becoming increasingly critical. They are no longer simply supporting areas of operations, but, so it is argued, fully-fledged warfighting domains.[40] The electro-magnetic spectrum, for example, exerts an increasingly ubiquitous influence on operations because of its centrality in navigation, communications, command and control, data networking, surveillance, and targeting.[41]

The conditions wrought by these changes create a battlefield logic that seems especially conducive in the future to manoeuvre warfare. On the one hand, attritional approaches are likely to carry increasing costs for Western armies. As Palazzo comments: 'On today's battlefield, if it can be sensed, it can be killed from afar, often with a single round'.[42] There is an essential problem, therefore, in applying attritional approaches in an environment in which adversaries have access to large quantities of accurate, long-range firepower.

---

[37] Multi-Domain Battle: Evolution of Combined Arms for the 21st Century, 2025–2040 (2017), 4–8, https://admin.govexec.com/media/20171003_-_working_draft_-_concept_document_for_multi-domain_battle_1_0.pdf.

[38] TRADOC Pam 525–92, 19–20.

[39] David Kilcullen, *The Dragons and the Snakes: How the Rest Learned to Fight the West* (London: Hurst and Company, 2020); Mikael Weissmann, Niklas Nilsson, Bjorn Palmertz, and Per Thunholm (eds), *Hybrid Warfare: Security and Asymmetric Conflict in International Relations* (London: I. B. Tauris, 2021).

[40] *The U.S. Army in Multi-Domain Operations 2028*, TRADOC Pam 525-3-1 (Training and Doctrine Command, 2018), 8–10.

[41] John G. Casey, 'Cognitive Electronic Warfare: A Move Towards EMS Maneuver Warfare', Over The Horizon: Multi-Domain Operations and Strategy, 3 July 2020, https://othjournal.com/2020/07/03/cognitive-electronic-warfare-a-move-towards-ems-maneuver-warfare/

[42] Albert Palazzo, 'Precision and the Consequences for the Modern Battlefield', Small Wars Journal, 19 August 2016, https://smallwarsjournal.com/jrnl/art/precision-and-the-consequences-for-the-modern-battlefield.

The application of mass, too, becomes highly problematic under conditions in which lethality has increased and in which Western armed forces have tended to get smaller. Indeed, the US Army has identified the need to develop approaches to warfare that will allow it to shift the balance of forces required for offensive success from the traditional 3 to 1 ratio, to 1 to 2.[43]

On the other hand, contemporary trends also seem to reinforce the salience of the key tenets of manoeuvre warfare. The growing importance to modern militaries of networking as a critical enabler reinforces the importance of system-based thinking, making systemic disruption of the enemy, by attacking their networks, potentially even more effective. For example, in breaking into an adversary's A2/AD complex, systemic disruption may be the best route. As one US officer has noted: 'We need to overwhelm an enemy's command and control, then we can penetrate and create a window for the joint force.'[44] At the same time, developments in AI, machine learning, and ISTAR (intelligence, surveillance, target acquisition, and reconnaissance) capabilities could allow '[a]ccounting and mapping adversary systems to the nodal level', providing exquisite detail on systemic enemy vulnerabilities.[45] Nontraditional domains provide new methods of exploiting these vulnerabilities increasing the tools available to surprise and shatter an opponent's cohesion. AI and machine learning, for example, could allow electronic warfare systems to identify, adapt, and attack vulnerabilities faster than the enemy can respond.[46]

New technologies for acquiring, processing, and disseminating information, linked to AI and machine learning provide 'game-changing'[47] opportunities to speed up war and generate greater tempo in future military operations. Since tempo is founded, amongst other things, on information superiority and speed of decision-making, the informationalization and greater automation of war increases the relevance of manoeuvre styles of warfare. Automated information and battle-management systems, along with greater integration reflected in developments in the 'internet of things', in tandem with longer range, faster, and more precise capabilities, allow an acceleration in our ability to identify enemy weaknesses and then to apply physical and/or non-physical means of attack. Satellites identify targets, cloud

---

[43] AFC Pam 71-20-1, 41.

[44] Kris Osborn, 'Army Pursues New "Combined Arms Maneuver Warfare" Attack Plan', *Fox News*, 18 September 2021, https://www.foxnews.com/tech/army-pursues-new-combined-arms-maneuver-warfare-attack-plan.

[45] AFC Pam 71-20-1, 19.

[46] Casey, 'Cognitive Electronic Warfare.'

[47] Paul Scharre, *Robotics on the Battlefield, Part II: The Coming Swarm* (Center for a New American Security, October 2014), 5. https://s3.amazonaws.com/files.cnas.org/documents/CNAS_TheComingSwarm_Scharre.pdf?mtime=20160906082059.

computing pools and shares the data, and AI then assesses the data and creates responses.[48] In essence, AI will accelerate the 'kill chain' linking sensors to shooters. The future, then, might be a 'hyperactive battlefield',[49] in which information superiority, and the capacity to leverage it, will give decisive advantages in military initiative. As Ardis and Keene have argued, in the future: 'Dominance in the information space is a critical capability that will enable the US Army to determine if, how, and when it will engage in conflict'.[50]

As the velocity of war increases, de-centralization becomes even more significant. There are clear difficulties in trying to apply centralized, sequential approaches in a warfare environment that has become even faster, more complex, and non-linear and which requires even greater dispersion and complex synchronization. Survivability demands that forces have the capability to mass effects, rather than mass physically. This demands effective networking, and the capabilities, logistics, and command philosophy to operate separately for days at a time. In these circumstances, 'intent-based mission command—enabled by a culture of trust and risk' will be critical to maintaining the tempo of operations.[51]

Reflecting these developments, an expanded view of manoeuvre is at the heart of debates on the concepts and doctrines required to fight future warfare. If manoeuvre is directed towards obtaining 'positions of advantage', these positions are no longer conceived of in an exclusively, or even predominantly, physical way. Information, electronic, and cyber manoeuvre are recognized as being increasingly central to success.[52] Cyber manoeuvre, for example, entails the 'application of force to capture, disrupt, deny, degrade, destroy or manipulate computing and information resources in order to achieve a position of advantage in respect to competitors'.[53] In achieving these positions of advantage, even within the non-physical domains, the emphasis is on 'rapid, focused, and unanticipated actions' as a way of shattering the adversary's cohesion.[54]

---

[48] Sydney J. Freedberg, Jr, 'SecDefs Multi-Domain Kill-Chain: Space-Cloud-AI', *Breaking Defense*, 22 November 2019, https://breakingdefense.com/2019/11/secarmys-multi-domain-kill-chain-space-to-cloud-to-ai/

[49] Kris Osborn, 'Army pursues new "Combined arms maneuver warfare" attack plan', 18 September 2020, https://www.foxnews.com/tech/army-pursues-new-combined-arms-maneuver-warfare-attack-plan.

[50] John A. S. Ardis and Shima D. Keene, *Maintaining Information Dominance in Complex Environments* (Strategic Studies Institute, October 2018), xiii.

[51] AFC Pam 71-20-1, 27.

[52] 'New Electromagnetic Maneuver Warfare Strategy Emerges, *Interference Technology*, 30 October 2014, https://interferencetechnology.com/new-electromagnetic-maneuver-warfare-strategy-emerges/

[53] Scott D. Applegate, *The Principle of Maneuver in Cyber Operations* (NATO CCD COE Publications, 2012), 185. https://ccdcoe.org/uploads/2012/01/3_3_Applegate_The Principle Of Maneuver In Cyber Operations.pdf.

[54] Casey, 'Cognitive Electronic Warfare'.

For these reasons, the current *zeitgeist* in military thinking is 'multi-domain operations'. Multi-Domain Operations (MDO) are founded upon the assumption that the challenges of future warfare can only be overcome by the orchestration of all of the domains of war, at every level of conflict, into a single effort. As one commentator has argued, there is a need to 'become much more attuned to forms of maneuver in all … realms, and until [we] develop an appreciation for and understanding of multi-domain maneuver, true innovation' will be lacking.[55] Published concepts for these kinds of operations have continued to apply the essential precepts of manoeuvre warfare, but with some modifications to reflect the changing context. The US Army's *Maneuver in Multi-Domain Operations* concept, for example, focuses on the idea of 'echeloned maneuver', which is: 'Army air-ground movement in depth supported by ground fires along with air, maritime, space and cyberspace generated effects to gain positions of advantage, penetrate adversary defenses, and conduct exploitation.'[56] The explicit purpose of this concept is to enable manoeuvre, which is regarded as the critical route to success.[57] In this, manoeuvre is conducted in the non-physical domains as well as physical, and it is applicable at all of the levels of war.[58] The aim is 'to achieve physical, temporal and psychological advantage over enemy forces'.[59] In MDO, manoeuvre is also expanded outside of the arena of direct armed conflict, since armed conflict is likely to be preceded by periods of hybrid activity which may create a 'continuous, dynamic, and simultaneous competition arena that elevates up to conflict in non-linear cycles'.[60]

At the root of these approaches remains an implicit commitment to the tenets of classic manoeuvre warfare. Thus, the purpose of 'echeloned maneuver' is to avoid the need to mass physically and to focus instead on massing effects; MDO 'enables independent maneuver of distributed formations' with a focus on agility, flexibility, seizing the initiative, attaining momentum, and controlling the tempo of operations.[61] It does this by creating a system that can call quickly on any sub-element, in any domain, at any level.[62] By employing all of the domains in a continuous, synchronized way, multi-domain

---

[55] Gay, 'Modernizing ISR'.

[56] AFC Pam 71-20-1, 8.

[57] Ibid.

[58] Ibid., 10.

[59] *The U.S. Army Functional Concept for Movement and Maneuver, 2020–2040*, TRADOC Pam 525-3-6, (Training and Doctrine Command, 2017), 8.

[60] AFC Pam 71-20-1, 10.

[61] Ibid., 8.

[62] Colin Clark, 'Gen. Hyten on the New American Way of War: All-Domain Operations, *Breaking Defense*, 18 February 2020, https://breakingdefense.com/2020/02/gen-hyten-on-the-new-american-way-of-war-all-domain-operations/.

manoeuvre seeks to create the maximum synergies for friendly efforts and to create multiple dilemmas for an adversary, maximizing the chances that they will have no effective response. The aims are 'convergence', the creation of 'simultaneous effects from all domains faster than the enemy',[63] and the systemic paralysis of the enemy, 'shattering the coherence of his military system'.[64]

This potential increase in the future relevance of manoeuvre warfare is unsurprising, one might argue, given the parallels between conditions today and those which first brought forth doctrines of manoeuvre warfare: perceptions of an urgent military threat from increasingly potent adversaries (Russia, China); defeats in unconventional warfare (Iraq, Afghanistan) which have created an appetite for reform and a desire to refocus on conventional operations; extant conflicts (Ukraine, Syria, Nagorno-Karabakh, hybrid war campaigns) from which lessons for the future can be drawn; and parallel conceptual developments, in the form of multi-domain approaches.

Alternative Perspectives

Uncertainty, however, permeates all of our thinking about the future. This uncertainty applies just as much to the relevance of manoeuvre warfare. We cannot be certain that the positive analysis presented thus far is correct. There are many who are sceptical regarding the role and value of manoeuvre warfare in future conflicts. The final part of this chapter focuses on three related criticisms of manoeuvre warfare's future: that we cannot assume that the future of warfare necessarily would make manoeuvre warfare more relevant; that even if it remains a relevant concept, it is not universally applicable; and finally, that perhaps a distinct style of manoeuvre warfare does not actually exist at all—it is thus an invented concept.

The first critique is that we cannot assume that the agents of change outlined earlier in this chapter will increase the relevance of manoeuvre warfare. This conclusion is based on a number of observations. One very general point is that we are simply very poor at all forms of accurate prediction in relation to war and international affairs. As the political scientist Philip Tetlock famously commented, research demonstrates that the predictions of experts have about the same level of accuracy as a monkey using a dartboard.[65]

---

[63] Tom Greenwood and Pat Savage, 'In Search of a 21st Century Joint Warfighting Conflict', *War on the Rocks*, 12 September 2019, https://warontherocks.com/2019/09/in-search-of-a-21st-century-joint-warfighting-concept/.

[64] AFC Pam 71-20-1, 20.

[65] Philp Tetlock, *Superforecasting: The Art and Science of Prediction* (London: Random House, 2016).

At the end of the 1990s, for example, whatever future was envisaged for Western armies, it certainly wasn't nearly two decades of stability operations in Iraq and Afghanistan. In relation to the specifics of future manoeuvre warfare, there are a wide range of challenges. It may be, for example, that our assessments of the nature and outcome of salient trends in future warfare are flawed. As the futurologist Christopher Coker has identified: 'The future we envision can only be an extrapolation of present trends taken to a logical and therefore often illogical conclusion.'[66] The future trajectory and implications of AI, for example, are a matter of vigorous contestation, with wide variations in the conclusions.[67] As one commentor has noted, 'there is still little clarity regarding just *how* artificial intelligence will transform the security landscape'.[68] It may be, therefore, that, if AI underperforms in relations to optimistic expectations, MDO cannot be delivered effectively and that we succeed only in adding to the 'fog of war' a kind of 'fog of systems'.[69] It may also be that, since warfare is a relational activity, the theoretical benefits delivered by MDO-type approaches are unrealizable in the face of enemy action and adaption, and the frictions of war. For example, if networks become increasingly vital to Western militaries, adversaries inevitably will target them as our centre of gravity. As one US officer has commented: 'Capabilities create dependencies, and dependencies create vulnerabilities.'[70] Nor can we presume in the end that it is we who will be able to obtain a clear advantage in arenas such as the information domain, especially given the heavy investments made by China and Russia in AI systems.[71]

The future may therefore be very different from that one that we predict. It may be one in which standoff firepower rather than manoeuvre is key.[72] Or it may be one in which attrition and mass, for example in the form of large

---

[66] Christopher Coker, *Warrior Geeks: How 21st Century Technology is Changing the Way We Fight and Think about War* (London: Hurst and Company, 2013), xix.

[67] See, for example Erik J. Larson, *The Myth of Artificial Intelligence: Why Computers Can't Think the Way We Do* (Cambridge, MA: Harvard University Press, 2021).

[68] Ben Garfinkel and Allan Dafoe, 'Artificial Intelligence, Foresight, and the Offense-Defense Balance', *War on the Rocks*, 19 December 2019, https://warontherocks.com/2019/12/artificial-intelligence-foresight-and-the-offense-defense-balance/.

[69] Franz-Stefan Gady, 'What Does AI Mean for the Future of Manoeuvre Warfare?' *International Institute for Strategic Studies*, 5 May 2020, https://www.iiss.org/blogs/analysis/2020/05/csfc-ai-manoeuvre-warfare.

[70] Williamson Murray, 'Technology and the Future of War', *Hoover Institution*, 14 November 2017, https://www.hoover.org/research/technology-and-future-war.

[71] See, for example, Elsa B. Kania, 'Chinese Military Innovation in Artificial Intelligence: Testimony before the US-China Economic and Security Review Commission Hearing on Trade, Technology, and Military Civil Fusion', *Center for a New American Security*, 7 June 2019.

[72] Steve May, 'The British Way of War—Balancing Fire and Manoeuvre for Warfighting', *The Wavell Room*, 9 May 2018, https://wavellroom.com/2018/05/09/the-british-way-of-war-balancing-fire-and-manoeuvre-for-warfighting/.

numbers of low-cost drone swarms, becomes more pre-eminent.[73] Nor is it clear if developments in such things as AI might not in the end favour the defender rather than the attacker.[74]

A second criticism of manoeuvre warfare is that it is presented as a concept of universal applicability when in fact it is difficult, or indeed dangerous, to apply in many circumstances. Three of these contexts are likely to have particular significance in the future: limited war; irregular war; and urban operations.

In limited wars of the future, manoeuvre warfare may be risky to implement, especially against nuclear armed adversaries, because it is escalatory: it is oriented towards the application of aggression, rapidity, deep offensive operations, and the annihilation of the enemy forces. It is a doctrine designed to produce decisive victory. Manoeuvre warfare, therefore, is not an approach that can be easily modulated in politically complex circumstances, because it is an approach that intrinsically threatens to impose high costs on an adversary.[75] It is certainly the case that MDO is less overtly focused on decisive military victory than doctrines of the 1980s and early 1990s, the concept noting that the purpose of operations is to 'achieve tactical, operational, and strategic objectives that support the return to non-crisis competition on favourable terms'.[76] Nevertheless, the whole construct of multi-domain manoeuvre is based on the simultaneous application of tactical, operational, and strategic-level actions against the whole depth of the enemy. Army corps areas of operation alone would extend up to 500 km deep.[77] Moreover, during armed conflict, friendly forces would still aim to manoeuvre in order 'to destroy or defeat enemy forces'.[78] When the enemy system is so extensive, both in density, capability, and geography, the system-focused nature of manoeuvre warfare, and the centrality of themes such as seizing the initiative, tempo, and risk-taking, carries intrinsic problems of escalation. Thus, there may be a whole range of scenarios in which the operational and strategic application of manoeuvre warfare might be regarded as dangerous, and in which more limited, defensive options might seem more applicable.

In relation to irregular warfare, many commentators have argued that the real future of armed conflict lies with so-called New Wars, in which armed

[73] Margarita Konaev, 'With AI, We'll See Faster Fights, But Longer Wars', *War on the Rocks*, October 29 2019. https://warontherocks.com/2019/10/with-ai-well-see-faster-fights-but-longer-wars/; Scharre, 'The Coming Swarm', 5.
[74] Garfinkel and Dafoe, 'Artificial Intelligence'.
[75] Malkesian, 'Airland Battle', 117.
[76] AFC 71-20-1, vi.
[77] Ibid., 45.
[78] Ibid., iii.

conflicts increasingly will be sectarian wars of 'state disassembly'.[79] Manoeuvre warfare will be less relevant because (a) most armed conflicts will not be state-on-state conflicts, and because (b) the irregular war used by Western adversaries will be an asymmetric strategy designed deliberately to avoid the application of powerful conventional capabilities.[80] Indeed, for some commentators, manoeuvre warfare is a route to a less effective military future because the post-Iraq and Afghanistan tilt back towards to conventional warfare is already leading to a loss of capabilities and skills for stability and counterinsurgency accumulated since 2001.[81] On the basis that, as Anthony Cordesman has noted, 'conventional wars never have a conventional ending',[82] even if manoeuvre warfare succeeds in the initial stages of a conflict, that success may be squandered, as it was in Iraq, because of a lack of capabilities to do what needs to be done afterwards.[83]

Finally, critics of the future relevance of manoeuvre warfare point to the problems posed by urban operations. Urban operations may well be an increasing feature of future warfare. They have certainly been an important feature of the current war in Ukraine. There are a variety of reasons why this might be the case: demographic trends (half the world's population lived in cities in 2007; by 2050 it will be two-thirds); the ways in which the physical and geographic density of urban environments might make these the battlegrounds of choice for some adversaries; and the decline in the size of military forces which makes a focus on urban areas, as decisive political and economic ground, more cost effective.[84] Thinking about wars over the last ten years, urban areas have often been key: as in Syria, in the fighting for Aleppo; or in Iraq, in Raqqa, the ISIS capital, or around Kyiv and in Mariupol in Ukraine. Experience in operations such as Fallujah demonstrates that urban operations are positional and attritional.[85] They are slow. The depth of the battle shrinks. The advantages conferred on defenders in urban environments requires that attacking forces have to mass physically in order to penetrate and occupy enemy positions. Combined arms become

[79] Mary Kaldor, 'Peacemaking in an Era of New Wars', *Carnegie Europe*, 14 October 2019, https://carnegieeurope.eu/2019/10/14/peacemaking-in-era-of-new-wars-pub–80033.

[80] Malkesian, 'Airland Battle', 117.

[81] Linda Robinson, Sean Mann, Jeffrey Martini, and Stephanie Pezard, *Finding the Right Balance: Department of Defense Roles in Stavilization* (Santa Monica, CA: RAND, 2018), 31–36. https://www.rand.org/pubs/research_reports/RR2441.html.

[82] Anthony H. Cordesman, 'Stability Operations in Syria: The Need for a Revolution in Civil-Military Affairs', Military Review (March 2017): 5. https://www.armyupress.army.mil/Portals/7/Army-Press-Online-Journal/documents/Cordesman-v2.pdf

[83] Applegate, 'The Principle of Maneuver', 185.

[84] Anthony King, *Urban Warfare in the Twenty-First Century* (Cambridge: Polity, 2021), 19–40.

[85] Ibid., 199–201.

more difficult. Factors such as tempo and the targeting of the enemy system become less effective because urban operations have a natural tendency to become more fragmented and localized anyway. Purely military considerations become complicated by interactions with the local population and the necessity to provide political, economic, and social means to consolidate success. Legitimacy becomes a central concern, with consequent constraints on military operations imposed by legal, ethical, political, media, and messaging considerations.[86] Recent experience also seems to show that the manpower requirements for such operations, as well as political conditions, mandate a significant reliance on local forces. These might include regular troops but might also include irregular militias and other proxies. In this case, the capacity for manoeuvre warfare becomes circumscribed by the weaknesses of these allied components.[87] Taking this perspective, a continued focus on manoeuvre warfare, therefore, is a doctrinal misstep. Indeed, the military sociologist Anthony King has gone so far as to assert that 'manoeuvre warfare is dead'.[88] Instead, this critical view argues that militaries need to reorientate themselves and consider the urban environment as 'a primary driver of capabilities', adapting their roles and structures accordingly.[89]

Given these contextual challenges, therefore, the limited applicability of manoeuvre warfare might in the future make it relevant only for a select range of circumstances.[90] However, for advocates of the continued relevance of manoeuvre warfare, these criticisms are easily answered. As one proponent of manoeuvre warfare has argued: 'Maneuverism is a frame of mind, not a prescription';[91] and for another, it is simply 'a thought process that seeks to pose our strengths against our adversaries' weaknesses'.[92] In addressing the specific critique that manoeuvre warfare is not applicable in urban operations, for example, one might make the point that becoming locked into attacking an adversary in urban environments is exactly an example of the failure to apply manoeuvrism. By definition, if we are confronting adversaries on the

---

[86] John Spencer, 'Square Peg, Round Hole: Maneuver Warfare and the Urban Battlefield', *Modern War Institute*, 11 March 2021, https://mwi.usma.edu/square-peg-round-hole-maneuver-warfare-and-the-urban-battlefield/.

[87] 'Is the Era of Manoeuvre Warfare Dead?', RUSI Podcast, 23 December 2020. https://rusi.org/podcasts/western-way-of-war/episode-30-is-the-era-of-manoeuvre-warfare-dead.

[88] Ibid.

[89] Joseph Bogan and Aimee Feeney, 'Future Cities: Trends and Implications', DSTL (February 2020), ii. https://assets.publishing.service.gov.uk/government/uploads/system/uploads/attachment_data/file/875528/Dstl_Future_Cities_Trends___Implications_OFFICIAL.pdf.

[90] Malkesian, 'Airland Battle', 120.

[91] Barnes, 'Maneuver Warfare'.

[92] Ricky L. Waddell, 'Maneuver Wafare and Low-Intensity Conflict', in Hooker, Jr, *Maneuver Warfare: An Anthology*, 119.

ground of their choosing, then we are failing to apply such basic tenets of manoeuvre warfare as avoiding enemy strengths and focusing on surprising and dislocating an adversary.[93]

But this defence of manoeuvre warfare leads us to our final critique: that manoeuvre warfare does not actually exist as a discrete style of warfare. Instead, it is an imaginary construct, fabricated and instrumentalized by Western militaries to service a variety of value-related and political functions. Whilst manoeuvre enthusiasts portray the concept as one that, in terms of rigour, 'meets the standards of contemporary social and political science',[94] sceptics argue instead that there is an elusiveness to the concept of manoeuvre warfare that makes it vague, fluid, and unfalsifiable. There is, for example, a basic lack of definitional clarity concerning what manoeuvre warfare is.[95] If manoeuvre warfare really is 'attaining positions of advantage' or 'avoiding enemy strengths and focusing on their weaknesses', then it is so general as to be meaningless as a distinct approach to warfare. Manoeuvre warfare then simply becomes common sense and 'anything that works': successful operations are successful because they are manoeuvrist; and failures occur because the defeated party was not manoeuvrist enough.[96]

Indeed, critics also attack the evidence base for manoeuvre warfare, accusing exponents of the selective use of history and of manipulating case studies to fit the manoeuvre argument.[97] This creates grotesque over-simplifications in our understanding of warfare. First, manoeuvre warfare is presented as the sole route to success in land warfare. But this ignores a wide range of literature that highlights other critical variables, including strategy and policy, command and control, and cohesion.[98] Stephen Biddle, for example, locates the key roots of tactical and operational success in land warfare in force employment—the relative competence of belligerents in 'modern system' land warfare, this being: 'a tightly interrelated complex of cover, concealment, dispersion, suppression, small-unit independent maneuver, and combined arms at the tactical level, and depth, reserves, and differential concentration at the operational level of war'.[99] Biddle argues that where the gap in the quality of force employment between belligerents is large, then rapid and decisive

---

[93] Barnes, 'Maneuver Warfare'.

[94] Richard D. Hooker, 'Part 1: The Theory of Maneuver Warfare', in Hooker, Jr, *Maneuver Warfare: An Anthology*, 3.

[95] Spencer, 'Square Peg, Round Hole'.

[96] Bolger, 'Maneuver Warfare Reconsidered', 21.

[97] Ibid., 22–29; Spencer, 'Square Peg, Round Hole'.

[98] Colin S. Gray, *The Strategy Bridge: Theory for Practice* (Oxford: Oxford University Press, 2010); Ryan D. Grauer, *Commanding Military Power: Organizing for Victory and Defeat on the Battlefield* (Cambridge: Cambridge University Press, 2016); Jason Lyall, *Divided Armies: Inequality and Battlefield Performance in Modern War* (Princeton, NJ: Princeton University Press, 2020).

[99] Stephen Biddle, *Military Power: Explaining Victory and Defeat in Modern Battle* (Princeton, NJ: Princeton University Press, 2004), 28.

successes are possible; but where the gap is small, attritional warfare is likely to result. Second, and building on the point just made, the idea of manoeuvre warfare builds a false distinction between manoeuvre and attrition. Context, including politics, terrain, force-to-space ratios, limitations in friendly forces, and, critically, the actions and capabilities of the enemy, may limit the ability to apply the prescriptions of manoeuvrism. Contemporary examples make this point. Fighting in Ukraine, urban operations in Syria, and the campaign against Islamic State have all illustrated 'the continued efficacy of positional and attrition warfare'.[100]

For these reasons, manoeuvre warfare may simply be imagined. For many, it certainly is not reflected in the reality of military force structures or operations. The US Army, critics argue, remains focused on firepower and attrition;[101] its concept of 'manoeuvre' actually still wedded to physical movement and fires.[102,103] Nor is MDO a specifically manoeuvrist concept.[104] The US Army still anticipates a future in which physical manoeuvre and closing with the enemy are critical.[105] But why, then, has manoeuvrism had such an impact on Western doctrines, and why might it continue to do so in the future?

One reason is because manoeuvre warfare is consonant with Western concepts of military skill. Essentially, attrition and positional warfare are seen as symptomatic of failure and incompetence, a view reflected in Winston Churchill's observation that: 'Battles are won by slaughter and manoeuvre. The greater the general, the more he contributes in manoeuvre, the less he demands in slaughter.'[106] Another is that manoeuvre warfare provides an apparent solution to the political and material limitations faced by Western armies: increasingly, they are too small to withstand prolonged attrition, and society is unwilling to sanction heavy losses. Explicitly, manoeuvre warfare is founded upon the idea that qualitative superiority can compensate for quantitative weaknesses and deliver success more rapidly and at lower

[100] Amos C. Fox, 'A Solution Looking for a Problem: Illuminating Misconceptions in Maneuver-Warfare Doctrine', *RealClear Defence*, 2 February 2018. https://www.realcleardefense.com/articles/2018/02/02/a_solution_looking_for_a_problem_113002.html.

[101] Brian Clark, Dan Patt, and Harrison Schramm, 'Mosaic Warfare: Exploiting Artificial Intelligence and Autonomous Systems to Implement Decision-Centric Operations' (Center for Strategic and Budgetary Assessments, 2020), iii. https://csbaonline.org/uploads/documents/Mosaic_Warfare.pdf.

[102] Leonhard, *The Art of Maneuver*, 235.

[103] Jeff Becker and Todd Zwolensky, 'Making Sense of Military Doctrine: Joint and Service Views on Maneuver', *War on the Rocks*, 3 July 2014, https://warontherocks.com/2014/07/making-sense-of-military-doctrine-joint-and-service-views-on-maneuver/.

[104] Michael Gladius, 'The Case Against Maneuver Warfare', War on the Rocks, 1 April 2019, https://smallwarsjournal.com/jrnl/art/case-against-maneuver-warfare.

[105] Shawn Woodford, 'Multi-Domain Battle and the Maneuver Warfare Debate', *The Dupuy Institute*, 20 February 2017, http://www.dupuyinstitute.org/blog/2017/02/20/multi-domain-battle-and-the-maneuver-warfare-debate/.

[106] *Greenhill Dictionary of Military Quotations* (Barnsley: Greenhill, 2020), 274.

costs.[107] Manoeuvre warfare also reflects a failure of imagination. It is a projection onto the future of the sorts of wars that we would like to fight. As General John R. Galvin notes: 'We arrange in our minds a war we can comprehend on our own terms, usually with an enemy who looks like us and acts like us.'[108] Thus, for hyper-critics, manoeuvre warfare is not a meaningful concept or doctrine: it is for Western militaries a religion, a panacea, and 'a solution looking for a problem'.[109]

Conclusion

Thus, the relevance in the future of manoeuvre warfare remains contested. For some, manoeuvre warfare will remain relevant because it has demonstrably stood the test of time. It is a valuable concept that is agnostic to the character of conflict. Indeed, it is our failure to understand and execute manoeuvrism that explains many of our recent defeats. In Afghanistan, for example, it has been argued that 'it was the insurgents who truly practised the Manoeuvrist Approach through initiative, surprise and the leveraging of their superior information'.[110] For others, these deficiencies become even more important because the changing character of conflict embodies trends that will make manoeuvre warfare even more powerful in the future. In particular, AI and machine learning promise to deliver the capabilities to dramatically increase the tempo of war. Concepts such as Multi-Domain Operations will allow us to harness these new possibilities and make manoeuvrism even more powerful.

On the other hand, critics have argued that we cannot presume that MDO-type futures will indeed be the futures that we get. Moreover, the efficacy of manoeuvre warfare may be challenged in the coming years by the salience of contexts such as limited war, and irregular and urban operations. At a more fundamental level, many challenge the existence at all of a distinctly manoeuvrist style of warfare. Indeed, on this basis we may need to separate the concept of 'relevance' from that of 'utility'. Manoeuvre warfare may continue to have relevance, in the sense that it will continue to form an important part of the lexicon of modern land warfare. This relevance is no guarantee, however, that manoeuvre warfare actually will work.

---

[107] Garfinkel and Dafoe, 'Artificial Intelligence'.

[108] Brian McAllister Linn, *The Echo of Battle: The Army's Way of War* (Harvard, MA: Harvard University Press, 2009), 4.

[109] Fox, 'A Solution Looking for a Problem'.

[110] Nick Reynolds, Performing Information Manoeuvre through Persistent Engagement (RUSI, June 2020), 4. https://rusieurope.eu/sites/default/files/20200611_reynolds_final_web.pdf.

3

Commanding Contemporary and Future Land Operations

What Role for Mission Command?

Niklas Nilsson

Introduction

Western land forces are undergoing adaptation in the face of new demands and possibilities surrounding armed conflict. This takes place in an operational environment that is increasingly conceived as an integrated conglomeration of threats, assets, and capabilities, beyond those provided by land, maritime, and air forces, also including space and cyberspace, as well as the electromagnetic spectrum and the information environment. The perceived complexity of conducting land operations in this environment is underscored by the solutions fielded in response, most of all the US Army's concept of multi-domain operations. Adding to this complexity is the ever-growing availability of information on all aspects of military activity, requiring considerable resources for collection and analysis, coupled with rapid developments in weapons systems, information, and digitized communications technology, satellite surveillance as well as artificial intelligence and automation.

The challenge posed by high-technological near-peer adversaries such as China and Russia, as well as the possibilities emerging from new technological innovations, have induced a drive toward the convergence of capabilities and synchronization of actions between military branches, other national services, as well as allied and partner states. These trends stem from the imperative of adapting to an emerging era of great-power competition across the conflict spectrum. This process will undoubtedly have significant implications for the current and future command of land forces. Yet exactly how these consequences will materialize remains an open question as the evolving operational environment places varying and frequently contradictory

Niklas Nilsson, *Commanding Contemporary and Future Land Operations*. In: *Advanced Land Warfare*.
Edited by Mikael Weissmann and Niklas Nilsson, Oxford University Press. © Niklas Nilsson (2023).
DOI: 10.1093/oso/9780192857422.003.0003

demands on command systems. The incentive to attain unprecedented levels of coordination between military and other resources available to Western states suggest that command systems must follow suit through increased centralization. Yet simultaneously, decentralized command structures that can develop capable tactical commanders prone to initiative are still considered to constitute a crucial component of manoeuvre warfare capability and the capacity to cope with the uncertainty that remains an unavoidable feature of war.

This chapter examines the concept of mission command, a decentralized command philosophy with adjacent methods and practices that is formally embraced by land forces across the West, in light of ongoing trends in the evolution of warfare and military operations. The chapter starts with a discussion of mission command in terms of a culture or command philosophy, and as a set of methods and practices of command. This distinction is important, since the view of mission command as a decentralized method allows it to be combined or replaced with other command methods, whereas a fundamental view of mission command as a culture is more rigid and inflexible. The chapter then discusses the role and future utility of mission command in light of developments in three broad areas that are of central importance to the evolution of military command in general. These are, first, general trends in the current and future operational environment with implications for the command of land operations, with a focus on the US Army's concept of Multi-Domain Operations (MDO). The second area is the ever-increasing demand for information management, and the daunting challenge it poses for any military command system. The third area concerns developments in information technology over the last decades and the more recent but very rapid shift toward artificial intelligence and automation, which together have opened new horizons, as well as vulnerabilities, to military command.

Mission Command: Culture or Method?

After the Second World War, the utility and effectiveness of mission command in German and earlier Prussian warfare was thoroughly analysed in militaries across the western world.[1] As Eitan Shamir has shown,

---

[1] Martin van Creveld, *Fighting Power: German and US Army Performance, 1939–1945* (Westport: Greenwood Press, 2007); Bruce I. Gudmundsson, *Stormtroop Tactics: Innovation in the German Army, 1914–1918* (New York and London: Praeger, 1989); Martin Samuels, *Command or Control? Command, Training and Tactics in the British and German Armies, 1888–1918* (London: Frank Cass, 1995); Bruce Condell and David T. Zabecki, *On the German Art of War: Truppenführung* (Boulder and London: Lynne Rienner, 2001).

these interpretations gave rise to various approaches to implement mission command practices, depending on the historical heritage and cultures of different militaries.[2]

The pragmatic and efficiency-based arguments for mission command draw on the Clausewitzian fog of war. The dictum of manoeuvre warfare puts a premium on speed in decision making and action (in accordance with Boyd's famous OODA-loop).[3] Since the battle is assumed to be unavoidably chaotic, higher commanders cannot expect to gain an adequate overview of events sufficiently quickly to identify fleeting windows of opportunity. Instead, lower commanders in direct contact with the battle should be allowed to decide and act, without seeking direct permission from their superiors. Thus, instead of commanding through direct and detailed orders, higher commanders are to communicate intent, the overall objectives that the mission is to fulfil. This has concrete implications for command practices, prescribing brief operational plans and orders defining what to achieve but not how to achieve it, in conformity with mission command principles. Subordinate commanders are expected to acquire a deep understanding of the mission's purpose, preconditions for accomplishment and limitations. Based on this understanding, they should then exercise initiative and creativity to fulfil the mission, even by means that contradict existing orders, should these become obsolete in the course of fighting.[4] In his historical study of command in war, Martin van Creveld identified the principles consistent with mission command to provide a timeless advantage for command systems organized to exercise them.[5]

For the purposes of this chapter, it becomes important to distinguish between a deeper understanding of mission command as an institutionalized leadership philosophy, that is, a distinct culture of leadership engrained in a military organization and embraced by the officers working in it, and a more superficial reading of the command methods and practices associated with mission command.

The cultural perspective on mission command draws on the antecedent German *Auftragstaktik*, emphasizing battlefield command practices as an effect of the socialization of the officer corps into a certain professional ethos emphasizing the centrality of responsibility and initiative, from the

---

[2] Eitan Shamir, *Transforming Command: The Pursuit of Mission Command in the U.S., British, and Israeli Armies* (Stanford: Stanford Security Studies, 2011).

[3] William S. Lind, *Maneuver Warfare Handbook* (New York: Routledge, 1985).

[4] B. A. Friedman, *On Tactics: A Theory of Victory in Battle* (Annapolis: Naval Institute Press, 2017).

[5] Martin van Creveld, *Command in War* (Cambridge, MA: Harvard University Press, 1985).

earliest stages of their career.[6] In this view, employing mission command as a leadership philosophy places far-reaching demands on a military organization beyond the ability to exercise command in battle. Mission command is enabled by a cultural environment defined by mutual trust and common understanding among superiors and subordinates, which in turn relies on professionalism and skill as well as a permissive approach to risk-taking and creative problem-solving.[7] The ability of junior officers to exercise initiative and judgement should not only be expected in certain situations, it should be actively encouraged and enabled by the organization, in all aspects of their professional work as well as in training and education.[8] This cultural or philosophical perspective depicts mission command as an all-encompassing practice where the leadership philosophy cannot easily be separated from the methods and techniques utilized for command, and where the advantage acquired from the decentralization of command cannot be achieved without also embracing and institutionalizing mission command as a culture. This has far-reaching consequences for the officer corps as a collective as well as the military organization in which they serve.[9] The cultural view of mission command places the human resource of highly trained and capable officers and investments into education and training at the centre of military capability.[10]

A narrower, managerial, understanding of mission command as encompassing a set of command methods posits a much more malleable concept, reducible to a question of locating the mandate to make decisions, and thereby to determining the appropriateness of centralized or decentralized command depending on operational circumstances.[11] In a methodological and pragmatic perspective, there is no direct contradiction between centralized and decentralized command in a military organization that

---

[6] Jörg Muth, *Command Culture: Officer Education in the U.S. Army and the German Armed Forces, 1901–1940, and the Consequences for World War II*, 1st edn (Denton: University of North Texas Press, 2011).

[7] Shamir, *Transforming Command.*

[8] Donald E. Vandergriff, *Adopting Mission Command: Developing Leaders for a Superior Command Culture* (Annapolis: Naval Institute Press, 2019).

[9] Niklas Nilsson, 'Practicing Mission Command for Future Battlefield Challenges: The Case of the Swedish Army', *Defence Studies* 20, 4 (2020): 436–452; Joseph Labarbera, 'The Sinews of Leadership: Mission Command Requires a Culture of Cohesion', in *Mission Command: The Who, What, Where, When and Why, an Anthology*, edited by Donald Vandergriff and Stephen Webber (CreateSpace Independent Publishing Platform, 2017). See also Joseph Labarbera, 'Planting the Seed', in *Mission Command: The Who, What, Where, When and Why: An Anthology, Volume II*, edited by Donald Vandergriff and Stephen Webber (CreateSpace Independent Publishing Platform, 2018); Peter. C. Vagnjel, 'Mission Comman', in *The Who, What, Where, When and Why: An Anthology, Volume II*, edited by Donald Vandergriff and Stephen Webber (CreateSpace Independent Publishing Platform, 2018).

[10] Donald E. Vandergriff, 'How to Develop for Mission Command: The Missing Link', in *Mission Command: The Who, What, Where, When and Why, an Anthology*, edited by Donald Vandergriff and Stephen Webber (CreateSpace Independent Publishing Platform, 2017).

[11] David S. Alberts and Richard E. Hayes, *Understanding Command and Control* (Washington, DC: CCRP Publications, 2006).

fundamentally embraces the principles of mission command. Rather, the allocation of decision mandates is determined by the need for coordination in order to resolve particular operational tasks.[12] The guidance provided by mission command as a philosophy nevertheless holds that decision mandates should be no more centralized than necessary to resolve the task at hand. Thus, mission command denotes a subsidiarity principle of military command.

The cultural and methodological aspects of mission command may be inseparable in theory. The ability to exercise mission command and utilize its full potential arguably requires the officer corps to become socialized into this way of being and acting in their organizational environment. Thus, fully institutionalizing mission command denotes making this philosophy an ingrained part of the culture of the military organization, implying a commitment that is far broader than conduct during combat, and that addresses human interaction within the organization relating to basically all aspects of military work, including peacetime tasks, education, and training.

However, mission command is frequently addressed in separation from its cultural side and it is easy to get the impression that it can be reduced to an issue of practices regarding decision-making and command. This is a methodological perspective on mission command, focusing on the speed and efficiency gained from the decentralization of authority. Mission command then becomes one option among several, suited to certain types of tasks but not others. Whereas the legitimacy of mission command flows from its ability to execute speedy and expedient decision-making in an operational environment defined by uncertainty and blurred situational awareness, the increasing technology-enabled ability to monitor the battlefield, and to communicate and command from a distance, has led several analysts to argue for a more limited role of mission command in future command systems.[13] We will return to the question of information management and situational awareness later in the chapter.

The question, then, is what place mission command has, or can have in contemporary and future military command. As Anthony King has observed, the age of the individual military genius, embodied in the example of Erwin Rommel, has passed. He and other legendary German generals of the time enjoyed maximum freedom of action, accomplishing loosely defined

---

[12] Michael Flynn and Chuck Schrankel, 'Applying Mission Command through the Operations Process', *Military Review* (March–April 2013).

[13] Andrew Hill and Heath Niemi, 'The Trouble with Mission Command. Flexive Command and the Future of Command and Control', *Joint Force Quarterly* 86, 3rd quarter (2017): 94–100. See also Robert R. Leonhard, *Fighting by Minutes: Time and the Art of War* (Westport and London: Praeger, 1994).

missions by rapid movement and manoeuvre and deep penetration behind enemy lines. Their corresponding contemporaries such as Jim Mattis or Stanley McChrystal operated in far more complex operational environments, on missions requiring consideration of a much larger set of variables; and in effect exercised a largely different form of command. Indeed, modern command has become a more collective exercise, supported by large numbers of professional specialists and enacted through trusted deputies.[14]

The distinction between, on the one hand, the philosophy and culture of mission command, and the implementation of mission command principles into a command system, on the other, have important implications for the command of contemporary and future land operations. Trends in military planning over several decades have sought to address the increasing complexity of military operations. This complexity stems in large part from the imperatives of managing increasingly abundant information flows and integrating new technologies with warfighting capabilities. In terms of command, these trends can be considered attempts to cut through the fog of war and to vastly improve situational awareness at the higher levels of command. In this light, the pragmatic argument in favour of mission command as an all-encompassing practice has increasingly become diluted and subjected to competing visions of future command.

Commanding Land Operations in the Contemporary and Future Operational Environment

Several significant events in the 2010s have prompted rethinking of the future requirements and utilization of Western land forces.[15] Russia's 2014 invasion of Ukraine and its subsequent standoff with NATO, as well as an increasingly assertive Chinese posturing in the eastern Pacific and beyond, have prompted US and NATO forces to reconsider the possibility that future military operations may involve combat against near-peer adversaries on a scale not conceived since the Cold War.[16] This has obviated the need to organize and equip Western land forces to execute military operations facing a radically different type of opponent than during the campaigns

[14] Anthony King, 'Mission Command 2.0: From an Individualist to a Collectivist Model', *Parameters* 47, 1 (2017): 7–19.
[15] Note that this chapter was written before Russia's full-scale invasion of Ukraine on February 24, 2022. It therefore does not consider the additional implications for Western land forces drawn from this war.
[16] Jim Mattis, *Summary of the 2018 National Defense Strategy of the United States of America: Sharpening the American Military's Competitive Edge* (Washington, DC: US Department of Defense, 2018); *Brussels Summit Communiqué*, NATO, 14 June 2021, https://www.nato.int/cps/en/natohq/news_185000.htm.

in Iraq and Afghanistan. Of course, this prospect involves challenges of a different magnitude from those faced during military operations in Iraq, Kosovo, Somalia, Afghanistan, or Libya. These near-peer adversaries are not only numerically strong; they also place much effort and resources into the development of advanced technology in order to achieve layered standoff capabilities, in the form of anti-access/area denial (A2/AD) systems, as well as advanced electronic warfare systems, unmanned air and land systems, and automated warfare capabilities. The evolution of anti-satellite technology and capabilities to conduct extensive hostile operations in cyberspace have added new dimensions to an already complex picture of how a future conflict between major powers might unfold.[17]

The recalibration of Western forces to face near-peer adversaries in the form of Russia, China, or high-technological regional powers like North Korea or Iran raises a large number of challenges. The execution of future, high-intensity land operations is envisioned to require extensive synchronization of forces in the land, air, and maritime domains for the convergence of effects. It also envisions integration of actions and defence in space and cyberspace, the electromagnetic domain, and the informational environment. Moreover, it has become increasingly recognized in western military thinking that much of the antagonistic competition in world politics takes place below the threshold of war—and that Russia and China in particular are increasingly refining strategies and tactics to pursue their interests vis-à-vis the West by means that will not trigger a military response. Whilst operating in the grey zone is by no means a new phenomenon per se, it has not been until recently that Western militaries have seriously sought to address doctrinally the problem of an increasingly blurred demarcation between war and peace, and identifying the demands placed on military forces in this operational context. The response to these challenges originating in and driven by the US Army is conceptualized as multi-domain operations (MDO)—integrating the designated five domains of land, sea, air, space, and cyberspace.[18]

Attempts to create synergies by integrating capabilities have a long history, and is indeed the purpose of joint operations. MDO nevertheless takes this thinking to new levels, both by including new dimensions of warfare and by envisioning ever-closer coordination between services as well as allies and

---

[17] Terrence K. Kelly, David C. Gompert, and Duncan Long, *Smarter Power, Stronger Partners*, Rand research reports RR-1359-A (Santa Monica: RAND Corporation, 2016–2017).

[18] James C. McConville, *Army Multi-Domain Transformation: Ready to Win in Competition and Conflict*, Chief of Staff Paper #1 (Washington, DC: Headquarters, Department of the Army, 16 March 2021); *The U.S. Army in Multi-Domain Operations 2028* (Washington, DC: TRADOC, US Army, 6 December 2018).

partners. The concept intends to generate unprecedented synchronization between activities in the five domains, presenting adversaries with constantly shifting dilemmas and converging the effects generated to sustain maximum damage to the adversary's will and cohesion.[19] Indeed, the concept of MDO can be characterized as addressing a complex problem with a highly complex solution.

MDO and Mission Command

Both the envisioned character of future major power confrontation and the conceptual response to the new operational environment present new specific challenges for Western land forces. Military confrontation between peer adversaries will imply dispersed high-intensity fighting in large theatres, which will complicate centralized planning and exercise of control. On the future battlefield, land forces will increasingly have to operate dispersed in order to improve survivability in the face of long-distance high-precision munitions and standoff fires. Improved capabilities in the electromagnetic spectrum as well as the cyber domain will increase the likelihood of interrupted communications. A2/AD capabilities will place limits on the possibilities of air support, implying that land, rather than air forces, might in some circumstances need to spearhead operations. All of these factors imply that the premium on forces proficient in the competent exercise of mission command, as a means for coping with uncertainty and building capacity for initiative and independent decision-making, will increase.

Yet simultaneously, the envisioned nature of MDO also implies drastically increased demands for coordination, synchronization, information processing, and situational understanding. It is far from certain that these requirements will be compatible with the decentralized vision of leadership implied by mission command.[20] As pointed out above, a methodological perspective on mission command does not preclude centralized decision making when prudent, that is, when the operational situation requires a high degree of coordination in order to obtain desired effects. However, the synchronization required in order to fulfil the potential of MDO suggests that high-level coordination will presumably be the rule rather than the exception. The unprecedented ambition to achieve convergence arguably puts a premium on

[19] Robert B. Brown and David G. Perkins, 'Multi-Doman Battle: Tonight, Tomorrow, and the Future Fight', *War on the Rocks*, 18 August 2017, https://warontherocks.com/2017/08/multi-domain-battle-tonight-tomorrow-and-the-future-fight/.

[20] Conrad Crane, 'Mission Command and Multi-Domain Battle Don't Mix', *War on the Rocks*, 23 August 2017, https://warontherocks.com/2017/08/mission-command-and-multi-domain-battle-dont-mix/.

centralization, delimiting the room for decentralized command. Moreover, whilst this will logically favour a further concentration of decision mandates at the higher levels of command, the MDO concept also envisions increased needs for horizontal coordination, beyond combined arms, underscoring the need to integrate capabilities and forces provided by services and agencies other than the military, as well as by other allies and partners.[21]

The acknowledgement and attention paid to competition 'below the threshold' also raises important questions regarding the command and control of military forces in general. The potential consequences of miscalculation and unwanted escalation in an ambiguous environment are arguably far larger today than 10–20 years ago, and are becoming comparable to the Cold War. Yet particularly after the Cuban missile crisis in 1962, robust safeguards were constructed to reduce the risk of unintended escalation between the two major power blocs. The current global strategic environment is far more ambiguous and uncertain, includes a larger number of state and non-state actors, and increasing fluidity between different means for aggression and retaliation. Against this backdrop, regardless of what roles and functions that land forces will fulfil in the grey zone, these will in all likelihood require very well thought through rules of engagement, which constitutes yet another motive for increased centralized control.

Diversity and Conformity

Another challenge pertains to US allies and partners. Realizing the full potential of MDO essentially presumes the capability to dominate and shape the future battlefield, which is extremely ambitious and relies on the formidable military and technological resources at the disposal of the US military. MDO is designed in response to perceived challenges to US military supremacy and to observed developments in contemporary antagonistic competition and warfare and no other military force, including near-peer competitors, possess corresponding means. Yet MDO also envisions close coordination with NATO allies and partners, whose capabilities are far more limited. For interoperability purposes, these must take account of the MDO concept in their development of doctrine and organizations. Allies and partners also differ considerably in their political strategic outlook and thus in their motivation and ability to expend the resources required to contribute to MDOs.[22] More-

[21] Mark Balboni, John A. Bonin, Robert Mundell, and Doug Orsi, *Mission Command of Multi-Domain Operations* (Carlisle: US Army War College, 2020).
[22] Jack Watling and Daniel Roper, *European Allies in US Multi-Domain Operations* (London: RUSI, 2019).

over, effective synchronization of allied forces will require the development of infrastructure and systems for providing a common understanding of the operational environment and an integrated infrastructure for information sharing and situational awareness, which is both demanding and controversial.[23] The question therefore is to what extent MDO as a construct can provide a one-size-fits-all solution and in a more general sense, to what extent the conceptualization of challenges and solutions devised by large military powers are also applicable and workable in smaller states.

Importantly, the technologically enabled information supremacy and control of the fragmented battlefield that is envisioned to enable MDO may not be available to smaller allies and partners located at the frontline in a confrontation with a peer adversary. Instead, these must expect to be inferior in terms of manpower as well as technology and will likely need to fight in a highly contested information environment where vertical as well as horizontal coordination is very difficult to attain. These conditions will place a premium on the competent execution of mission command in the 'deep' sense, as a baseline for command that allows military leaders to accomplish missions independently of directions from higher echelons, for extended periods of time.[24] With the increasing focus on MDO and related concepts, and their inherent incentives for synchronization, there is a risk that the capability for independent tactical initiative and decision-making may become degraded, particularly in smaller militaries.

The Informational Challenge

A central problem of contemporary command is the challenge associated with collecting, processing, and acting upon information. Whilst information availability has always been a prominent concern for military decision makers, this has historically been an issue of scarcity—information about the opponent's intentions, strength, movements, etc., and even about the status and location of own forces, has typically been limited and difficult to obtain. However, with the advent of information technology (IT), big data, and cloud computing, and the numerous technologies facilitating intelligence gathering and digital communications for military use, the amount of information available to military decision makers has vastly increased. Command systems

[23] Joseph Soeters and Irina Goldenberg, 'Information Sharing in Multinational Security and Military Operations. Why and Why Not? With Whom and with Whom Not?', *Defence Studies* 19, 1 (2019): 37–48.
[24] Niklas Nilsson, 'Mission Command in a Modern Military Context', *Journal on Baltic Security* 7, 1 (2021): 5–15.

thus risk becoming overwhelmed as the amount of information available vastly exceeds the capacity to process it.[25] It has therefore been suggested that the information problem has gradually become inversed; if previously an issue of access and availability, the problem is now one of prioritization, analysis, and interpretation among multiple data streams, implying that "'the fog of information" is replacing "the fog of war"'.[26] The problem of information management has taken a very concrete expression in the organization of command systems, reflected in an exponential growth in the number of specialist staff in modern headquarters, required in order to operate various command systems and process information.[27]

Of course, the increased demand for information management capacity is in part an outcome of the changing nature of military operations in recent decades. The coalition campaigns in Afghanistan and Iraq in large part consisted of counterinsurgency missions in highly sensitive and mediatized environments. Friendly or civilian losses have potentially been very costly, not only in terms of human life but also in terms of the strategic preconditions for these campaigns, risking the erosion of local support as well as political and public acceptance for these operations back home.[28] It has therefore become increasingly important to analyse and assess the consequences of actions across the tactical, operational and strategic levels. These conditions have increased the demands for operational situational awareness and comprehensive analysis. However, it has also been pointed out that these circumstances have inhibited action in uncertain situations, and contributed to risk aversion.[29]

When it comes to establishing situational awareness, the central challenge is to develop a relevant understanding of the situation by establishing which information is relevant and consequential in a vast flow of data from a myriad of sources, a majority of which may very well be irrelevant, inaccurate, misleading, or false. Indeed, the complexity of decision-making in the contemporary information environment has opened new opportunities for perception management and deception available to opponents

[25] Anthony King, *Command: The Twenty-First-Century General* (Cambridge: Cambridge University Press, 2019).
[26] Mie Augier, Thorbjorn Knudsen, and Robert M. McNab, 'Advancing the Field of Organizations Through the Study of Military Organizations', *Industrial and Corporate Change* 23, 6 (2014): 1417–1444.
[27] Jon R. Lindsay, *Information Technology and Military Power* (Ithaca, NY: Cornell University Press, 2020).
[28] Rupert Smith, *The Utility of Force: The Art of War in the Modern World*, 1st Vintage Books edn (New York: Vintage Books, 2008).
[29] Ad L.W. Vogelaar and Eric-Hans Kramer, 'Mission Command in Dutch Peace Support Missions', *Armed Forces & Society* 30, 3 (2004): 409–431; Jim Storr, 'A Command Philosophy for the Information Age: The Continuing Relevance of Mission Command' *Defence Studies* 3, 3 (2003): 119–129.

as well as friendly forces. The speed and multitude of information flows, as well as the dependence on systems for information management, have enhanced the opportunities to target the opponent's situational awareness by the dissemination of distorting information and narratives. The challenge thus has less to do with the amount of information available than with the ability to identify the information critical to making decisions in the midst of the vast amount of noise that constitutes the information environment. As an effect, information management has become increasingly time- and resource-consuming, as well as technology dependent, whilst any deficiencies in this capability become a source of vulnerability.[30]

The Information Environment and Mission Command

Regarding the prospect of practising mission command in a military operational environment demanding increasing coordination, and a military decision process increasingly dependent on the capacity to manage information flows, the key question is to what extent it will be possible to delegate decision mandates in such an environment. The complexity of commanding contemporary land forces in high-intensity warfare creates a contradiction between the need for centralized control and decentralized mandates to decide and act. As Van Bezooijen and Kramer note, whilst decentralized command is suitable for networked military operations, this requires low interdependencies between networked units.[31] Conversely, vertical synchronization of domain effects as well as horizontal self-synchronization would seemingly require more interdependence between units. In other words, the accumulated effects drawn from coordination conflicts with the speed and efficiency stemming from autonomy.

The distribution of information within an organization is closely interconnected with the possibility of distributing decision mandates.[32] Thus, expedient tactical decision-making requires that the level of command possessing the most accurate situational understanding should be granted mandates to decide and act. Proponents of mission command argue that this situational understanding necessarily rests with commanders in immediate contact with the situation on the ground. In principle, the commander directing the actual combat will be best placed to identify fleeting opportunities

---

[30] Christopher Paul, *Improving C2 and Situational Awareness for Operations in and Through the Information Environment* (Santa Monica: RAND Corporation, 2018).

[31] Bart van Bezooijen and Eric-Hans Kramer, 'Mission Command in the Information Age: A Normal Accidents Perspective on Networked Military Operations', *Journal of Strategic Studies* 38, 4 (2015): 445–466.

[32] Alberts and Hayes, *Understanding Command and Control*, 82–83.

and make the right decisions at the right time to exploit them. If decision mandates are placed with higher commanders, deferring to these in order to obtain necessary permissions will consume time and provide for a slower and unnecessarily static decision process.

The counterargument is that command in modern land warfare requires the management of a virtually unlimited amount of information in order to achieve a sufficient level of situational understanding to make decisions on appropriate courses of action and assess their potential consequences. Moreover, land operations are only one dimension of the operational environment which, aside from encompassing all three service arms, with the emergence of MDO also includes the space and cyber dimensions. Therefore, significant decisions in battle are thought to become ever more complex and therefore increasingly beyond the independent capability of tactical commanders.[33] Indeed, some analysts have argued that these emergent realities of military command have contributed to the obsolescence of mission command in the traditional sense, and that speed and accuracy in decision-making require ever-closer interdependence between different levels of command, rather than allowing subordinates autonomy.[34]

However, the fact that the evolution in warfighting has necessitated an evolution also in command systems does not negate the continued relevance and advantages of decentralized decision-making. The future utilization of mission command ultimately depends on the acquisition and dissemination of situational awareness. If information is exclusively accumulated and processed at the top levels of command and stays there, this exacerbates the tendency towards centralized decision-making. If, on the other hand, information is shared, disseminated and allocated where it is most acutely needed, this equips tactical commanders with the means to assess developments on the battlefield in light of the overall operational situation, and to act on opportunities as they present themselves.[35]

One alternative, or middle ground, that has been suggested is flexible (flexive) command, whereby the military organization can shift between centralized and decentralized command depending on the demands of the situation. In this perspective, centralized command is appropriate for situations in which higher command levels possess supreme situational understanding and are therefore positioned to coordinate effects across domains.

---

[33] Jesse Skates, -Multi-Domain Operations at Division and Below-, *Military Review* (January–February 2021): 68–75.

[34] King, 'Mission Command 2.0: From an Individualist to a Collecivist Model'.

[35] Sonia Lucarelli, Alessandro Marrone, and Francesco Niccolò Moro, eds, *NATO Decision-Making in the Age of Big Data and Artificial Intelligence* (Brussels: NATO HQ, 2021), 40–41.

Conversely, command is envisioned to become decentralized in uncertain situations, where higher command lacks sufficient overview.[36] Whilst flexible command appears to be an attractive alternative, at least in theory, allowing for adaptation to the situation at hand, there is also a tendency to present decentralization as a sort of reserve alternative, to be applied in situations where the expected informational supremacy is lacking. Indeed, this view ascribes a questionable 'normality' to certainty, and overconfidence in the premise that things will go according to plan. Moreover, a truly effective utilization of flexible command would require forces to spend equal amounts of time exercising both, as well as the transition between them. Given the overall thrust of MDO towards integration and coordination, this seems unlikely in practice. Instead, forces will likely spend more effort exercising the 'normal' scenario, which will consume a lot of time given its complexity. Moreover, the 'shift' between circumscribed and open decision mandates requires a cognitive shift that needs serious preparation. As Finkel has argued, the development of flexibility in military forces requires a considerable integrative effort in education, training, and exercises. Otherwise, forces are likely to function in the way they are most used to functioning—and if centralized command is predominating and normalized in peacetime, this will be the case also in war.[37]

Command, Communication, and Technology

The capacity for communication is central to the coordination of any military operation. Yet maintaining this capacity constant grows ever more important in highly synchronized and integrated operations. It is today possible to exert command from far out of theatre, including from a different continent, for example from a command central in the Pentagon in direct contact with theatre commanders in Iraq or Afghanistan. The more centralized a command structure becomes, the more it will depend on functioning communications to direct subordinated units. Whilst communication capabilities have evolved substantially over the last two decades, and have obviously been adapted for the type of operations undertaken, they will nevertheless have important limitations in a confrontation with a peer adversary. In this regard, communications will almost certainly be contested, which has not been the case in combat against low-technological opponents. Electronic warfare

---

[36] Hill and Niemi, 'The Trouble with Mission Command'.
[37] Meir Finkel, *On Flexibility: Recovery from Technological and Doctrinal Surprise on the Battlefield* (Stanford, CA: Stanford University Press, 2011).

capabilities, cyber warfare, and the prospective use of electromagnetic pulse weapons imply that the reality of units having to pursue objectives without direct guidance from higher command will be a normal precondition during operations, rather than an exception. Command and control (C2) nodes will be likely priority targets for attack, kinetic or otherwise, whereas overseas communications are vulnerable to attacks on vital infrastructure (such as undersea cables or satellite communication infrastructure). Thus, whilst the MDO concept will imply an increased dependency on constant communication, this also highlights a key vulnerability of centralized command.[38]

As noted in the previous section, access to information and situational awareness is key to the location of decision mandates, and thus a decisive factor to the degree of command centralization. The rapid advances in information technology and digitalization in recent decades have provided for an extensive transformation of the means by which information can be acquired, processed, and disseminated within a military organization. Ideally, this accessibility should facilitate flexible command by providing different command levels with a common situational understanding, enabling both centralized coordination toward a common purpose at the operational level and initiative at the tactical level. Integrated communication systems should also facilitate horizontal coordination within the organization, allowing units to self-coordinate and assume responsibility for appropriate action. Yet for this integrated situational understanding to emerge, and for it to be compatible with the decentralized decision-making envisioned in mission command, information must reach those levels of command where decisions are to be taken. This has proven difficult to achieve and conversely to its potential, the implementation of advanced digitalization and the centralized accessibility to weapons systems and intelligence, have frequently worked in favour of centralized decision-making practices in the form of detailed command and tendencies towards micromanagement.[39]

The introduction of technology permitting real-time updates on battlefield developments, including GPS-tracking, drones, and long-distance satellite communications, have enabled commanders at higher echelons to closely monitor and steer subordinates in detail. And when these abilities become available, they have also demonstrably been utilized in this way.[40] For example, it has been recorded during exercises that units equipped with Blue

[38] Miranda Priebe, Douglas C. Ligor, Bruce McClintock, Michael Spirtas, Karen Schwindt, Caitlin Lee, Ashley L. Rhoades, Derek Eaton, Quentin E. Hodgson, and Bryan Rooney, *Multiple Dilemmas: Challenges and Options for All-Domain Command and Control* (Santa Monica, CA: RAND Corporation, 2020), 12.

[39] Lucarelli et al., *NATO Decision-Making in the Age of Big Data and Artificial Intelligence*, 40.

[40] Thomas S. Sowers, *Nanomanagement: Superior Control and Subordinate Autonomy in Conflict: Mid-Level Officers of the U.S. and Britisk Armies in Iraq (2003–2008)* (London: LSE, PhD Thesis, 2011).

Force Tracker are much more likely than others to receive direct orders from higher command levels.[41]

In a more profound sense, the increasing reliance on technology for information processing and decision support builds on a problematic assumption regarding command, namely that the key challenge for military decision-making is to accumulate, interpret, and understand sufficient amounts of information in order to make optimal decisions and then communicate these decisions to subordinates. This assumption is problematic for at least two reasons. First, it suggests that military decision-making requires a certain amount of available information. This is of course true to an extent, yet it also puts a premium on information supremacy that is likely to inhibit action in uncertain situations. Whilst the purpose of sensor systems and data processing capacity is to eliminate uncertainty to the extent possible, uncertainty will always remain an element in military decision-making. Therefore, efforts to oversee and control developments on the battlefield risks diverting attention from the remaining necessity of being able to take advantage of fleeting opportunities in complex and uncertain situations. The key question is what degree of situational awareness is sufficient to make an informed decision, and what investment of time is acceptable in order to achieve it.[42]

Second, the informational requirement also risks inhibiting initiative at the lower levels of command. Whilst situational awareness at the top requires subordinate units to constantly relay information upwards in the decision-making system, it simultaneously becomes increasingly unlikely that supreme levels of command will accept independent action without direct approval. The necessity of clarifying the big picture implied by the integrated battlefield requires the capacity to assess the consequences of actions on such a broad scope that tactical commanders will unlikely be able to independently identify opportunities when they appear, or anticipate the consequences of acting on them. There is thus a risk that a technology-driven command centralization will reduce the sense of ownership of the mission at the tactical level, and the sense of responsibility to take initiatives in its favour.[43]

In sum, the assumption that the answer to uncertainty is simply more capacity to process information, and priority being given to developing technologies for this purpose, risks contributing to constructing command structures that are skewed towards highly centralized and detailed command practices.

---

[41] Augier, Knudsen, and McNab, 'Advancing the field of organizations through the study of military organizations', 1431.
[42] Paul, *Improving C2 and Situational Awareness for Operations in and through the Information Environment*, 30.
[43] Lucarelli et al., *NATO Decision-Making in the Age of Big Data and Artificial Intelligence*.

Future Technologies

The rapid pace of development in artificial intelligence (AI) also poses new questions as to the features of contemporary and future command. AI for military use is developing along multiple trajectories, and ostensibly has the potential to fundamentally change the nature of military operations. Michael Raska argues that it has become relevant to speak of a sixth wave of Revolutions in Military Affairs, AI-RMA, which 'differs in the magnitude and impact of human–machine interactions in warfare, in which algorithms increasingly shape human decision-making, and future combat is envisioned in the use of AI-enabled autonomous weapons systems'.[44]

Several analysts point out that an AI-driven arms race is underway, in which states and other actors compete in the development of AI technology for military use. As these technologies become deployable on the battlefield and in command structures, they are envisioned to drastically reduce the time required for planning, executing, and responding to military action and will thus become increasingly crucial in offensive and defensive operations alike.[45]

Military organizations are to different degrees placing hope in AI-enabled systems for decision-making support as the answer to the challenge of information management. These systems are envisioned to provide situational awareness at unprecedented speed, relying on algorithms and machine learning. Technology that has in many cases been in civilian use for a long time, are predicted to increasingly become employed to sift through vast quantities of data for relevant information. These systems will prospectively be capable of detecting indicators of change in the strategic and operational environments in order to predict adversary action, and calculate risk and probability pertaining to different courses of action at a fragment of the time required by humans to perform the same tasks, whilst unaffected by human factors such as groupthink, confirmation bias, stress, anger, or fear.[46] Yet reliance on AI-enabled command structures and decision support also have built-in vulnerabilities since the purported situational awareness that is expected to stem from them can be distorted through the intentional input of misleading information, causing these systems to divert commanders' attention from

[44] Michael Raska, 'The Sixth RMA Wave: Disruption in Military Affairs?', *Journal of Strategic Studies* 44, 4 (2021), 456–479.
[45] Jürgen Altmann and Frank Sauer, 'Autonomous Weapon Systems and Strategic Stability', *Survival* 59, 5 (2017): 117–142; Kareem Ayoub and Kenneth Payne, 'Strategy in the Age of Artificial Intelligence', *Journal of Strategic Studies* 39, 5–6 (2016): 793–819; Balboni et al., *Mission Command of Multi-Domain Operations*, 21.
[46] Altmann and Sauer, 'Autonomous Weapon Systems and Strategic Stability'.

significant developments or produce useless or harmful recommendations for action. Whilst human decision makers are by no means immune to deception, distraction, or information overload, the speed with which AI-enabled systems will be able to react and communicate presents new risks of manipulation, misinterpretation, and escalation.[47]

Similarly to the above discussion on information and communication in military command, the increasing utility of AI-enabled command systems may pose a challenge to current practices of mission command. As military decision-making becomes increasingly dependent on AI-enabled decision support, as a consequence of the increasing demands for information processing in a complex operational environment, this will likely also contribute to a further (perceived) reinforcement of situational awareness, and therefore also a concentration of decision-making, up the chain of command. These systems will in all likelihood be more prevalent at the operational level of command than at tactical levels, thereby motivating and allowing for more detailed command from the top. Moreover, the risks involved in relying on automated decision-making, or action based on decision support emanating from these systems, constitute arguments for additional controls and safeguards, which will likely be concentrated to higher levels of command.[48]

Automated Warfare and Robotics

Yet the increasing military utilization of AI and automated systems also potentially holds more profound implications for the future exercise of command. The envisioned fielding of automated robotic systems such as networked missile systems, self-driving vehicles, and UAV swarms removed from direct human control may be a distant prospect, due at least as much to ethical concerns as to technological limitations. However, the prospect of future human–machine interaction that goes beyond the enhancement of human capabilities raises additional questions regarding the utility of mission command. This is inherently a human-centric concept devised and employed as a solution to the social complexity of coordinating large numbers of people towards a common purpose, which may be subject to change over time and subjective in terms of defining success and failure. Machines, in contrast, function according to algorithmic logic. As long as humans remain in charge of the operations of robotic systems, they can be viewed as a piece of machinery. However, the prospect of developing systems that can be granted

[47] Paul, *Improving C2 and Situational Awareness for Operations in and through the Information Environment*, 96, 99; Altmann and Sauer, 'Autonomous Weapon Systems and Strategic Stability'.
[48] Lucarelli et al., *NATO Decision-Making in the Age of Big Data and Artificial Intelligence*, 40–41.

increasingly high levels of autonomy raises the possibility of employing them to perform less well-defined tasks in line with commander intent, reminiscent of mission command, but with much faster reaction times than humans.[49]

Given the limitations to these technologies, this will for the foreseeable future only be applicable to non-complex and logical environments. So far, states have been reluctant to employ AI-systems with autonomous kinetic effects, except for performing very narrowly defined tasks (such as the Phalanx close-in defence system).[50] However, in a future where military technology increasingly relies on autonomous AI and robotics, the interaction and interoperability between humans and machines may require a fundamental rethinking of command practices. This is particularly true for the advent of AI-enabled command systems, developed to empower military commanders, but which may nevertheless interfere with human decision-making. As one analyst has suggested, on a future multi-domain battlefield where a premium is put on an extreme capacity for processing information and acting on it, and where antagonistic actors possess the resources and capabilities for employing increasingly sophisticated AI-enabled solutions to these problems, the exclusively human-centric practice of mission command may become redundant.[51]

There are additional problems associated with the development and adoption of new technologies in relation to the continued applicability of mission command. For example, the centrality of mutual trust and common understanding in the philosophy of mission command is potentially at odds with the envisioned teaming of humans and machines. Even if autonomous systems will in theory become capable of performing increasingly complex tasks, it is questionable whether humans in charge of these systems will endow sufficient trust in these capabilities to take advantage of their full potential.[52] Yet another concern stems from the risk that an exaggerated reliance on technology, for example through the large-scale integration of robotics with army forces, would require extensive retraining and reconceptualization of tactics in order to utilize these systems, which could simultaneously result in a loss of skills in operating without relying on technology.[53]

[49] Robert J. Bunker, *Mission Command and Armed Robotic Systems Command and Control: A Human and Machine Assessment*, Land Warfare Papers 132 (Arlington, VI: The Association of the United States Army, 2020), 6.
[50] Altmann and Sauer, 'Autonomous Weapon Systems and Strategic Stability'; Bunker, *Mission Command and Armed Robotic Systems Command and Control: A Human and Machine Assessment*.
[51] Bunker, *Mission Command and Armed Robotic Systems Command and Control: A Human and Machine Assessment*, 13.
[52] Lucarelli et al., *NATO Decision-Making in the Age of Big Data and Artificial Intelligence*; Bunker, *Mission Command and Armed Robotic Systems Command and Control: A Human and Machine Assessment*.
[53] Jai Galliott, 'The Limits of Robotic Solutions to Human Challenges in the Land Domain', *Defence Studies* 17, 4 (2017): 327–345.

Conclusion

As land operations are perceived to become increasingly complex endeavours, and the solutions devised to coordinate them add to this complexity, this will both exacerbate existing challenges for military command systems and raise new ones. As we enter an era of great-power competition across the conflict spectrum, western militaries, and particularly land forces, must develop the ways and means to make the most of the resources at their disposal, through convergence, synchronization, and the adoption of new technologies. However, the means developed to address this complexity tend to provide an exaggerated picture of the possibility of controlling the operational environment and mitigating uncertainty.

Whilst the envisioned drive towards domain convergence will provide numerous incentives for the centralization of command systems, the approach simultaneously attaches considerable importance to horizontal coordination and networking within military organizations. These solutions will rely heavily on communications technology and bandwith, which may or may not be accessible in the event of high-intensity conflict. Developing these command systems and the training to master them will require considerable time, effort, and resources. There is a risk that this will come at the expense of capabilities to exercise decentralized autonomous command when the operational situation so requires, or when it becomes a necessity in the face of interrupted communications or technology failure. Adapting for future land operations will require truly agile and flexible land forces trained to operate within flexible command systems capable of quickly shifting between tight coordination and open decision mandates. Decentralized mission command will continue to be an important part of these systems. Therefore, neglecting mission command would be highly problematic and would fundamentally imply a loss of capacity for dealing with uncertainty. Regardless of all efforts and resources expended to reduce uncertainty, this will remain a prominent feature of warfare in the foreseeable future, as it has in the past, which will only be exacerbated by the complexity of the emerging operational environment.

4

Combat Logistics in the Twenty-first Century

Enabling the Mobility, Endurance, and Sustainment of NATO Land Forces in a Future Major Conflict

Christopher Kinsey and Ronald Ti

> The line between disorder and order lies in logistics.
> **(Sun Tzu)**

Introduction

What Sun Tzu understood 2,500 years ago is still true today: in warfare, logistics is critical to warfighting and without it defeat is more likely than victory. Unlike warfighting components such as tanks or artillery which exert an immediate, direct, and visible effect on the battlefield, logistics is less prominent and often overlooked. Nevertheless, effective combat logistics[1] has been decisive throughout the history of warfare and this holds true today. The latest tank may be equipped with the most advanced weapon systems and might easily defeat its opponents, but without fuel it will last, on average, 6 to 8 hours on the battlefield before it comes to a halt. It is the logistic system that provides ammunition and fuel as well as supporting the personnel operating that tank which is critical to its successful utility. The same situation applies to all military equipment. Without combat logistics, nations would not be able to fight, because their capacity on the battlefield would be severely restricted.

This chapter will reinforce the enduring importance of combat logistics in modern land warfare and will emphasize its enduring importance today. At

---

[1] This chapter intentionally uses the term 'combat logistics' throughout. Whilst this phrase is not discretely defined in NATO logistic doctrine, its sense is widely understood by logisticians. By the deliberate use of the term 'combat logistics', this chapter is highlighting its major focus on military logistics in the land domain.

Christopher Kinsey and Ronald Ti, *Combat Logistics in the Twenty-first Century*. In: *Advanced Land Warfare*.
Edited by Mikael Weissmann and Niklas Nilsson, Oxford University Press. © Christopher Kinsey and Ronald Ti (2023).
DOI: 10.1093/oso/9780192857422.003.0004

the same time, new and emerging challenges to providing combat logistic support on the battlefield continue to appear. These include the effect of both unmanned air and ground systems, both of which may be partially enabled by artificial intelligence as well as newer technologies such as 3D printing. Combat logistics matters, and if it is unable to fulfil its key roles of enabling movement, strengthening endurance, and providing sustainment, overall military operations are unlikely to succeed. So, in relation to future land warfare, this chapter seeks to answer this question: in the face of change, what needs to be done by military logisticians so that they can continue to provide essential combat logistics in future major conflicts?

This chapter discusses new and emerging threats to operational and tactical military logistic systems, together with how NATO is preparing to deal with Russian threats to its combat logistic support. It will selectively highlight some important initiatives being undertaken to address these perceived threats. The ultimate aim is for NATO to both deliver and sustain effective, resilient combat logistics that enables the mobility, endurance, and sustainment of its coalition forces. Put bluntly, if NATO cannot fulfil these tasks, it will lose. The conclusion to this chapter will round off the discussion and points presented.

The following scheme will be followed. First, the initial section will present a short summary of the nature, scope, and principles of combat logistics. The emphasis here is on explaining important enduring principles together with different aspects of combat logistic systems. It will discuss general principles underpinning the character and scope of combat logistics and link these to NATO's situation as a multi-national coalition, whilst emphasizing the complexity in which NATO combat logistics operates, especially in the current volatile, ambiguous, uncertain, and complex strategic environment.

Following this discussion on principles, the chapter will then highlight current applications. It will illustrate its arguments using the current example of how the North Atlantic Treaty Organisation (NATO) is preparing to deliver effective combat logistics as it once again prepares for high intensity, high lethality, major joint operations in Europe. In the present circumstances (2021), this will most likely take the form of an attack by Russia on an eastern European NATO member nation triggering Article V[2] of the NATO treaty[3].

---

[2] Article V is provision within the NATO Treaty that refers to collective defence. It states that external aggression exercised upon a single NATO member is considered to be an attack on the entire alliance. In distinction to Article 51 of the UN Charter on which it is based, it also imposes a binding collective defence obligation on all parties to the Treaty. See: 'The NATO Treaty, 4 April 1949', NATO, 10 April 2019, https://www.nato.int/cps/en/natolive/official_texts_17120.htm.

[3] The treaty which created the military alliance in 1949 is more correctly referred to as: 'The North Atlantic Treaty'. It is also sometimes referred to as the 'Washington Treaty'. This chapter will use the informal term 'NATO Treaty' as a synonym for: 'the original treaty which created NATO in 1949'.

This is commonly referred to as an 'Article V attack' and would trigger collective defence obligations on all thirty NATO member states. In the event of such a conflict, effective NATO combat logistic integrity would be absolutely critical, particularly as the Alliance prepared for conflict.

The Character and Scope of Military Logistics

Before the specific discussion on the application of combat logistics to the issue of a Russian attack provoking NATO Article V, it is first necessary to examine, in general terms, the character and scope of combat logistics. Whether one is analysing combat logistics from the perspective of a single military organization, for example the German Army (or 'Bundeswehr'), or from a coalition perspective, such as NATO's, certain characteristics of combat logistical systems remain constant, and apply to both perspectives in equal measure. The same also applies to the scope of military logistics, in that the content of its political, social, economic, and technological reach remains the same whether one is examining a single country or a coalition.[4]

In one of the few seminal works on combat logistics, *Supplying War*, the author Martin van Creveld gives equal importance to logistics and strategy.[5] According to Lynn, the book 'shifted logistics from a supporting role to centre stage, convincing soldiers and scholars alike that throughout modern history, strategy has rested upon logistics'.[6] The notion that in the end it is logistics alone that shapes strategy is captured in van Creveld's statement that 'strategy, like politics, is said to be the art of the possible; but surely what is possible is determined not merely by numerical strength, doctrines, intelligence, arms and tactics, but in the first place, by the hardest facts of all: those concerning requirements, supplies available and expected, organisations and administration, transportation and arteries of communication'.[7] Ultimately, combat logistics is about planning and carrying out the movement and maintenance of air, sea, and land forces.[8] This is already a challenging task but is made even more difficult when planning and executing operations in coalitions such as NATO. Enabling the mobility, endurance, and sustainment of multiple land

---

[4] Noting that in the case of coalitions, the content covers multiple state militaries and not merely one.
[5] Martin Van Creveld, *Supplying War: Logistics from Wallenstein to Patton*, 2nd edn (New York: Cambridge University Press, 2004).
[6] John A. Lynn, *Feeding Mars* (Oxford: Westview Press, 1993), 9.
[7] Van Creveld, *Supplying War*, 1.
[8] *NATO Logistics Handbook* (Brussels: NATO Logistic Committee Secretariat, Logistic Capabilities Division, NATO HQ, 2012), 20.

forces during future major conflicts in an operational environment that today faces many diverse challenges that did not exist during the Cold War will push the organization to its intellectual and material limits. These challenges will be outlined in the next section. But none of this is new, even regarding coalition forces. As Moore and colleagues remind us, 'the practice of logistics, as understood in its modern form, has been around for as long as there have been organised armed forces'.[9]

Whether considering a single military force or, as per this chapter's example NATO, combat logistics is a vital component of overall combat power. The reason for this is simple, as combat logistics alone 'determines what military force can be delivered to an operational theatre, the time it will take to deliver that force, the scale and scope of forces that can be supported once there and the tempo of operations'.[10] Combat logistics in the fullest sense refers to more than the immediate equipping, deployment (mobility), sustainment, and endurance of warfighting units in war, but also extends to the ability of NATO countries to manage their defence industrial base and commercial supply lines to meet future military requirements. This is a crucial area for NATO concern since combat logistics is the key enabler to carrying out NATO campaign plans for any future Article V operation.

According to Lonsdale, even though the character of war is changing with the introduction of new technologies, the nature of war remains the same.[11] This is also true for combat logistics. Its inherent nature, which is to do with the 'movement of force … and the sustainment of personnel, weapons systems, and other support requirements to achieve tactical, operational, and strategic objectives' has not changed since the era of ancient warfare.[12] Because of the close link between strategic intent and its expression in logistics, combat logistics also faces another inherent problem in the formulation and execution of strategy. In the case of NATO, however, the problem is made more challenging in that meeting the logistical requirements of any strategy will involve coalition partners not only agreeing on the said strategy but also agreeing on how to support it.

A very important point regarding NATO must be made here. Each NATO member state is responsible for its own logistics, especially first and second line support. There is actually no overall 'NATO combat logistics'

---

[9] David M. Moore, Jeffrey P. Bradford, and Peter D. Antill, *Learning from Past Defence Logistics Experiences: Is What Is Past Prologue* (London: RUSI, 2000), 1.

[10] Christopher Kinsey and Matthew Uttley, 'The Role of Logistics in War', in *The Oxford Handbook of War*, edited by Julian Lindley-French and Yves Bowyer (Oxford: Oxford University Press, 2012), 401.

[11] David J. Lonsdale, 'Strategy', in *Understanding Modern Warfare* edited by David Jordan et al. (Cambridge: Cambridge University Press, 2008), 16.

[12] William G. Tuttle, *Defense Logistics for the 21st Century* (Annapolis: Naval Institute, 2005), 1–2.

organization. What exists in NATO are various initiatives at the overall NATO level that seek to improve the interoperability of coalition logistics. These initiatives include standard doctrine and guidance, and centres such as the Multinational Logistic Coordination Centre in Prague. The aim of all of these coalition-wide logistic initiatives is to improve interoperability and attempt to produce something like a consistent level of combat logistic support over all of the individual national logistic components that together comprise what this chapter labels as 'NATO Logistics'.[13]

Hence, NATO decision makers need to agree on a set of logistical choices which reduces their strategy or strategies to a coherent and practical logistical system. This, in turn, involves making certain assumptions about two vital issues. First, NATO decision makers will need to decide contingencies where NATO forces might be deployed in support of policy objectives surrounding Article V. Second, the same decision makers also need to determine both the necessary logistical capabilities to achieve the required operational tempo(s) and the military warfighting capabilities to perform the operation(s) if such contingencies occur.[14]

What this entails is NATO decision makers deciding on a set of cascading choices, in an environment dominated by a lack of data and considerable ambiguity, that turns their appraisal of the existing strategic environment into a judgement on what is necessary for NATO to respond in the event of Article V being triggered. This, in turn, will mean ranking member state national interests, whilst at the same time assessing potential contingencies which will in all likelihood require military action. Next, the decisions taken by NATO decision makers will need to ensure the efficient conversion of combat logistics into policy, strategy, force posture, and other important military capabilities. These decisions, moreover, will also be influenced by their perception of the international political environment, the diverse national and strategic cultures of NATO member states, as well as other technological and economic interests. Each one of these conceptual and practical steps is an inherent aspect of policy and strategy preparation and design in an environment where reasoned conclusions link NATO member state interests, perceived threats, and military capabilities.

The resultant choices derived from this process then inform NATO decision makers about the most appropriate combat logistic system to meet

---

[13] The principle of interoperability amongst NATO members is underwritten by Article III of the NATO Treaty. Where Article V refers to collective defence, Article III refers to cooperation between the nations and preparation, before a crisis.

[14] Kinsey and Uttley, 'The Role of Logistics in War', 403.

the demands of a NATO Article V operation.[15] NATO logisticians will fundamentally seek to achieve the optimal interface between NATO operational commanders and a combat logistics system that is able to provide mobility, endurance, and sustainability. An example of these kinds of operational challenges facing NATO logistic support is the need to repair and recover combat assets and still provide ongoing information regarding availability and status, whilst still responding to sudden and urgent logistic demand signals from Commanders in an unpredictable, ever-changing battlespace.[16]

An important aspect of essential information flow is for NATO commanders to maintain the feedback loop that ensures continuous assessment of the combat logistics system for potential weaknesses and vulnerabilities to enemy action, whilst at the same time seeking to identify and exploit weaknesses in the enemy combat logistical system.

The Applications of Combat Logistics

The applications of the principles of combat logistics described above will now be discussed, illustrating arguments with selected examples from current NATO combat logistic practice. The discussion will be framed within the context of a hypothetical NATO Article V defensive operation in Europe against the Armed Forces of the Russian Federation, currently NATO's most likely and most dangerous peer/near-peer adversary. Selected issues affecting NATO logistics that remain to be fully addressed will also be discussed at the conclusion of this section.

Challenges Arising from the Post-Cold War Environment

Wider geopolitical changes, such as the inclusion of former Soviet states into NATO, have resulted in greatly expanded NATO borders—which still need to be defended. The most obvious challenge for NATO combat logisticians has been the greatly increased NATO Joint Operational Area along with the additional logistic burden this imposes. Differences in the practice and theory underpinning ex-Warsaw Pact states are also an important and often overlooked factor: these will be discussed at the end of this chapter. Other reasons

---

[15] David J. Foster, 'Air Operations and Air Logistics', in *Perspectives on Air Power: Air Power in Its Wider Context*, edited by Stuart Peach (London: The Stationery Office, 1998), 220.
[16] Foster, 'Air Operations and Air Logistics', 223.

for changes in the post-Cold War environment are complex and wide in scope and range from socio-political factors arising within individual NATO member states, to those resulting from global 'megatrends' extending over the past four decades. These include the effects of globalization and the prevalence of 'free-market' economics. A particular feature of the latter has been the rise of privatization and the downsizing and sale of state-owned enterprises. Both factors have led directly to increasing global commercialization within military logistics. Other no less important factors include the reduction of defence spending amongst NATO states since the 1990s. Consequently, by contrast to the situation prevailing during the Cold War, current logistic reinforcement and sustainment of NATO forces has become considerably more challenging.

However, not everything has changed as a result of the 'Fall of the Wall'. One example of an issue pre-dating the end of the Cold War which constrains combat logistic operations in Europe today, is the presence of ongoing legal and procedural obstacles to cross-European border movement. The problem persists despite status of forces agreements and negotiated memoranda of understanding. Customs procedures to clear cargoes, especially Dangerous Goods, often applied inconsistently, are an important factor. Although not absolutely critical, the requirement for the correct cross-border paperwork, varying between countries and often requiring paper forms available only in the host nation language, does impose a layer of delay should the need for a rapid deployment arise in an Article V situation. As a recent group of experts (including a very recent former NATO Supreme Allied Commander Europe (SACEUR)) have commented, in the case of paperwork required in a crisis:

> timescales for completing the required paperwork are likely to be of the same order of magnitude as timescales for the movement itself and legal and procedural delays may have operational impact.[17]

The following sections will discuss the applications of combat logistics and provide examples of how NATO is optimizing combat logistics through improved logistic command and control (leading to enhanced Coalition interoperability), better logistic movement, endurance, and sustainment through improved Reception, Staging, and Onward Movement (RSOM), and increased capability in delivering combat logistics over the 'last tactical mile'. It should be noted that these three factors roughly align themselves with the strategic, operational, and tactical levels of war respectively.

---

[17] Ben Hodges, Tony Lawrence, and Ray Wojcik, *Until Something Moves—Reinforcing the Baltic Region in Crisis and War* (Tallinn: International Centre for Defence and Security, 2020), 14.

Improved Command, Control, and Coordination Enhances NATO Combat Logistic Interoperability

As described previously in the section on principles, NATO is a multinational coalition that must be prepared to potentially command, control, and coordinate up to thirty different member state forces in combat operations—and win. The requirement to successfully deliver combat logistics to enable the movement, sustainment, and endurance of this force is directly related to optimizing logistic command, control (C2), and coordination. To this end, NATO command has recently identified the need for specific logistic C2 structures in order to improve NATO logistic command arrangements. Following endorsement by NATO Defence Ministers in February 2018, two new NATO Joint Force support commands were created specifically to coordinate combat logistic functions in NATO's main 'rear areas' of Continental Europe and the Euro-Atlantic land/maritime space. Reporting directly to the NATO SACEUR, both Joint Commands are tasked with a range of supporting functions, particularly with the movement, coordination, and force sustainment across Europe and the Atlantic.[18] The European-based entity, NATO Joint Support and Enabling Command (JSEC),[19] is based in Ulm, Germany, and achieved its initial operating capability (an important milestone in reaching full functionality) in September 2019. The Euro-Atlantic entity, NATO Joint Forces Command, Norfolk (JFC-NF), is based in Norfolk USA. Both JSEC and JFC-NF will coordinate the response deployment of NATO 'follow-on forces' should an Article V attack occur on NATO. JSEC focuses on 'intra Europe' support, whilst JFC-NF is primarily concerned with the movement of US and Canadian forces across the Atlantic Ocean. These Headquarters will also have responsibilities in providing rear area security for deploying forces and their national combat logistic supporting elements in transit. In the event of an Article V crisis, individual NATO nations may also request overall logistic support from either of the two support commands described above to augment and coordinate their own mobility, supply, and sustainment efforts. Both commands will focus on the principal mobility activity of RSOM, which will be further discussed below.

---

[18] The Brussels Summit declaration which announced these initiatives is at paragraph 29 of the Joint Declaration of 11–12 July 2018, see: *Brussels Summit Declaration*, NATO, 11 July 2018, https://www.nato.int/cps/en/natohq/official_texts_156624.htm#29.

[19] For a recent NATO Review article discussing JSEC, see: Sergei Boeke, 'Creating a Secure and Functioning Rear Area: NATO's New JSEC Headquarters', NATO Review, 13 January 2020, https://www.nato.int/docu/review/articles/2020/01/13/creating-a-secure-and-functional-rear-area-natos-new-jsec-headquarters/index.html.

Prior to the 2018 initiative that created these new strategic-level commands, a number of additional combat logistic initiatives had already been initiated at the operational level. A key initiative directly related to a Coalition interoperability project has seen the establishment of a NATO Joint Logistic Support Group (JLSG) Headquarters (which is co-located with JSEC in Ulm).[20] This deployable HQ has been established specifically to coordinate overall combat logistics at the operational level for a NATO Response Force. In summary, the initiatives described here seek to improve combat logistic capability by providing better overall NATO command, control, and coordination of Coalition combat logistic assets. Creating an overall combat logistic command and coordination framework is linked to another key factor in effective combat logistics, which is the enablement of mobility, endurance, and sustainment of forces, also known by the acronym: 'RSOM'.

Reception, Staging, and Onward Movement

RSOM refers to the receiving, staging, and onward movement of NATO military forces which have proceeded from more distant locations (particularly from across the Atlantic) and are in transit *en route* to the Article V Joint Operational Area.[21] Given the importance of reinforcement and deployment of forces, RSOM is a major operational-level task. Inherent RSOM tasks are implicit in its name: in essence, 'reception' gathers forces in transit, 'staging' prepares those forces specifically to the respective theatre of operations. This includes activities such as preparatory training, or specific equipment allocation. This may include the issuance of protective equipment if, for example, there is an identified chemical warfare threat. Finally, 'Onward Movement' refers to movement forward directly into the NATO Joint Operational Area (JOA) after these forces have transited and completed staging according to the specific requirements of the operation. 'RSOM' is hence a collective term that groups a number of diverse logistic activities that include the marshalling, theatre preparation, and movement of NATO forces. The range of essential tasks includes not only obvious ones such as directly facilitating road, sea, and air movement, but also a range of less apparent, but no less important, movement enablement tasks. These often include critical but less

---

[20] This US Army article gives a good summary, dated January 2020, of the NATO JLSG concept and functions, see: Aaron Cornett, 'Multinational Operations: JLSG Offers Effective Role with Allies, Partners, U.S. Army, 16 January 2020, https://www.army.mil/article/231676/multinational_operations_jlsg_offers_effective_role_with_allies_partners.

[21] *Logistics Handbook* (Brussels: NATO Logistic Committee Secretariat, Logistic Capabilities Division, NATO HQ, 2012), 73.

prominent procedural, physical infrastructure, and host nation support and coordination issues. The issue of legal and procedural obstacles in enabling cross-border movement permissions has been previously discussed.

Host Nation Support (HNS) and coordination is another critical input into interoperability and RSOM.[22] An ongoing issue related to cross-European theatre movement is the need to optimize coordination between multiple military and civilian national agencies. Unlike the Cold War when HNS and coordination was more seamless,[23] a major deployment of NATO forces:

> (now) requires the mobilisation of civilian strategic transport assets, and the infrastructure to receive and re-deploy those forces on arrival on European soil. (this) is not just about *military* preparedness; *civil* preparedness is equally important ["italics in original"].[24]

Whereas during the 'Cold War' HNS and coordination required coordination with what was in essence a functional 'single point of contact' in the form of European governments and their agencies (effectively functioning in the absence of privatization as single points of contact), current NATO–Host Nation preparedness now requires both much wider and more complex interaction amongst NATO, national militaries, and both governmental and private sector bodies.[25]. These interactions focus on issues such as local Host Nation procurement and supply, and the harmonization of national legislation. The latter includes HNS-NATO activity on issues such as improved liaison and deconfliction of national customs procedures, particularly in synchronization/ harmonization of NATO member state regulations. These may be in important logistic areas involving the transport of Dangerous Cargo such as ammunition, pyrotechnics, and fuel. The goal of such joint initiatives lies ultimately in enabling NATO Combat Logistics and the RSOM process.

As mentioned previously, a further post-Cold War trend has been the increasing use and integration of commercial contracted logistic firms by NATO member militaries. Consequently, the range of HNS support activities now routinely includes the procurement of logistic services from Host Nation

---

[22] *Logistics Handbook* (NATO Logistic Committee Secretariat), 107–112.

[23] The reasons for this are complex, but a principal one was the primacy of governments as single points of contact resulting in less numerous actors than today.

[24] Jonathan Hill, 'NATO: Ready for Anything?', NATO Review, 24 January 2019, https://www.nato.int/docu/review/articles/2019/01/24/nato-ready-for-anything/index.html.

[25] Ongoing preparedness in this area is also being undertaken jointly between NATO and the European Union (EU) through EU initiatives such as the 'military mobility' project undertaken as part of PESCO, the 'Permanent Structured Cooperation' program. For an introduction see: Permanent Structured Cooperation (PESCO), accessed 19 May 2022, https://pesco.europa.eu/project/military-mobility/

commercial firms.[26] Whilst cost savings and increased efficiencies might result from these arrangements, these nevertheless increase the complexity of logistic activity by introducing additional actors.

Bridging the 'Last Tactical Mile'

It's no longer the 'last logistic mile'-it's now the last 1,000 logistic miles …
Brigadier Robert Wilhelm, Deputy Commander, Bundeswehr Logistic, Budapest Command, quoted at the 2021 Central European Logistics Conference, Budapest, 22 March 2021.

The above quotation refers to a battlespace where linear concepts of 'fronts', 'frontlines', and 'rear areas' are now outdated. Increased Russian indirect offensive strike capabilities which extend well beyond the range of conventional indirect fire by artillery have significant implications for the delivery of combat logistics nearer to the 'forward edge of battle' (noting that the term itself reflects obsolete linear battlefield ideas), or, in the context of this discussion, the 'last tactical mile'.

This section will discuss the challenges of delivering combat logistics with the focus on providing combat logistics in the 'last tactical mile'. The term 'last tactical mile' is itself anachronistic, and likely originates from early twentieth-century trench warfare. In the First World War, static trenchlines were dug such that the 'last mile' before encountering the wire tangle of 'no mans' land' contained support, communications, and reserve trenches, all of which often extended rearward for—literally—one Imperial mile.[27] The 'last tactical mile' persists today as an ill-defined term, and is commonly (and vaguely) understood to refer to that part of the land battlespace which is in close physical proximity to the enemy and therefore highly vulnerable to direct enemy action. As defined by the United Kingdom's Autonomous Last Mile Resupply System (ALMRS) project, the so-called 'last tactical mile' may actually extend up to 30 kilometres behind the notional 'frontline'.[28]

---

[26] NATO overall procurement is undertaken by a number of agencies according to the respective level. For strategic and operational level procurement, the NATO Supply and Procurement Agency (NSPA) is generally the lead agency. See: NATO support and procurement agency (NSPA), accessed 19 May 2022, https://www.nspa.nato.int/about/nspa.

[27] Or about 1.6km in the metric system.

[28] The United Kingdom Autonomous Last Mile Resupply System (ALMRS) project was first open to competition in mid-May 2017 and is an ongoing project within the UK's Defence Science and Technology Laboratory (DSTL). See: Competition Document: Autonomous Last Mile Resupply, last accessed 19 May 2022, https://www.gov.uk/government/publications/accelerator-competition-autonomous-last-mile-supply/accelerator-competition-autonomous-last-mile-resupply.

Current Russian land-based surface to surface missile systems deployed in Europe have ranges extending far beyond the 'traditional' 15 km range of most conventional First World War artillery. These trends in the range and accuracy, particularly of Multiple Rocket Launcher Systems (MRLS) are alarming, and with ranges typically over 100km, 'the long range heavy MRL has a major Anti-Access/Area-Denial capability if present in an opponent's arsenal. Today's ... [MRLS] ... are now capable of attacking targets at ranges that in the 1990s were the purview of short-range ballistic missiles ...'.[29] Given this situation, taking the 'last tactical mile' as a synonym for the area of modern-day logistic vulnerability means that combat logistic units are now subject to direct enemy action in an area extending rearward up to the one thousand kilometres referred to in the quotation. The present situation is returning military thinkers back to older ideas such as Boulding's 'Loss of Strength Gradient' theory, which relates combat power directly to its proximity to the 'last tactical mile'. Put simply, modern 'anti-access/area denial' systems, by their range, are restoring the importance of proximity back to the modern battlefield.[30] The 'transaction costs' of moving logistics through the 'last tactical mile' in terms of losses and damage from enemy action, together with the increased range and effectiveness of indirect fire weaponry has re-established 'proximity' to the 'front line' as a key determinant of modern combat logistics.

The result is that the practice of combat logistics has become more dangerous, and considerably more manoeuvre restrained. Whereas in previous world wars (aside from aerial attack), combat logistic units were placed further 'back' in 'rear' echelons removed from the 'frontline' of a linear battlespace, this is no longer the case, as the 'rear' has been functionally erased. The notion of a 'rear area' where logisticians would historically ply their trade in relative safety has ceased to exist on the modern battlefield.

Given that these so-called 'rearward' combat logistic personnel and systems often operate unarmed and unarmoured logistic vehicles only compounds their vulnerability. In a recent comment by a former senior British Commander,[31] the NATO alliance faces a '360 degree threat' consisting of enhanced indirect fire and air threats from both manned and unmanned platforms which have hitherto never been faced by NATO forces. This will result

---

[29] John Gordon IV et al., *Comparing US Army Systems with Foreign Counterparts Identifying Possible Capability Gaps and Insights from Other Armies* (Washington, DC: RAND Corporation, 2015), 113.

[30] Kieran Webb, 'The Continued Importance of Geographic Distance', *Comparative Strategy* 26, 4 (2007), 295–310.

[31] Brigadier (ret.) Ben Barry, formerly British Army, Senior Fellow for Land Warfare at the Institute for International Strategic Studies, quoted at FINABEL Land Forces Modernisation Seminar, Brussels, 23 March 2021.

in the dispersal of NATO combat logistic units for reasons of survival, as a concentration of combat logistic assets into bases (as has been the case in the recent Afghanistan operation) or even into smaller, but still discrete, logistic support hubs presents targeting opportunities for Russian indirect fire or air attack assets. In the modern battlespace, this will present commanders with even greater challenges in maintaining the integrity of their units on the battlefield. The necessary dispersal of combat logistic services to counter these threats and its potentially disruptive effects on logistic command and control, will 'disaggregate' logistic formations. This 'logistic disaggregation', will become a prominent feature of combat logistics on the future NATO Article V battlespace.

Unmanned Aerial Systems

Unmanned aerial systems (UAS) are now widespread for surveillance use, and increasingly, weaponization. Currently, the hitherto least developed capability for UAS use has been in combat logistics, especially in battlespace delivery. As a recent commentator has put it:

> Compared to the ability of loitering UAVs to reconnoitre terrain, or weaponising an autonomous platform with Hellfire missiles, logistic drones and machines don't appear to be sexy. There's not much hype around drones for resupply, in stark contrast to the sensation of drones for the kill chain. But the kill chain is useless without a resilient supply chain.[32]

Further … 'while many military projects have toyed with the unmanned, or reduced manning, model of resupply, the pressure is on civilian freight and distributor companies to pioneer the hardware, software and systems to make unmanned resupply relevant to a competitive world of business logistics'.[33] UAS are being employed increasingly in global logistics, with a 2016 analysis by a prominent international business advisory firm suggesting that 'drones' in commercial logistics currently occupy only $USD13 billion of market share out of a potential $USD127.3 billion market.[34] Whilst Combat Logistics is currently lagging behind this trend, work is proceeding in

[32] Jacob Choi, 'Autonomous Resupply: Drones for the Supply Chain reinforce the Kill Chain', Australian Army Research Centre, 30 September 2017, Available at: https://researchcentre.army.gov.au/library/land-power-forum/autonomous-resupply-drones-supply-chain-reinforce-kill-chain.

[33] Choi, 'Autonomous Resupply'.

[34] Michal Mazur, Adam Wisniewski, and Jeffery Mc Millan, *Clarity from Above: PwC Global Report on the Commercial Applications of Drone Technology* (London: Price Waterhouse Coopers, 2016), 40.

some NATO states, with, for example, the UK Ministry of Defence currently evaluating a 180 kg lift UAS (the Malloy T-400).[35]

However, quite aside from unarmed, logistic use, an expanding technological threat in the modern land battlespace,[36] with especial relevance to Combat Logistics, is the increasing presence of weaponized unmanned aerial systems, which are employed increasingly as 'loitering munitions'. Their deployment as weapons, whilst retaining their lethal 'real-time' surveillance function, has substantially increased the threat to combat logistic forces. The ever-improving range and persistence of UAS is further reason why the 'last tactical mile' has now been pushed further out from the notional 'front-line' maximizing their threat to 'rearward' combat logistic units. In addition, the smaller size of most tactical UAS means that they can remain below the detection threshold of conventional ground-based air defence radar, further adding to the threat. Finally, despite current research, there is a current lack of readily available effective weaponry against UAS for combat logisticians, often carrying only small arms.

The other factor affecting NATO combat logistics is more widespread Russian use of UAS operating at multiple levels over the tactical battlespace. In contrast to virtually all other NATO forces, Russia currently deploys UAS directly from lower echelon, tactical levels, for example, unlike equivalent NATO forces where UAS are generally deployed from centralized units (often at formation or higher divisional levels). The net result of such a preponderance of UAS across echelon levels within Russian organizational charts has produced a substantial reduction in the 'reconnaissance–target acquisition–targeting–battle damage assessment' loop. This has not only increased the rapidity by which Russian indirect fire assets acquire targets and execute fire missions, but also the speed at which these assets can adjust or re-direct targeting. The result is to increase risk even more for NATO combat logistic units.

Russian Offensive Doctrine

The current Russian operational 'way of war' has direct impacts upon NATO combat logistics and this situation is exacerbated by the fact that NATO has simply not faced a peer/near-peer adversary since its creation. The recent

---

[35] Personal communication to the author from the UK MOD desk officer responsible for the UK ALMRS and Project Theseus, 5 March 2021.

[36] Seen especially in the recent 2020 Armenia–Azerbaijan conflict. For a recent (December 2020) commentary from a reputable US think tank, see: Shaan Shaikh and Wes Rumbaugh, 'The Air and Missile War in Nagorno-Karabakh: Lessons for the Future of Strike and Defense', CSIS, 8 December 2020, https://www.csis.org/analysis/air-and-missile-war-nagorno-karabakh-lessons-future-strike-and-defense.

Russian offensive in Eastern Ukraine has provided key insights into Russian offensive doctrine.[37] An important factor is the preponderance of offensive fire assets within Russian organizational structures which include rocket missile units. For example, when compared to the organizational structure of the average US Stryker Brigade, the current equivalent Russian motorized/mechanized brigade holds substantially greater amounts of offensive support assets in terms of both conventional gun and rocket artillery.[38] This doctrinal and historical Russian emphasis on concentrated, intense area fires, compared to more discriminatory precision strike modes, when coupled with the Russian air threat, presents hitherto novel and unaccustomed threats to NATO combat logistics.[39] This is exacerbated by the greatly expanded zone of logistic vulnerability previously described.

The logistic disaggregation resulting from all of these factors will have profound effects on NATO combat logistics in a future Article V conflict with Russia, especially as certain combat logistic services cannot easily disperse in response to novel air and indirect fire threats from a peer/near-peer adversary such as Russia. The result is that combat logistic functions will need to become less centralized on the battlefield (however many 'miles' this extends' backwards') to survive. The 'de-centralization' of logistics on the battlefield is somewhat analogous to the so-called 'ink spot' deployments practised in recent conflicts in Afghanistan. In the case of logistics, however, the difference is that whereas in the latter case, 'ink spot deployment' was employed to improve control over territory, this dispersal is now necessary for survivability. Hence both situations employ the same method, but for quite different reasons.[40]

Emerging Technology and Combat Logistics

This section will briefly discuss selected emerging technologies of logistic relevance. These not only have the potential to bridge the 'last tactical mile' space, but to enable combat logistics to also function effectively *within* this increasingly restricted tactical battlespace in spite of threats. The technologies that will be discussed include 3D printing, also known as Additive

---

[37] Peter B. Doran, *Land Warfare in Europe: Lessons and Recommendations from the War in Ukraine* (Washington, DC: Centre for European Policy Analysis, 2016).

[38] Brigadier (ret.) Ben Barry, formerly British Army, Senior Fellow for Land Warfare at the Institute for International Strategic Studies, quoted at FINABEL Land Forces Modernisation Seminar, Brussels, 23 March 2021.

[39] Phillip A. Karber, 'Lessons Learnt from the Russo-Ukrainian War: Personal Observations', 8 July 2015, https://prodev2go.files.wo.

[40] Cyrus Hodes and Mark Sedra, 'Chapter Four: International Military Support', *The Adelphi Papers* 47, 391 (2007): 46.

Manufacturing (AM), Robotic and Autonomous Systems (which may be enabled by Artificial Intelligence (AI), and Unmanned Ground Vehicles (UGV) for combat logistics.

Additive Manufacturing

Additive Manufacturing builds up the desired object with raw material in powder form. These may be metals, composites, or plastics. Digital three dimensional designs are used to guide the process. It is termed 'additive' because it differs from the 'subtractive' process described in the original 3D printing concept first described in 1981.[41]

Additive Manufacturing is being actively researched and trialled specifically in combat logistics settings. Recent deployed field projects have successfully produced smaller components in 3D printing mobile facilities entirely contained within a standard 40 foot shipping container.[42] AM has the potential to address certain supply issues by functionally concentrating the production and distribution of 3D printed items into a single combat logistic node located *within* the battlespace itself, at either 'first' or 'second' lines of logistic support.[43] This offers the potential for combat logistic personnel to produce a range of logistic items for immediate consumption in real time. For example, when considering a spare part two general alternatives exist. One alternative is for a conventionally manufactured part to be supplied, which has been produced in a distant node and transported through a logistic distribution system from that remote location. The other alternative is to 3D print that part *in situ* using a 3D printing plant (which is wholly contained within a standard 40 foot shipping container) and manufacture that part *de novo* in the forward area. The obvious advantage of AM is that the manufactured item does not have to undergo onward forwarding and transport. In addition, stockpiling and warehousing is minimal. The supply chain is effectively shrunk, together with the reduction in the risks of delay, mis-delivery, and vulnerability to attack compared to a conventional manufacturing supply chain. The value proposition behind AM is simple: it is better to carry raw material capable of (conceivably) being turned into 1,000 different parts than to source and carry those 1,000 parts. AM has '(the) enormous potential

---

[41] This website from US multinational 3M gives a brief history of 3D printing till today's AM application: 'The History of 3D Printing: 3D Printing Technologies from the 80s to Today', Sculpteo, accessed 19 May 2022, https://www.sculpteo.com/en/3d-learning-hub/basics-of-3d-printing/the-history-of-3d-printing/.

[42] For an example of a recent field deployment, see: Marines, Engineers Conduct a first-of-its-kind 3D printing Exercise, Marines, 26 August 2019, https://www.marines.mil/News/News-Display/Article/1943919/marines-engineers-conduct-a-first-of-its-kind-3d-printing-exercise/.

[43] Matthew Wood, 'Reintroducing Manufacturing to Army's Supply Chain', *Australian Army Journal* 16, 1 (2020): 101–113.

in assisting Army in eliminating much of the "iron mountain" synonymous with 20th century logistics'.[44]

Robotic and Autonomous Systems (RAS)

Robotic and Autonomous Systems[45] are a rapidly evolving capability in modern warfare. Mixed Human–Robot teams are already well established in the area of explosive ordnance disposal with robots deployed routinely alongside human operators. Current military logistic research focuses on human-led applications with two overall aims. These are 'the opportunity to achieve greater combat power within its planned budget by increasing its physical and non-physical mass (coupled with) … the opportunity to fundamentally alter the structure of Defence from a force of a few large and expensive platforms to one of many small and cheap platforms'.[46]

Current RAS research specifically related to combat logistics can be seen in three applications. These are Machine Learning enabled by semi-autonomous AI, human 'Leader-follower' systems, and human augmentation. Briefly summarized here, Machine Learning harnesses AI technology for various potential logistic applications. These include machine sensing for maintenance and capability life cycle applications, or applications such as data analysis of logistic supply demands to optimize logistic routing or route finding.[47] 'Leader-follower' robotic systems in logistics include experimental transport systems where, for example, as one manned truck may lead and control a number of 'follower' unmanned robotic trucks (which themselves will be enabled partially by semi-autonomous AI direction systems). The essential component is that teaming occurs of human operators/controllers with robotically enabled systems.[48] Current human augmentation research focuses on the development of exoskeletons which are 'worn', and which greatly augment the ability of individual logistic human operators. These enable an individual soldier to carry a significantly heavier logistic load per (human) unit. This capability will enable a combat soldier to carry a load far exceeding 25–35 kg on foot in remote warehousing or storage situations.[49]

---

[44] Matthew Ng, 'Additive Manufacturing, Taking the Iron Mountain out of Logistics', Australian Army Research Centre, 24 August 2018, https://researchcentre.army.gov.au/library/land-power-forum/additive-manufacturing-taking-iron-mountain-out-military-logistics.
[45] For a concise definition of RAS, see *Concept for Robotic and Autonomous Systems* (Canberra: Australian Defence Force, 2020), 8.
[46] *Concept for Robotic and Autonomous Systems*, 9.
[47] *Concept for Robotic and Autonomous Systems*, 36.
[48] *Concept for Robotic and Autonomous Systems*, 29–31.
[49] Gordon, *Comparing US Army Systems with Foreign Counterparts*, 116.

In a situation where logistic vehicle fleets may be too detectable, an augmented logistic operator may well be able to move loads that currently require
machinery (which generate noise, sound, and a heat signature, all of which
can be targeted).

Unmanned Ground Vehicles (UGV)

Finally, robotic Unmanned Ground Vehicles show promise as a viable alternative to conventional logistic vehicles in a greatly restricted 'last tactical
mile'. In an Article V situation where NATO would not enjoy freedom
of movement in a restricted tactical battlespace, the use of smaller, less
detectable, stealthy swarms of UGV's may ultimately represent the only viable
alternative to conventional logistic transport.[50]

AM, Robotics, and AI in Future Force Structures

It would be appropriate to make a few comments at this point regarding the
future prospects of these technologies in support of logistics. First, information, particularly regarding development, is restricted: virtually all research
in these areas is classified work being undertaken by nations. This lack of
information coupled with the ongoing and exponential development of computing power, makes the trajectory, scope, and degree of these changes very
complex and difficult to assess. Given these factors, predicting a 'realistic
timeframe' for implementation with any confidence is challenging. Secondly,
particularly in the logistic area, much of this enabled technology will be
in the form of 'black boxes' which in essence will 'bolt on' to existing systems: the utilization of logistic system machine learning is a prime example
of this. This raises a host of secondary issues around intellectual property
and the sharing of critical Original Equipment Manufacturer (OEM) technology. This will become an issue of increasing prominence, especially since
much research and development, particularly in the USA and the UK, in AM,
robotics, and AI is driven through public–private partnerships.[51] Thirdly,
there are specific issues that relate directly to the actual nature of the capability itself. For example, Unmanned Ground Vehicles might well operate

---

[50] The Estonian-produced MILREM UGV is currently (2021) undergoing trials as part of the UK
MOD's Project Theseus. This unit has already been trialled on operations in Mali with the Estonian
Defence Force. See the company's own website: The THeMIS UGV, Milrem Robotics, accessed 19 May
2022, https://milremrobotics.com/defence/.

[51] With both the UK ALMRS and the US RAS strategy (which posits a fully autonomous unmanned
aircraft by 2040) being prime examples.

semi-autonomously, but due to limitations in their size and payload, will exist on the future battlespace as multiple units. This 'swarm' of UGV's creates not only additional imposts on logistic command and control, but also increased vulnerabilities if the control systems are interdicted electronically, for example through the 'hacking' of Command and Control networks. Unlike, say, a human-operated truck, a UGV has no accompanying human operator, thus leaving no option if semi-autonomous systems are somehow neutralized. Lastly, it is a historical truism that military bureaucracies procure new equipment and technology far more readily than they can properly assess and integrate novel capabilities and derive optimal operational applications. Thus, whilst AM, Robotics, and AI are all poised to become increasingly important, there are also significant organizational challenges to their wide employment and integration into any force structure. Perhaps the greatest challenge they represent lies in how competently (or not) militaries as technical and cultural institutions, can incorporate 'lessons learnt' derived from analysis of both the 'known' and 'unknown unknowns' of this evolving and novel technology.

The Persistence of Warsaw Pact Legacy Systems on NATO Logistics

This section will now highlight a current strategic issue that goes to the heart of NATO's efforts to engender interoperability with the goal of optimizing movement, endurance, and sustainment. NATO membership has increased considerably since the original 1949 group of twelve and there are currently (2021) thirty NATO member states. The most recent increase has occurred since 2004 with significant expansion into former eastern Warsaw Pact states. This issue concerns the persistence of Warsaw Pact era legacy logistic systems at the national level of these former Communist, but now NATO, member states, despite almost two decades of NATO membership. The root causes of this situation lie in these states retaining highly resilient and deeply embedded legacy systemic and cultural factors which can be difficult to clearly discern. Despite these states gaining (nominally) 'full' NATO membership status, in most cases their military logistic systems continue to function quite differently from their counterparts in the West.[52] The situation is complex,

---

[52] 'West' is used here as a synonym for western European, non-Warsaw Pact states, generally west of the Cold War West–East German border.

but in essence, one of the fundamental causes lies in systemic and organizational cultural differences between 'Western' concepts of logistics versus Ex-Warsaw Pact legacy Communist-era 'Eastern' concepts of logistics. There are two key elements of this. The first is the persistence of Communist-era command and management structures which are diametrically opposed to western structures, and which fundamentally shape how logistics is delivered. In the words of one experienced Eastern bloc commentator '(ex-Warsaw Pact) legacy command concepts ... (have) instead ... been "grafted" to the new democratic paradigm, resulting in unclear chains of command, while allowing continued overcentralized decision making'.[53]

The second element is related to the theoretical concept of logistics itself. Logistics as practised in former Warsaw Pact states, has its ultimate origins in Soviet Military Economic Science.[54] One of the outcomes of this ideology for combat logistics is a disconnection between higher goals and operational tactical application. The reasons again are complex, but in this instance much is due to the subordination of logistics to higher national production. This places logistics as subordinate to 'supply side' production and not 'consumer side' demands, as is the case with operational, Western logistics. This is an approach which betrays much of its Soviet legacy. The result is that because logistics is determined by supply side considerations which are not directly related to operational demand and requirements, the process of logistics itself no longer acts as the vital link between national strategic intent and expressed operational/tactical effect. This nexus between logistics and strategy has been identified by some commentators as 'timeless' (perhaps unconsciously echoing Clausewitzian ideas of conflict possessing certain 'enduring' features). Such a 'logistic-strategic nexus' is a reciprocal one where 'grand strategic plans influence the general shape of the military logistic system, while future strategic options are circumscribed by the logistical system of the day'.[55] With a legacy former Warsaw Pact logistic system, NATO member states cannot as readily establish the logistic–strategic nexus so essential for translating strategic aims into operational success.

Cultural and attitudinal differences in command, control, and delegation, have become apparent when attempting to integrate former Warsaw Pact states with Western NATO member systems. Differing attitudes to delegation are an example. For example, a western-based NATO military will enable unit

---

[53] Thomas-Durell Young, 'Can NATO's "new" Allies and Key Partners Exercise National-level Command in Crisis and War?', *Comparative Strategy* 37, 1 (2018): 18.

[54] Thomas-Durell Young, 'The Challenge of Reforming European Communist Legacy Logistics', *The Journal of Slavic Military Studies* 29, 3 (2016): 354–355.

[55] Mark Erbel and Christopher Kinsey, 'Think again—Supplying War: Reappraising Military Logistics and its Centrality to Strategy and War', *Journal of Strategic Studies* 41, 4 (2015): 5.

staff at relatively low tactical levels to determine a logistic estimate[56] to guide sustainment for an operational brigade size echelon. Alternatively, in an eastern, former Warsaw Pact military, this function will typically be performed by higher logistic staff located outside the unit who are often responsible for multiple subordinate units. In this instance, authority is highly centralized, not 'devolved' to the lower unit, and in accordance with Communist-era directive command models. This contrasts with widespread application of the 'mission command'[57] model in western NATO militaries. When comparing mission command philosophies in the West with 'dictatorial, directed, central command' in former Warsaw Pact-now NATO member states- it is clear that not only are they quite dissimilar, but philosophically quite inimical.[58] These are substantial differences which have the potential to become significant obstacles to effective coalition interoperability in wartime. This includes logistic command. As one experienced commentator has written:

> these legacy command concepts compromise the ability of governments to respond quickly and effectively in periods of escalation and war, but by avoiding fully adopting Western command concepts (and retiring their legacy counterparts), they leave their countries at risk of not being able to respond in a timely fashion to threats to their interests, and indeed their own national security.[59]

The discrepancy between 'Western' demand-driven logistics and 'Eastern' command directed logistics affects virtually all key enablers of Combat Logistics previously discussed in this chapter. The fact that these differences are deeply embedded within the organizational cultures of the respective organizations responsible for logistics in these eastern member states not only makes them highly persistent, but very resistant to change. Whilst alignment to achieve interoperability and robust combat logistic systems is ongoing, what is now readily apparent is that these challenges are impervious to short-term modification and will require medium- to long-term organizational cultural change in order to achieve the ultimate goal of effective interoperability.

[56] A logistic estimate is a general term that describes the planning by tactical and operational level logistic staff principally to advise the commander about the logistic support required to undertake any action.

[57] Mission command is a philosophical approach to command widely practised in western militaries which emphasizes delegation of authority and responsibility to subordinates in closest proximity to the scene of action, coupled with ongoing overwatch, and if necessary direct intervention by the Commander as necessary.

[58] Thomas-Durell Young, 'Legacy Concepts: a Sociology of Command in Central and Southern Europe', *Parameters* 47, 1 (2017): 33.

[59] Young, 'Can NATO's "new" Allies and Key Partners Exercise National-level Command in Crisis and War?', 18.

Conclusion

This chapter has argued that combat logistics is critical to the conduct of war. Without it achieving the key roles of enabling movement, strengthening endurance, and providing sustainment, military operations are very likely to fail. This will be especially so in the case of any future NATO collective defence against Russia, which has been the focus of this chapter. In examining how NATO is preparing to deliver effective combat logistics in the event of a Russian attack, the article points to a number of important initiatives the organization has taken to ready itself. These include establishing two new commands, JSEC and JFC-NF, specifically oriented toward logistic delivery. As the chapter points out, these organizations are responsible for controlling and coordinating troops deploying to the operational space, with a major focus on the mobility activities of receiving troops, preparing them for operations, and then organizing their onward movement. At the same time, complicating this picture is the integration of contracted logistic firms that are often utilized through HNS. Providing force protection to these contracted logistic firms is another challenge NATO needs to address. With the disappearance of the frontline and rear bases and the 'last tactical mile' now effectively 'the last 1000 logistic miles', targets that would have been relatively safe had the Cold War turned hot are now no longer so. This means logistic firms moving NATO supplies are likely to be targeted by a long-range Russian strike and thus will need NATO to provide force protection. However, probably the biggest challenge NATO faces with respect to organizing combat logistics is addressing the lack of logistic capabilities from former Warsaw Pact countries that are now full NATO members. As the chapter notes, this is an organizational and cultural challenge that will take time to resolve.

But it is not all bad news for NATO combat logistics. Some of the challenges mentioned above may be mitigated through emerging technologies and in particular robotics, autonomous aerial and land systems, artificial intelligence, and 3D printing.[60] Further, NATO logistic capabilities are likely to rely heavily on these technologies to enable the mobility, endurance, and sustainment of NATO land forces in any future conflict with Russia. It is imperative, therefore, that NATO continues to improve on these technologies, as well as develop new ones, not only to stay ahead of Russian combat

---

[60] An approach highlighted in the recently released (16 March 2021) UK MOD Integrated Review. See: Global Britain in a Competitive Age: the Integrated Review of Security, Defence, Development and Foreign Policy, Cabinet Office, 16 March 2021, https://www.gov.uk/government/publications/global-britain-in-a-competitive-age-the-integrated-review-of-security-defence-development-and-foreign-policy/global-britain-in-a-competitive-age-the-integrated-review-of-security-defence-development-and-foreign-policy.

logistic development, but to ensure it is able to counter any military surprises from Russian forces in the event of war. For example, countering new Russian missile and cyber technology, as well as threats from technologies that have not yet been developed, and aimed at undermining NATO combat logistics. At the same time, developing new technology will also enable NATO itself to threaten Russian combat logistic capabilities. Finally, insufficient emphasis on sound, resilient combat logistics is more likely to be a critical element in the operational failure of a future NATO collective defence than any strategic misconception. Put simply: combat logistics matters.

5

The Command of Land Forces

Jim Storr

Introduction

June, 1999. NATO's Allied Command Europe Rapid Reaction Corps entered
Kosovo as a result of Operation Allied Force. The initiating operation order
arrived from Headquarters, Allied Forces Southern Europe (in Naples) after
the force had arrived in Kosovo. It filled two lever arch files and was used as
a doorstop.[1,2]

Coalition forces deployed to Saudi Arabia and Kuwait in 2002 and 2003 to
invade Iraq under a four-phase plan. Phase Three covered offensive operations. Phase Four would be 'post hostilities'. Every operation order down to
battalion or battlegroup level showed no detail for Phase Four.

In 2009, the headquarters of the British 19 Brigade planned to seize the
crossings of the Shamalan Canal in Helmand, Afghanistan, because intelligence assessments indicated that the canal was impassable. Unfortunately, it
wasn't, and the Taliban defenders withdrew across it easily.[3] It is easy to tell if
a waterway is impassable: just go and measure it. A year later, Headquarters
11 Brigade made the same mistake on the same canal.[4]

NATO corps operation orders regularly run to 750 pages. Theatre-level
orders typically run to well over a thousand.[5] Nobody reads them. A NATO
corps headquarters can employ about 450 people, but the joke runs that only
about a hundred work there. Officers reckon that perhaps 20–25 per cent do
any useful work.[6]

---

[1] A major general, personal communication.
[2] I have known hundreds of generals. I have known dozens of them fairly well. The material for this
chapter is based on interviews with, discussions with, and remarks by hundreds of senior officers over
many years. My first notes were made in 1991, but some observations refer to events in the 1980s. However, for reasons which should seem obvious, I do not reveal my sources. Thus many references here will
effectively be anonymous.
[3] Theo Farrell, *Unwinnable: Britain's War in Afghanistan 2001–2014* (London: Penguin Random House,
2017), 257–8.
[4] Farrell, *Unwinnable*, 310.
[5] A lieutenant colonel, personal communication.
[6] A major general and a lieutenant colonel, separately, personal communications.

Jim Storr, *The Command of Land Forces*. In: *Advanced Land Warfare*. Edited by Mikael Weissmann and Niklas Nilsson, Oxford
University Press. © Jim Storr (2023). DOI: 10.1093/oso/9780192857422.003.0005

Such problems are symptoms of a systematic malaise which infects the command of NATO and coalition land forces. Repeated questioning of middle-ranking and senior officers found *unanimous* agreement with three simple propositions:

- that modern headquarters are too big;
- that the orders they produce are too long; and that
- they take too long to produce them.

That is, *unanimous* agreement. All those questioned pointed to the same problems, and to the same areas where improvements could be made. So, why do such problems exist, and how should they be overcome? Overcoming them would remove significant obstacles to land force effectiveness.

The accepted meanings of terms related to 'command' are problematic. 'Command' is defined by NATO as 'the authority vested in an individual for the direction, coordination and control of military forces'; together with the exercise of that authority (that is, 'to command').[7] So, very simply:

'command' is 'direction, coordination and control';

thus

'command *and control*' is 'direction, coordination, control *and control*'?

Clearly that is ridiculous. Furthermore, 'control' is normally considered to be a level of authority over assigned forces which is somehow less than that of (full) command.[8] That also renders the term 'command and control' largely meaningless. NATO does not define 'command and control', but most western armed forces appear to use it, often without thought. Additionally, NATO has no definition for 'control' in the sense of oversight, supervision, and coordination.

Command is considered here to consist of three major functions: the making of decisions; leadership; and the control of subordinates.[9] Thus 'command' here is what some readers might think of as 'command and control'.

Command is essentially a human activity. Collective human activities tend to be complex and poorly understood. There can be huge differences between how similar groups perform: think of sports teams. As we shall see, in this

---

[7] AAP-6, *NATO Glossary of Terms and Definitions*. Various editions.
[8] Ibid.
[9] Originating in *ADP Command*, 1–5; and carried forward in several subsequent high-level doctrine publications.

instance the key problem is not what it seems. The way ahead is, however, relatively clear. But it is likely to meet considerable organizational resistance.

This chapter looks, first, at the *purpose* of command. It then considers the main features of the overall command system. That is: the *products* it generates; the *processes* it uses; its *structures* and *systems*; and the *people* within it.[10]

Purpose

Three major aspects of purpose should be considered. They are: the purpose of command systems; the role of purpose within military operations; and whether a command system is fit for purpose.

The word 'command' suggests 'order' and 'control'. 'To order' has the sense of 'to direct'. To control is to regulate; to set bounds or limits.[11] This supports the description of command as consisting of decision-making, leadership, and control. Command directs everything else. Above the level of a company, the conduct of conflict and waging war *is* command. So is some of what occurs at lower levels. By the company level, we should include (for example) military intelligence and signal companies. The purpose of command systems is simple. It is to assist the commander in the execution of command: no more, and no less.

Historical sources often describe military activities which appear to have been pointless. That is, they seemed to lack operational or strategic utility. That perception is made worse where it is linked to loss of life: the suggestion that individuals, or numbers of soldiers, died for no purpose. That is an aspect of the wider narrative of the futility of war.

Wars and armed conflicts should serve valid political purposes. Thus purpose should stem from the grand strategic level, and there should be a continuous thread of purpose from there downwards to the activities of every single soldier. If no thread of purpose exists, the activity in question is, by definition, without sensible purpose. Here we define the strategic level as the national and political direction of the conflict. We define the operational level as that of the theatre and campaign, and the tactical level as the conduct of battles and engagements within a theatre. We shall see that

[10] The argument and structure of this chapter follows that of Jim Storr, *Something Rotten: the Command of Land Forces in the Early 21st Century* (Havant: Howgate Publishing, 2022), which discusses the subject at much greater length. For practical reasons much of the subject material is omitted here; not least most of the 600 or so references. I wish to thank both the Swedish National Defence University and Howgate Publishing for their consideration. Three reviewers contributed to the book, and hence this chapter. A fourth reviewed a draft of this chapter. I am grateful to them all.
[11] *The Concise Oxford English Dictionary*.

command processes can systematically, and inadvertently, break that golden thread of purpose.

Are command systems fit for purpose? Command systems only exist to assist the commander in the execution of command, and the purpose of military operations should flow down the chain of command. Therefore fitness for purpose should largely be equated with mission achievement. Do command systems support commanders in achieving missions? If so, they are fit for purpose. An alternative construct could be to suggest that command systems should assist the commander in 'winning' or 'losing'. Unfortunately, winning and losing are subjective and notoriously poorly defined.[12]

Command systems must function in the environment of war and conflict. That environment is, and long has been, dominated by complexity. War is unutterably complex. This is not a novel problem: soldiers have struggled to master complexity in war for at least two centuries. They have often succeeded. Conceptually, the way to master complexity is through familiarity with its particular conditions. So, much of commanding military operations in complex environments should centre on familiarity with those conditions, and on methods which allow commanders and staffs to develop that familiarity.

War and conflict are also lethal and adversarial. Massive advantage can be gained through pre-emption, which gains the initiative. Successful pre-emption should be exploited through speed of action. That helps retain the initiative, and contributes to surprising and shocking the enemy (or adversary) in subsequent operations. Thus command systems which can decide and act very quickly are particularly effective.

War and conflict are also unpredictable. Therefore there is little point planning operations several days in the future when the tactical situation is likely to change as a result of operations conducted today. Clearly that needs to be balanced by having a plan for the campaign, or even the war, as a whole. However, very few headquarters should ever need to plan beyond the next operation. For a division in war, that may mean planning no more than 24 hours ahead.

Soldiers can deal with complexity. The key problem is unfamiliarity, not complexity. Even the most astonishingly complex situations can be mastered if they are, or become, familiar. Once that is achieved, commanders can make decisions and act intuitively, fast, and effectively. As we shall see, that is not an alternative to formal planning methodologies. It is an improvement.

---

[12] See for example Jim Storr, *The Hall of Mirrors* (Warwick: Helion Publishers Ltd, 2018), 256–8.

Thus the purpose of command systems should be to assist commanders achieve the desired military outcome of the endeavour. That exposes several real-world problems. They include the possibility that the desired military outcome is poorly described, or unachievable; that the command system does not in practice assist commanders; that the golden thread of purpose breaks; or, worse, that the command system routinely breaks it as a matter of course.

Products

The primary output of a command system is the orders which headquarters issue to their subordinates. We now consider the requirement for those orders, their content, and how that should be revised.

Corps orders may be 750–1000 pages long. Some *battalion* orders have been over two hundred. But in Afghanistan in 2009, Regional Command South produced an order which ran to 120 pages. An Afghan Army brigade received a copy, analysed it, and produce their own order. It was two pages long. The apparent complexity of modern conflict does not require long, wordy, woolly, imprecise orders. Division-level operations in Iraq and Afghanistan were sometimes conducted by Coalition forces based on orders just 12–15 pages long.[13]

Orders should balance the need to succeed in the given mission with the need to ensure that the operation is relevant to its overall purpose. That is, the orders should form an explicit link in the golden thread of purpose. How the command intends to achieve a given mission is largely a matter of tactics, and not discussed here. How the golden thread of purpose is spun, however, is critical but largely overlooked.

What is needed is a nested set of orders, particularly from the theatre level downwards. The initiating directive may be quite short. For example, the initiating directive for Operation Overlord, the Normandy landing in 1944, was five pages long. (The sixth page showed the distribution of the 24 copies.) It had six annexes. All but two were single-page tables. The other two were two pages long.[14] There then followed an unimaginably vast amount of paperwork.[15] But after the landing, orders were typically very short. General Leslie McNair, the commander of US Army Ground Forces, had directed that '[f]ield orders should be oral or in message form for all elements of

---

[13] A major general, personal communication.

[14] *Operation OVERLORD*, 9 March 2021, https://www.ibiblio.org/hyperwar/ETO/Overlord/Overlord-SHAEF-Dir.html.

[15] See, for example, Headquarters 1st Canadian Army War Dairy for December 1943 to March 1944 13 August 2021, https://heritage.canadiana.ca/view/oocihm.lac_reel_t6676/5?r=0&s=5.

divisions and frequently for the corps'. They should typically be 'a few lines long'.[16] When the Second French Armoured Division (*Deuxième Division Blindée*, '2DB') liberated Paris on 24 August 1944 the divisional order was one page long.[17] Wehrmacht orders were typically even shorter.

The nesting of orders should start with the initiating theatre directive and operation order. Thereafter every subordinate headquarters should concern itself only with the current operation and what to do next. Planning too far ahead is a gross waste of time and effort. Thus, in general war, a battalion needs to know what to do in a few hours; a division tomorrow; and so on. With timelines so short, orders can be very short. Only theatre headquarters should ever write campaign plans.

Commanders are currently encouraged to focus their individual input on the formulation of a narrow part of the order known as 'the commander's intent'. That is not enough. Commanders should drive planning. The content of the Situation and Execution paragraphs should describe *the commander's view* of the situation, and *how he plans* to achieve the mission given to him by his superior. Anything else smacks of staff-driven planning. That tells us that the commander is not in charge.

If the initiating order for Overlord can be just five pages long, why should *any* order be much longer? 'An order should contain all that a subordinate needs to know to be able to execute his mission—and nothing more.'[18] The proper characteristics of an order are timeliness, clarity, simplicity, and brevity. But today orders tend to be excessively formulaic, both in structure and in language. Linguistic precision means more than the precise meaning of individual words. In writing orders, planners should focus on the precise use of language to convey meaning, rather than on formulaic assumptions of completeness.

The most important tool in ensuring that the golden thread of purpose is not broken is the structure of mission statements. The commander should write the mission of each of his principal subordinates, in the form of an instruction. The order, and if appropriate the mission statement, should explicitly describe the intended purpose. Subordinates should not change that mission without exceptional reason. Alternative processes, such as those generally used today, often break the golden thread of purpose. They typically do so in small, cumulative, and ultimately critical ways.

[16] Kent Roberts Greenfield, Robert R. Palmer, and Bell I. Wiley, *The United States Army in World War II. The Army Ground Forces. The Organisation of Ground Combat Troops* (Washington, DC: Center of Military History, United States Army, 1946), 378.
[17] 'S'emparer de Paris. Ordre d'operation pour la journée du 24 août 1944', in *La liberation de Paris*, edited by Jean-Pierre Bernier (1984).
[18] Bruce Condell and David T. Zabecki, eds, *On the German Art of War: Truppenführung* (Boulder, CO: Lynne Rienner Publishers, 2001), Para 73.

When looked at in detail, much of the content of current orders is either duplication or counterproductive. Practically any order, after an initiating operation order, can and should be reduced to one page or less. It will normally have to be accompanied by a graphic, and often by a couple of short annexes. For that to happen, the main requirement is that the initiating headquarters, and those receiving the order, are broadly familiar with the operational scenario. Planning and preparation prior to deployment should therefore focus on generating and sustaining that familiarity. In 1944 2DB's order could be one page long not because the division was familiar with seizing national capitals; but because commanders and staff were very familiar with working together as a team.

Processes

Products, such as orders, result from processes. However, command processes include not just the planning but also the conduct of operations. We must also consider how to *train* headquarters to plan and conduct operations.

Eisenhower wrote that 'in planning for battle I have always found that plans are useless, but planning is indispensable.'[19] That is clearly not literally true. But what did Eisenhower actually mean?

Current planning methods include NATO's 'Comprehensive Operational Planning Directive' (COPD) at the theatre level and the Military Decision Making Process for the tactical level. The COPD includes the 'Comprehensive Preparation of the Operational Environment'. The US Marine Corps has its Marine Corps Planning Process; and so on.

All such methods are lengthy, explicit and (in practice) collective. Planning involves several people; often dozens. Two British brigades, observed recently, involved 40–45 people each. They took 10–12 hours to plan a fairly straightforward, conventional warfighting mission. Such planning involves a number of discrete, explicit techniques. Examples include intelligence planning of the battlefield (or environment), review of concept drills, and wargaming. Because those techniques are explicit and collective, they take a long time. Because they are collective, they are also consensual. That typically results in mediocre, lowest-common-denominator plans.

Planning does not need to involve so many people, nor take nearly as long. They did not during the Second World War nor the Cold War. Two officers who served as chiefs of staff of a British armoured brigade in the 1980s could

---

[19] Dwight D. Eisenhower, quoted in *Allied Joint Doctrine for the Planning of Operations (UK Joint Doctrine)*, Allied Joint Publication-5. OPERATIONS. Edition A Version 2 with UK national elements, May 2019, 1–2.

plan and write orders for very similar operations in 30–40 minutes.[20,21] That was common in many armies. It is generally accepted that a force which can decide and act faster will normally beat a slower force. So why should armies choose to, or make themselves, decide and act slowly?

Battalions should be able to plan for a new mission in a familiar environment in an hour. Corps should be able to do so in four hours. In practice they should be *required* to do so, and therefore trained and exercised in doing so. Such speed was entirely normal in the Second World War and persisted in places throughout the Cold War. The inability to do so today is largely a result of low expectation, coupled to a mistaken fascination with complexity. To that should be added collective, explicit processes, and a lack of familiarity with the scenario. The latter can be overcome quite simply. The former requires far fewer planners and moving away from explicit, structured planning methods.

Psychologists have studied how experts make decisions in real-life situations. When not constrained by explicit process, experts tend to mentally explore a problem until they recognize it as something with which they are familiar, and to which they can envisage a solution. They then mentally adapt that potential solution to the problem at hand. That is rarely conscious: the decision maker just 'knows what to do'. At brigade level doing that, and writing down the solution (in this case as a set of orders), might take half an hour or so. There may need to be some technical input from others (such as time and distance calculations, or an artillery fireplan). For platoon-level situations the process might be almost instantaneous. However, most importantly: for experienced decision makers, naturalistic methods typically produce better results than explicit processes do. They are also *much* faster.

Thus the way forward for planning is threefold. First, one person should make the plan; advised where necessary by a few others. Secondly, explicit process should largely be abandoned. Lastly, training should develop decision makers' overall expertise and their familiarity with the particular scenario.

Turning from planning to conduct, conflict (and battle) is not a stage play which will follow a closely-worded script. It simply is not. So the conduct of operations is not the same as executing closely-synchronized plans made in advance. Not least, 'the enemy gets a go, too'. Therefore the adversary's

---

[20] A lieutenant general, personal communication.
[21] A colonel, personal communication.

(and other parties') actions will seriously disrupt closely-synchronized plans made in advance. Subordinates' actions must be coordinated, but top-down synchronization constitutes over-control. It generally does not work well.

Some of the techniques currently used in planning would be better used in collective training. That is, outside major command post exercises (CPXs). They include cloth model, sand table, or table-top exercises currently used as 'review of concept' tools'. They can, and should, be used to teach subordinate commanders how to coordinate actions *between each other*. They would also teach staff how to coordinate subordinates' actions from above, but *only where strictly necessary*. That would result in far greater decentralization, which is a very powerful tool for coping with complexity.

Battle is not a stage play. Therefore CPXs should not be acts of theatre, which they currently tend to be. The emphasis is currently on working through the processes, and to some extent on presentation. CPXs should be conducted as wargames: two-sided, free-play, and in real time. If they were, staff would soon learn to think and act much faster, to trust subordinates (and therefore decentralize), and to abandon most explicit processes.

Much collective training is currently conducted as CPXs supported by computer simulation. Such exercises are themselves generated by a very process-driven methodology. Nothing significant is allowed to go wrong, not least because that would wreck the closely-written Main Events List (MEL). Many shortcomings result from that. Not least, the tempo is driven by the MEL, not the speed at which the better side can operate. Long, procedural CPXs allow little time for other individual or small-group training, so professional expertise is often paper thin.[22]

So, for example, in some armies officers could not coordinate their actions in time and space. That resulted in explicit, top-down synchronization.[23] Cloth model, sand table, or table-top exercises, and tactical exercises without troops, should receive far more attention.

The overall focus should not be on staff processes but on making decisions very quickly, then translating those decisions into action against the enemy much more quickly than at present. The current separation of operations into 'planning' and 'execution' is artificial and unhelpful.

---

[22] The chief of staff of a British brigade admitted as much, about his own professional knowledge, after an exercise in 2021. Roughly half of the staffs of two brigade headquarters present agreed with him.
[23] For example, see Lt Col John F. Antal, 'It's Not the Speed of the Computer that Counts: The Case for Rapid Battlefield Decision-making', *Armor*, May–Jun 1998, 12–16.

Structures

To repeat: commanders were unanimous that current headquarters are too big. Three aspects of structure are relevant: the organization of the force as a whole; the size of the command post; and its internal rank structure. All three need attention.

The management of an organization is roughly dependent on the square of the number of major subordinates.[24] So, for example, a division with five combat brigades is about three times more complex than one with three. Historically, commanders have often struggled when the effective span of command of a formation is materially greater than two.[25] However, in Iraq we saw corps commanding six principal subordinates. In Afghanistan divisions commanded up to eight brigades.

Above a certain size, divisions need brigade- or group-level staffs, led by brigadier generals or colonels, for many of their major functions (such as engineering, maintenance, or logistics). In smaller divisions the commander of the relevant battalion functions as the staff adviser in the headquarters, and staffs are much smaller.

During the Second World War, staffs were very small. For example, a British divisional headquarters contained a total of forty-eight staff officers. German staffs were smaller; American staffs were slightly larger. Staffs grew very slowly: they were roughly double that size by 1990–1991. By 2020 they were perhaps ten times that size.

Rank representation was also very low. That British divisional headquarters had one brigadier (the artillery commander), one colonel (the chief medical officer), and seven lieutenant colonels. Today there might be four brigadiers, four colonels, and many more lieutenant colonels. Officers of such ranks, and specialists, have a negative impact. The net result is too many levels between the junior staff and the commander. Amongst other problems, the commander tends to be told what he wants to hear, not the objective facts. The presence of senior officers tends to lead to more process, slower decision making, and longer orders. The negative consequences of the presence of numerous senior officers greatly outweigh the perceived advantages. The establishment of assistant and deputy formation commanders is particularly damaging. Both staff size and unnecessary rank representation cause major difficulties. Both should be reduced considerably.

Overlarge headquarters are less effective (measured by, for example, timeliness or quality of output.) They are also less robust. Their lack of mobility, and vulnerability, are obvious and (in practice) ignored.

---

[24] Jim Storr, *The Human Face of War* (London: Continuum, 2009), passim and 128.
[25] Storr, *The Human Face of War*, 119–20.

The end of the Cold War and the move to conflicts of choice increased the availability of staff officers. Additionally, new NATO partner countries from eastern Europe were keen to contribute staff. The net result was a massive increase in the availability of officers to staff increasingly large headquarters.

There are strong relationships between command post size, coherence, and effectiveness. Put simply, effectiveness falls considerably with increasing size. These issues were studied extensively 20 years ago and more, but the findings were overlooked. Simple mathematical analysis suggests that larger staffs are much more complex and harder to manage.[26] Typical consequences are more internal structure and formal process. Thus the greatly enlarged headquarters which have developed since the Cold War are unbelievably and unnecessarily bureaucratic. Additionally, larger staffs tend to demand more information and then use it inefficiently.

Staffs should be scaled so that command posts can plan; conduct operations; and move. They must be able to do those things around the clock for months on end. However, very few dedicated planners are needed. Staffs scaled to the *minimum* needed to meet those requirements (with a sensible amount of double-hatting) can be very small indeed.

The way ahead would see slightly more, much smaller, corps headquarters. Divisions should be smaller, thus requiring less rank representation in their headquarters. Those headquarters would also be much smaller. Armies would typically have slightly more, but smaller, divisions.

Brigade and battalion (or battlegroup) headquarters should also be reduced. The target might be fifteen or twenty officers in a brigade headquarters, leading up to perhaps a hundred at corps. However, staff would also be needed to bridge the gap between formation and land component (or national land contingent) headquarters. A dedicated staff group designed to do that might contain as many as ten officers. There should be very, very few staff officers in any headquarters above the rank of major.

The cost savings would be dramatic. However, there would be consequences for staff selection and training, which will be considered later under 'people'.

Systems

This section considers the tangible elements of the command system. It largely focuses on communication and information systems (CIS). After 20 years of using digital CIS widely on operations, we should really question

[26] Storr, *The Human Face of War*, 150–1.

what impact it has had. In practice much of that impact is negative. That is largely a consequence of the way that CIS was introduced.

Humans aggregate information all the time. For example, they would not generally notice a group of perhaps two dozen trees; they would see, and report, a wood. Furthermore, trials have repeatedly shown that information quality and quantity make very little difference to the outcome of decision-making. The personality of the decision maker, and his skill and experience, had far more impact.[27]

Interviews with senior commanders up to theatre level showed that very little information is required for decision-making: if the assessment of the situation is appropriate and relevant.[28] 'Relevance' includes being aggregated to the right level. Hence it seems that, for skilled commanders, communications provision is generally adequate: any shortcomings are typically organizational or procedural. In practice any apparent demand for more information is a sign of poor commanders, poor processes, and over-large staffs. Command systems should not be information-*intensive*. They should be information-*sensitive*. That is: sensitive to critical items of information; aggregated to the appropriate level; and timely.

Command systems should not be seen as technical, but as *socio*-technical, entities. The way that the people interact, both verbally (through speech and text) and non-verbally has a major and largely unexplored impact on command effectiveness. That partly explains why, when developing its first-generation digital systems, the British MoD found *no* credible evidence that digital IT would have any positive impact on operational effectiveness.[29] Wider studies found that the impact of digital technology is not neutral: it is negative, unless the organization considers the business which it digitizes. Improvements to effectiveness usually come from analysis of the business (prompted by digitization); and only indirectly from the technology itself.[30] That still seems to be true.

The introduction of digital CIS into land forces resulted in over-command, information drag (not least, delays whilst waiting for systems to update), information overload, and information management overheads. It would, however, be entirely possible to redesign the use of CIS to hold and pass less data, but rely far more on information aggregated to the appropriate

---

[27] The relevant trials were first conducted by the Swedish armed forces. They were repeated, with the same results, on a group of forty-five qualified Royal Navy Principal Warfare Officers in 2001.

[28] The relevant interviews, some conducted over 20 years ago, included a future British CDS and a future DSACEUR.

[29] I was responsible for monitoring that research from 1997 to 1999, and then contributed to it from 1999 to 2001.

[30] The late Graham Mathieson, DERA Portsdown West, personal communication. Mathieson was a highly perceptive, very experienced, and internationally respected senior analyst.

level of abstraction. (Figuratively, 'woods, rather than trees'.) That would support both expert decision-making and the development of decision-making expertise. It would result in better and faster use of information, and therefore increased operational effectiveness.

In broad terms, CIS experts did not collect particularly perceptive information requirements. Therefore they did not develop good information services. But why did they not collect good information requirements? Was it because they were arrogant, and thought that they already knew what was needed? Or did operations staff not take the time to tell them? Or did they not know enough about socio-technical systems to know which questions to ask? Did *anyone*? Did people listen to the few people who did? Whatever the answer, this area is very much a human issue. It is not primarily technical.

The way that CIS was fielded is a major driver in the design of current command posts. By accident rather than design, western armies have created the best military targets in the world. That is: static, poorly-protected, high-value command posts. Practically every major command post in Iraq and Afghanistan was attacked; some of them several times. One was hit by a FROG rocket.

Open-plan command posts are fundamentally flawed. They interrupt the concentration of expert planners. Planning becomes collective, consensual, and slow. It results in lowest-common-denominator plans. Open-plan command posts also interrupt Current Operations staff working hands-on in near-real time. Reconfiguration on a cellular basis would largely overcome those shortcomings. Doing that, together with a different approach to the use of IT and better processes, would *improve* situational awareness; both within and between command posts. That would improve effectiveness considerably. Not least, it would largely remove the need for time-consuming, inefficient, face-to-face meetings.

Command systems are much more than just the CIS which people use. The introduction of digital IT is now largely taken for granted. However, it has largely been responsible for (or contributed to) several negative consequences. They are often not obvious, or accepted as simply 'the way things have to be'. They are not.

People

There is considerable variation between individuals. There is also considerable variation between armies. Although all armies have corporals, captains, and colonels, the processes which produce them vary considerably.

Accepting such differences, this section looks at the main characteristics of the human component of the command system. That is: career structure and progression; selection; and education and training.

The broad requirement for a career structure should be to fill command appointments and to provide small numbers of talented staff. It should not be to produce bloated hierarchies, nor to fulfil inappropriate ambitions. There are considerable differences between different armies' career structures. Some appear to be better than others.

It is an honour, a privilege, and an almost sacred responsibility to command soldiers. It is the primary duty of an officer. It is what he or she should do most, as their careers progress. If an officer does not see it that way, she or he has chosen the wrong career. If officers see time in command as merely steps to be climbed on the way up the greasy pole of promotion, the army is selecting the wrong people and giving them the wrong incentives.

Many armies appoint far too many middle ranking and senior officers. In 2007, Swedish officers pointed out that the Finnish Army had less than half as many field and general officers as the Swedish Army did, and that that seemed to be beneficial. In 2008 the US Army had proportionately 3.33 times as many colonels as the US Marine Corps. It is hard to believe that the impact of that is positive.

There are typically three platoons per company, three to five companies per battalion, and so on. That implies a requirement for fewer officers at successive ranks. Every subaltern cannot expect to command a battalion, let alone a brigade. Should they all expect to command companies, as they do in some armies? Longer command tours would generally be beneficial, and would reduce the numbers promoted to higher ranks. Command tours of less than two years tend to allow officers to escape many of the consequences of their decisions. Reducing numbers at successive ranks, selecting for talent, and satisfying sensible career aspiration is not easy. It requires careful management over decades.

The process which eventually selects generals also selects captains and colonels. It seems that most armies broadly select and promote the right people. However, some of the wrong people are promoted whilst others, just as capable or more so, are overlooked.

In the interwar period, the Wehrmacht was cautioned to avoid promoting officers with 'sordid ambitions'.[31] Today we would say 'excessively careerist'. *Officers* who are clearly driven by career advancement should be weeded out.

---

[31] German General Staff Project # 6. *Training and Development of German General Staff Officers.* Vol III. Operational History Branch, Historical Division, European Command 1948. Interview with General of Infantry Kurt Brennecke, former director of the Wehrmacht School for Commanding Generals, P30.

Armies whose officers are clearly driven by career advancement need to take a good look at their reward structures. Their selection processes are probably not sensitive enough.

There is good evidence of preference for stereotypes. There appears to be an evolved preference for taller, more authoritative, square-jawed men in leadership roles.[32] It seems more likely that they are preferred, than that they make better commanders.

Academic research can identify the psychological profile of successful senior officers in a given army. It can also demonstrate that selection processes can fail to promote some of those who fit that profile, whilst promoting some that do not.[33] Furthermore, the profile might indicate the most successful, but not the best. Authoritarian characters will tend to thrive in highly organized hierarchies, such as armies in peacetime. Unfortunately, highly authoritarian senior officers will tend to fail, and fail catastrophically, in war.[34] In some armies there seems to be a tendency to prefer extremely self-confident senior officers over their more competent peers. Analysis of those in command at the end of long wars suggests that the charismatic, the authoritarian, and the inept generally disappear from the higher ranks.

From a base of having enough subalterns to command platoons, armies must train officers for command and staff posts at successive ranks. As we saw above under 'structure', staffs should be small, highly trained, and relatively junior. That implies delivering the bulk of individual training and education to captains and majors; and focusing intensive training on a small number of carefully-selected majors.

Reflection suggests, first, that in practice officers just cannot be given too much instruction. Secondly, in most armies they do not receive enough. Since the end of the Cold War, the British and American Armies have reduced the amount of training given to their captains. Instruction for the best British majors has been reduced alarmingly. In many armies syllabus time has been diluted considerably by introducing 'Defence Studies' into the curriculum. Defence Studies are a nicety, not a necessity. The overall result is that very few officers, if any, are trained to the point where they can be expected to make good, rapid, naturalistic decisions. They genuinely do not 'just know what to do'. That is not their fault. The fault lies with the training and education system.

[32] 'Bartleby' in *The Economist*, 9 September 2019, 64. The use of the word 'men' was deliberate.

[33] Richard Sale, 'Towards a Psychometric Profile of the Successful Army Officer', *Defence Analysis* 8, 1 (1992): 3–27.

[34] Norman Dixon, *On the Psychology of Military Incompetence* (London: Random House, 1976), passim. Modern historians sometimes disagree with historical aspects of Dixon's work, but generally agree with his insight into the authoritarian character, and authoritarians, in high command.

To summarize, there seems to be little wrong with the human raw material. However, armies tend to promote too many officers to middle and upper ranks. The emphasis in peacetime selection is somewhat misplaced. Amongst the higher ranks it may prefer the overconfident over the highly competent. It can allow authoritarians to prosper, and sometimes promotes the dutiful but dull. Armies should focus their command and staff training on fewer officers, train them better, and focus the scope of that training more narrowly. The reason for doing so is simple: to improve the overall effectiveness of the command system.

Summary and Conclusions

Commanders were unanimous that western headquarters are too big, and take too long to produce orders which are too long. There is a real Gordian Knot of products, processes, structures, systems, and people. It needs to be cut through in order to improve effectiveness.

Command systems only exist to support the commander in the exercise of command. Conducting operations should be a closely-integrated process of planning and execution. Separating planning and execution, or seeing them as consecutive, is an error. Every headquarters should be a link in a chain that creates, and maintains, a golden thread of purpose from the grand strategic level down to the actions of every soldier. Command systems must function in the complex, lethal, and adversarial environment of conflict. Not least, they should be able to decide and act very quickly.

Once a campaign is under way, orders should be very short: perhaps a page of text and a schematic, supported by a few short annexes. Mounting an entirely new divisional-level operation, for example, might need as much as a dozen or fifteen sides of text.

That will require major reductions to processes. Commanders should play a far bigger role in planning. If a commander does not drive the planning, he does not own the plan. If he does not own the plans which his headquarters creates, he is not in charge.

Prior to hostilities, planning should concentrate on making the commander and staff deeply familiar with the environment and dynamics of the coming campaign, at a level relevant to the command. Battalions will not know precisely where and who they are going to fight after the first day. However, they should be familiar with the terrain and enemy in general, and how to fight and win in those conditions.

Command systems should support naturalistic decision-making. That requires highly-talented planners who are thoroughly familiar with the environment, the capabilities of their forces, and the team they work with. Plans should be made by one man, with input from a few others. The resulting order should be written by one man: often the same individual. It should be routine for battalions to plan and issue orders in an hour. Corps may need as much as four hours.

That requires much leaner structures. The staff should mostly be very junior but extremely well trained. There should be very few people in a command post above the rank of major.

The CIS within command posts should be trimmed down. Its functionality should be reduced to support rapid, naturalistic planning. Fewer staff would need smaller command posts, which could then become more mobile and more survivable.

Smaller, more junior staffs should reflect the structure of their armies. Many western armies produce far too many senior officers. Selection should focus on promoting the genuinely competent, weeding out the merely self-confident or charismatic and the authoritarian. Individual, residential training and education should be more focused on fewer, more junior officers.

If you are a battlegroup commander reading this, you have a stark choice. You can spend ten or a dozen hours with your staff planning and writing a set of orders. Or you can produce a set in an hour; debrief it; repeat that a few times; and reflect. By the end of the day, you and your people will be much, much better off. Adjust the times, and the same applies to brigades; or divisions; or corps.

History tells us that it is simple to beat a 'dinosaur' army: do things which are simple, violent, but above all quick. If you belong to a western army, you can continue as you currently do. You have good people. You might not have the right ones in the right places. The best of them are not trained well enough. You use too many of them, in headquarters which are too big and too busy. They produce orders which are too long, and take too long to do so. The CIS they use often gets in the way. You can continue that way if you wish. But if you do, somebody will beat you. Wake up.

6

Tactical Tenets

Checklists or Toolboxes

B. A. Friedman and Henrik Paulsson

Introduction

Tactical theory is about thinking about tactics. Many of the classic works on warfare, from Sun Tzu to today, reflect how people and institutions think about combat more than they reflect actual events, which are always mired in the muddy and muddled reality of human conflict. Yet, understanding how military forces intend to fight should combat occur is an important aspect of military affairs. Tactics is the practice of combat.

Whilst tactics are generally seen as the sole province of practitioners, they should also be of interest to politicians and other policymakers that oversee them, and to academics in the history, strategic and security studies, and military science disciplines. Although the focus of the latter two is generally, and should be, on strategy, tactics should not be ignored as strategic possibilities are bounded by tactical realities. The only strategic goals that can be accomplished are those which can be tactically achieved. Thus, a common set of tactical ideas, a theory of tactics, is as useful and important as strategic theory, and would foster a healthier civil–military discussion between those that employ war, those that fight wars, and those that study wars. These common sets of tactical ideas are termed here tactical tenets, core concepts that are useful tools for theoretical and practical analysis.

In this chapter we discuss first what tactics is, its history, and how different types of tactics exist. Next, we discuss what tactical tenets are, and how these are not checklists, in that they follow law-like rules as principles, but rather toolboxes of useful heuristics. These tenets are useful both for the practitioners of tactics in the field, but also for analysing military forces, their doctrine, and historical engagements. We use these tenets to contrast the US Army from the Marine Corps as two services with distinct tactical approaches despite being from the same country and both fighting as ground forces.

B. A. Friedman and Henrik Paulsson, *Tactical Tenets*. In: *Advanced Land Warfare*. Edited by Mikael Weissmann and Niklas Nilsson, Oxford University Press. © B. A. Friedman and Henrik Paulsson (2023). DOI: 10.1093/oso/9780192857422.003.0006

What Is 'Tactics'?

The word tactics comes from the ancient Greek word 'art of arrangement'.[1] At its core, it is the art of arranging military forces to defeat an opposing force. Arrangement implies more than one agent or object, so this is not about the actions of two individuals fighting each other. Art implies creativity, which is less about rules and more about inspiration.

The point of this artful arrangement is combat. Specifically, it is about winning in combat. It is less about why the combat is occurring, that is, the purpose of the war or strategy, and more about how to win the engagement at hand. The tactician, by profession, is less concerned about the war and more about the battle. Because of this focus on combat, tactics refers to the interaction of opposing military forces; those within rifle or cannon shot of each other. In most cases, tactics is not concerned with scale. Whether the forces are composed of just a few soldiers or entire fleets and armies, tactics is what they do to defeat each other.

Tactics is easily confused with techniques, procedures, and doctrine. Frequently lumped together as if they were synonyms or even one word, there are important distinctions between them. Tactics refers to the entire range of possible ways military forces can be employed in combat. For most military forces, doctrine is the codification of preferred and tested tactics, along with preferred techniques and procedures for the equipment and units of a specific military institution. This is an important distinction as a theoretical discussion of tactics must look beyond the specifics of any one military institution or even one time and place, and rather must examine tactics as a constant phenomenon across history The relationship between tactics, as a constant, contrasts with techniques and procedures; this is akin to the Clausewitzian constant *nature* of war compared to the spatially and temporally dependent *character* of war; this chapter covers tactics and the tenets stemming from it.

The History of Tactical Theory

A history of how tactics have changed over time would occupy many pages and, indeed, many books have been written on that subject, such as Archer Jones' *The Art of War in the Western World*.[2] This section will instead examine changes in how practitioners and theorists thought about tactics, based on

---

[1] Brett A. Friedman, *On Tactics: A Theory of Victory in Battle* (Annapolis: Naval Institute Press, 2017), 16.

[2] Archer Jones, *The Art of War in the Western World* (Champaign: University of Illinois Press, 2001).

available discourses on the subject. Whilst most landmark works of military theory focus on strategy and occasionally statecraft, few ignore the battlefield entirely. We can therefore glean predominant ideas about warfare even from extant sources that focus more on war as a whole. There have been three major paradigms, or regimes, of tactical theory, although there are certainly exceptions. These three tactical regimes can be called virtue tactics, linear tactics, and modern tactics.

Virtue tactics predominated until the pre-modern period focused on the character and virtue of leaders and combatants themselves. Thinkers were less focused on technological superiority or the arrangement of forces, focusing more on the moral factors and courage displayed both collectively and individually. Virtue tactics thus focused on the human capital of the battlefield rather than weaponry, although this clearly did not preclude innovation in weapons and force design. Tactical leadership focused on the general fighting in or around the ranks displaying virtue through personal example. Mechanical tactics, beginning in the pre-modern era with its focus on rationality, went in the opposite direction, focusing on geometrically-based manoeuvres and rules almost to the point of denying human agency entirely (which, as we will see, at least one writer did). Tactical leadership was depicted as a general surveying the entire battlefield and acting as a puppet master, moving pieces on a chessboard rather than being a piece himself. The modern regime, beginning with Romanticism and specifically Carl von Clausewitz, can be viewed as a synthesis of the two previous regimes. Rules of thumb, or rather principles, are important in guiding the tactics, as is technology and employment, but moral factors and virtue cannot be discounted entirely. Tactical leadership, whilst performed behind the lines by a directing general, still requires the general to have virtue as he or she seeks to overcome the friction of institutional inertia and the human factors of combat to achieve victory.

Virtue Tactics

The earliest written depictions of war all stress the virtue of the participants as a determining factor in victory. This includes Homer, whose epics focus on individual virtue, as well as Plato and Thucydides who began to depict collective as well as individual virtues in war.[3] For the ancients, virtue also compelled nations to war. However, for Plato at least, it was the negative virtue of avarice that compelled nations to go to war. Whilst we know a great

---

[3] James L. Cook, 'Plato: Virtue and War', in *Philosophers on War*, edited by Eric Patterson and Timothy J. Demy (Middletown: Stone Tower Press, 2019), 39.

deal about ancient and early medieval tactics through works of history, we can learn about the ideals of virtue tactics more through works that resemble manuals, such as Sun Tzu's *The Art of War*, than through works of history. Whilst *The Art of War* is by far the most famous, a clearer picture of virtue tactics emerges from a wide view of such works, including other Chinese works such as T'ai Kung's *Six Secret Teachings* and *Three Strategies of Huang Shih-kung*, the *Arthashastra* of India, *De Rei Militari* by Vegetius, a Roman military manual complete with excerpts from earlier works, and *Strategikon*, written or compiled by the Byzantine Emperor Maurice.

These works tend to focus on individual and collective character. It is not necessarily superior technology or employment of forces that wins, but the moral superiority—in terms of courage, fortitude, and discipline—of one side or the other. This applies to the tactical leader, whether a general, king, or emperor, as well as the soldiers themselves, although to a lesser degree. *The Art of War*, for instance, discusses generals almost as if they must be god-like figures: 'The general is the supporting pillar of the state. If his talents are all-encompassing, the state will invariably be strong.'[4]

This focus on discipline makes sense because, prior to the modern era, soldiers and sailors were poorly paid, if they were paid at all. They were frequently conscripts, sometimes even slaves, or part-time militia. Such men had little reason to stick around and fight if things started to look bad. Discipline, even harsh discipline, was necessary to even maintain an army to fight.[5] Militaries that developed a more professionalized core of career soldiers—most notably the Romans—tended to be vastly superior to other forces. The Ottoman janissaries were, for a time, one exception to this trend.

The height of this view was the chivalry of the high Middle Ages. Military writing during this time was almost entirely focused on the moral character and martial prowess of the mounted, noble knight. Employing other arms, like infantry, was either ignored or taken from Vegetius, who continued to be popular. Even coordinated cavalry tactics went undiscussed as it was assumed that many would fight as individuals. Medieval military leaders believed that the faith, courage, and devotion of the nobility would win the day. This attitude surely led to hasty and ill-considered tactics such as the headlong cavalry charges in battles like Crécy in 1346 and Agincourt in 1415, where French knights were mowed down by peasants wielding longbows despite courageous action.[6]

---

[4] Ralph D. Sawyer, *The Seven Military Classics of Ancient China* (New York: Basic Books, 1993), 161.

[5] Beatrice Heuser, *Strategy before Clausewitz: Linking Warfare and Statecraft, 1400–1830* (London: Routledge, 2018), 58–59.

[6] See John A. Lynn, *Battle: A History of Combat and Culture* (Boulder, CO: Westview Press, 2003), ch. 3 for attitudes about chivalry and military affairs.

Tactical leadership was personal. The general, who was usually also or primarily a political figure such as a king or consul, led from the front with his or her voice and personal example. This was not just a recommendation: generals such as Alexander the Great and Gaius Julius Caesar were known for using personal, front-line leadership to inspire their men to greater efforts, a method which frequently worked. Conversely, the death of a leader could cause the collapse of entire armies, such as the Battle of Cunaxa in 401 BC where the coalition of Cyrus the Younger fell apart mid-battle when Cyrus himself was killed.[7] The focus on the character of leaders and soldiers continued into the Renaissance. Niccolo Machiavelli, for example, advocated less reliance on mercenaries and the formation of a standing militia based on the presumed greater motivation and devotion.[8]

It is commonly believed that Eastern military theory focuses on deception, stratagems, and intelligence whilst Western military theory focuses on direct battle, especially infantry battle, and open confrontation. This belief is based mostly on Sun Tzu's *The Art of War* and Kautilya's *Arthashastra*, both of which focus on indirect fighting and the use of espionage. This is less a function of martial practice and more a function of their audience and subject matter: both works were written by and for members of the political classes and thus focused more on statecraft than the actions of military professionals. Warfare in both ancient China and India featured as much direct infantry combat as anywhere else and even other, lesser-known Chinese works of military theory feature greater focus on concepts like concentration of force and more focus on direct battle. Far from being an Eastern ideal, Sun Tzu's focus on deception and subversion was seen as immoral by later generations of Chinese scholars.[9]

In fact, Western equivalents show a remarkable similarity, rather than dissimilarity, to Eastern thought. Vegetius' *De Rei Militari*, far and away the most influential work of military theory in the west prior to Clausewitz, also advocates and focuses on surprise attacks and ambushes: 'An able general never loses a favorable opportunity of surprising the enemy ... [M]ilitary skill is no less necessary in general actions than in carrying on war by subtlety and stratagem.'[10] Parallels between Eastern and Western military works are eerily similar during certain time periods. For example, military works by Byzantine writers such as *Strategikon* by Maurice and *Taktika* by Leo IV

---

[7] See Xenophon's *Anabasis*.

[8] Niccolo Machiavelli, *The Art of War*, trans. Ellis Farneworth and ed. Neal Wood (Cambridge: De Capo Press, 1965), 16–19.

[9] Edward L. Dreyer, 'Continuity and Change', in *A Military History of China*, edited by David A. Graff and Robin Higham (Lexington: University Press of Kentucky, 2012), 21.

[10] Flavius Vegetius Renatus, *On Roman Military Matters*, trans. John Clarke (St Petersburg: Red and Black Publishers, 2008), 87.

make many of the same recommendations as Chinese military writers like the Tang Dynasty general Li Jang. Major differences tend to be in the realm of civil–military relations, not tactics.[11] Conceptions of divergent 'Eastern' and 'Western' ways of warfare are simply not supported by the historical record or military theory.

Linear Tactics

During pre-modern times, military practitioners and thinkers in Europe began to see tactics in purely rational terms, especially as the Enlightenment fostered an environment of scientific and philosophical inquiry. In military affairs, the development of gunpowder and the consequent need for mathematics and engineering to better exploit it drove a premium on geometric tactics, especially when it came to siege and fortifications.

These trends produced linear tactics, also termed linear warfare by John Lynn.[12] Military thinkers believed tactics were linear in both senses of the word: that lines and angles were the key to success and that utilizing the 'correct' tactical arrangements would predictably and automatically produce victory. The English word 'martinet', meaning a strict and unbending enforcer of detailed rules, comes from this period and is named for a French inspector general.[13] Thinkers of this school were so focused on divining rules through maths and science that some, such as Prussian thinker Georg Heinrich von Berenhorst, went so far as to almost deny human agency entirely.[14]

This mechanistic view of tactics began with the science of sieges. With the development and early industrialization of artillery, tacticians needed to learn and gain expertise in the mathematical calculations necessary to employ cannon, and the science of fortification and defence had to keep pace. Military thought began to be led by engineers, the most famous of whom was Sébastien Le Prestre de Vauban, a French general and expert in siege warfare and fortification.[15]

The apex practitioner of this tactical regime was Frederick the Great. As King of Prussia, Frederick made the Prussian Army famous for its detailed and precise execution of battlefield manoeuvres, and its ability to drill served

---

[11] David A. Graff, *Medieval Chinese Warfare, 300–900* (London: Routledge, 2002), 254–255.

[12] Lynn, *Battle*, 114.

[13] Ibid., 116.

[14] Anders Engberg-Pedersen, *Empire of Chance: The Napoleonic Wars and the Disorder of Things* (Cambridge, MA: Harvard University Press, 2015), 38–39

[15] See Henry Guerlac, 'Vauban: The Impact of Science on War', in *Makers of Modern Strategy*, edited by Peter Paret (Princeton, NJ: Princeton University Press, 1986), ch. 3.

it well on both the parade ground and in combat, so long as Frederick himself led it. Such a focus on discipline and drill produced an army that was an instrument that was only as good as the one holding it, and that someone should be an enlightened ruler familiar with the rules and laws of warfare. The soldiers themselves needed no such enlightenment. The general could adapt rules to specific situations, whilst subordinates should merely execute without question.[16]

The search for rules and predictability was not limited to kings and generals but also theorists, most famously Antoine-Henri Jomini, a Swiss officer and thinker who initially served with Napoleon's armies but later defected to Russia. Jomini spent his career searching for such general rules and especially whatever system of principles that might explain Napoleon's success. The summation of his life's work, usually published as *The Art of War*, is one of the most influential works of military theory of all time; it is arguably more influential than Clausewitz's *On War*, a book specifically written to argue against these linear visions of warfare.

Echoes of the linear school persist to this day. Linear tactics produced centralized command and control arrangements, inflexible and detailed adherence to doctrine, and an overarching focus on the destruction of opposing forces in direct, almost formal confrontations in battle. Much of the warfare during the industrial era, especially the First and Second World Wars, adhered to this school even as military organizations that eschewed them saw success.

Modern Tactics

Although Jomini and Clausewitz were contemporaries, the Prussian Clausewitz saw little value in the Swiss writer's assertions that war could be systematized to the point of predictability. Instead, from the other side of the Napoleonic Wars, Clausewitz saw the domination of intangibles like morale and probability. Unlike virtue tactics, however, the intangible human factors could be cultivated through training, education, and discipline. Clausewitz melded centuries of military thought together into one coherent system. Although most of his metaphors came from the then nascent science of thermodynamics, *On War* presents a theory of war as a complex adaptive system of interrelated agents. Where virtue tactics saw gods or honour as an agent, and linear tactics removed agency in favour of laws and rules, Clausewitz

---

[16] See R. R. Palmer, 'Frederick the Great, Guibert, Bülow: From Dynastic to National War', in *Makers of Modern Strategy*, ed. Paret, ch. 4.

stressed multiple agents and their interactivity as the nature of combat. For Clausewitz and his successors, the science of warfare matters, but so does the art of creative tactical employment and moral forces. War cannot be reduced to mere mathematical rules, and even rules could be discarded by a genius.

His most comprehensive depiction of tactics comes not in *On War*, which is focused more on defining war as a phenomenon, but rather a little-known work called *Guide to Tactics*.[17] This work stresses probability and interactivity at the lowest levels. Clausewitz describes the difference between infantry combat and artillery combat in terms of probability. Where Jomini saw rules and predictability, Clausewitz saw probabilities and chance. Presaging his later focus on the intangible elements of combat, Clausewitz describes a number of ways to win engagements. Only one involved the physical destruction of the opponent's forces, the rest all involved mental effects produced by different situations.[18]

Whilst linear tactics and even echoes of the virtue tactics persist even today, subsequent theorists have followed Clausewitz's example in stressing interactivity, especially the French Army officer Ardant du Picq whose book *Battle Studies* also stressed intangibles like unit cohesion, and US Air Force officer John Boyd who took Clausewitz's concept of friction and turned it around. Where Clausewitz was focused on overcoming friction, Boyd was focused on inflicting friction on the opponent by making better decisions faster than the opponent.[19]

The intangibles of human perception, decision-making, and organizational cohesion are becoming even more important with the rise of information warfare, which includes such emergent weapons and techniques as psychological warfare, influence operations, electronic warfare, and cyber warfare, among others.[20] These techniques can be used to increase the fidelity and accuracy of the information available to the tactician, but can also be used to corrupt and manipulate the perception, and thus the decision-making, of the opponent.[21] Whilst information has always been a factor in combat, these emergent technologies have now brought it to the forefront of all military operations.

[17] Olivia Garard, *An Annotated Guide to Tactics* (Quantico: Marine Corps University Press, 2021).

[18] Garard, *An Annotated Guide to Tactics*, 30–32.

[19] See Ardant du Picq, *Battle Studies*, trans. and ed. Roger J. Spiller (Lawrence: University Press of Kansas, 2017). For John Boyd, see Ian Brown, *A New Conception of War: John Boyd, the U.S. Marines, and Maneuver Warfare* (Quantico: Marine Corps University Press, 2018).

[20] Michael Raska, 'The sixth RMA Wave: Disruption in Military Affairs?', *Journal of Strategic Studies* 44, 4 (2021): 456–79.

[21] Antoine Bousquet, *The Eye of War: Military Perception from the Telescope to the Drone* (Minneapolis: University of Minnesota Press, 2018).

These three tactical regimes are distinct, yet also overlap in time and militaries today favour one regime over another, or more likely employ a combination of them. Modern theorists, especially those in the 'manoeuvrist' school, see compelling concepts in Sun Tzu, for example.[22] Some military forces, just like theorists, focus more or less on one regime or another, or employ a combination of them.

Importantly, the viewpoint that there is a 'Western' and an 'Eastern' or 'Oriental' way of warfare is not supported by the historical record of military history or tactical theory.[23] Western tactical manuals are far more positive about stratagems and deception, allegedly a focus of Eastern tactics, than is commonly presented. Eastern tactical ideas are far more similar to Western ones than is commonly believed. Indeed, it is remarkable that in conducting a search for how to win, disconnected Eastern and Western thinkers came to quite similar conclusions. The sources we have from the East tend to be works focused on politics and statecraft, and so discuss more espionage and spycraft than the Western military classics. However, espionage was far from unknown in the West even if it was not always depicted in military narratives. Claims of a Western 'way of warfare' focused on direct battle and Eastern 'way of warfare' focused on deception and subversion simply do not hold water.

Purpose of Theory

Although most of the above mentioned works of theory claim to provide battle-winning advice, as do a great number of modern works, providing answers is not the purpose of theory. Rather, theory provides for structured thought and communication. Clausewitz supplies what is probably the best and most famous vision of the purpose of theory, especially in regards to tactics. Theory, Clausewitz writes,

> [B]ecomes then a guide to him who wishes to make himself acquainted with war from books; it lights up the whole road for him, facilitates his progress, educates his judgment, and shields him from error. ... It should educate the mind of the future leader of war, or rather guide him in his self-instruction, but not accompany him to the field of battle; just as a sensible tutor forms and enlightens the opening mind of a youth without, therefore, keeping him in leading strings all through his life.[24]

---

[22] *MCDP-1 Warfighting* (Washington, DC. US Marine Corps, Department of the Navy. 1997).

[23] See especially Beatrice Heuser, *The Evolution of Strategy: Thinking War from Antiquity to the Present* (Cambridge: Cambridge University Press, 2010), and Lynn, *Battle*.

[24] Carl von Clausewitz, *On War*, trans. J. J. Graham (New York: Barnes and Noble, 2004), 82.

In other words, theory cannot provide answers on the battlefield, but it can help practitioners ask the right questions before they get there, help them communicate with each other through standardized concepts and ideas, and think through a planning process. Military theory is thus cognitive scaffolding, a structure built to facilitate the building of another structure. Tactical theory uses this scaffolding to produce plans for combat itself.

Cognitive scaffolding can not only assist in the building process but also provide a foundation for objective assessment. Most military organizations encourage their members to read widely, especially in military history. Simply reading military history is certainly not a waste of time, but it is also not quite the professional military education that can directly contribute to the tactician's purpose: making the difficult decisions necessary to succeed in combat. Tacticians must instead analyse military history, rather than just read it, and tactical theory is one lens through which to do so, with wargaming, staff rides, and tactical decision games being other common methods. Combined, these add up to quite a powerful toolbox which can then be used to strengthen a tactician's ability.

From Principles to Tenets

Tactical theory is often distilled into principles of war. These have gone through many permutations over the centuries, including debates about whether they apply to war and strategy or solely to tactics (which would mean they are more properly called principles of battle). As a teaching heuristic these lists are especially effective, the rote memorization of whatever list is chosen is not difficult to train and inculcate in troops, although the memorization demand does foster a belief that they are a checklist. They are also taken more or less as principles that must be adhered to to a greater or lesser degree by many different tacticians and thinkers. Here they will be referred to as tenets, as the term 'principles' implies that they are more rigid than indeed they can be in real life. We argue here that they are an analytical tool, be it for the tactical leader or for the sake of comparing doctrines, and thus act as tools to be used when appropriate—be it in officer education or on the battlefield.

There are many versions of 'the principles of war', far too many to list here. The list below of tactical tenets, rather than principles, serves here instead as a core set of concepts relevant for tactics. They are broken down by physical tenets (mass, manoeuvre, firepower, and tempo), mental tenets (deception, surprise, confusion, and shock) and moral tenets, with only moral cohesion in it.[25]

---

[25] Friedman, *On Tactics*.

Mass can be defined as an advantageous concentration of combat power in space and/or time.[26] It is sometimes depicted as a recommendation to concentrate forces at a singular or 'decisive' point to maximize the weight of their capabilities. Instead, it should be viewed as a decision the commander has to make on whether to concentrate forces or disperse them and when. Both concentration and dispersion can be advantageous at different times and in different situations. A guerilla force, for instance, needs to both disperse to avoid being targeted and to concentrate in order to conduct successful operations. The concentration of mass can also be used to counter an opponent's advantage in, for example, manoeuvre. One example of the use of mass as a tool in and of itself is the Soviet Army in the Second World War. The Soviets countered the German manoeuvre-based advantage through massed assaults, overwhelming the smaller force.[27] Whilst closely related to the tenet of firepower, another way of looking at mass is to attempt to concentrate the effects of weapon systems in space and time without necessarily concentrating them in one place.

Manoeuvre can be defined as attacking an enemy force from a position of comparative advantage.[28] This could be physical, such as approaching from a rear or a flank. It could also be cognitive, as in attacking in an unexpected way or from an unexpected direction. There are also more direct aspects of manoeuvre, such as penetrating attacks used to disjoint an enemy line or position. The essence of manoeuvre is that it is facilitated through an asymmetric application of mass. Examples of the successful use of manoeuvre include many of the more famous battles in history. Alexander the Great, Frederick the Great, and Napoleon all used manoeuvre as their favoured tactical approach.[29] With the advent of mechanized armies the German 'blitzkrieg' operations in the Second World War, Israel's attack in the 1967 war and the latter half of the 1973 war, and the American operations in both 1991 and 2003 all highlighted how modern and mobile forces can defeat—often rapidly—seemingly superior forces.[30]

Firepower is the application of long-range, missile weapons and supporting arms, such as artillery and close air support.[31] Beyond the destructive nature of firepower, it can also be used to facilitate manoeuvre by fixing enemy forces in place whilst another unit manoeuvers around or against them. One example of where firepower sticks out is the battle of Khe Sanh in 1968. The North Vietnamese forces had hoped to recreate the defeat of

---

[26] Ibid., 38.
[27] Ibid., 37–38.
[28] Ibid., 26.
[29] Ibid., 28–30.
[30] Ibid., 31–33.
[31] Ibid., 48.

Dien Bien Phu by surrounding the defenders and placing heavy artillery on the hills surrounding the base. However, the American artillery and close air support was so overpowering that the Vietnamese suffered upwards of 30,000 casualties over the three-month siege.[32]

Tempo is the ability to control the pace of combat to your advantage and the disadvantage of the opponent.[33] It is frequently called 'speed'. Whilst moving faster than the opponent, either physically or cognitively, is usually advantageous, sometimes it is not. In some situations, keeping the pace of fighting slower and outlasting your opponent can be an advantage. Mass, manoeuvre, and firepower all affect the ability of forces to move, and therefore to change or sustain tempo. Thus, these four physical tenets interact in highly dependent ways.

At the most basic level, a more manoeuvrist approach prioritizes manoeuvre, whilst a more attritionist approach prioritizes firepower; these are however rarely, if ever, black-and-white concepts, and better thought of as a balance between mass, manoeuvre, firepower, and tempo. Commanders must usually decide how much firepower is necessary for a mission. Too much firepower will make a manoeuvre difficult to execute or sustain logistically. Too little and the manoeuvre will strike with too little combat power to be effective. Added to this is the question of tempo, as both manoeuvre and heavy firepower approaches entail strengths and weaknesses. This stresses the need to act in a combined arms manner and achieve a good balance between the four physical tenets.

Combinations of the physical tenets produce mental effects when applied against an adversary force, and these mental tenets interact and function interactively with each other. The first mental tenet, *deception*, is the manipulation of the enemy's understanding or perception of the situation in order to achieve an advantage.[34] A concentration of forces or even firepower may deceive the enemy that an attack is imminent in that place when it is planned to happen elsewhere. A deceived opponent will still react to events but will do so based on inaccurate information. A successful deception involves a balance of letting the opponent gather enough intelligence on your forces that they believe in the deception, but not so much that they identify it as a ruse.

Surprise is perhaps the most potent of all the tenets. Surprise in combat is the act of presenting your enemy with a situation or capability for which they are not mentally prepared.[35] The essence of surprise is to cause the opponent

[32] Ibid., 53–54.
[33] Ibid., 57.
[34] Ibid., 65.
[35] Ibid., 70.

to be unable to react to events in time. The British Army officer Jim Storr even argues that surprise—combined with the follow-on tenets of confusion and shock—is the single most important factor of achieving victory in combat, with a successful surprise attack having a greater impact than a force ratio of ten to one.[36]

Confusion is a state of mental overload or disarray that makes it difficult both to react to events and understand the situation.[37] It can be produced by the application of mass and firepower in rapid or inconceivable ways. This tenet harkens back to Sun Tzu's recommendations to act in ways that the enemy cannot ascertain. One good way to cause confusion is to destroy or disrupt command and control nodes. A successful such attack would cause losses to both communication equipment and leadership, leading to confusion.

Shock is a state of psychological overload caused by the sudden, unexpected, or successive action of the enemy.[38] This tenet is the 'shock effect' of the cavalry charge or 'tank fright' in modern warfare. The mere act of attacking with certain forces intimidates and shocks the opponent. The two classic examples of German and Israeli assaults against French and Egyptian forces in 1940 and 1967, respectively, highlight just how powerful shock is. Not only did they achieve strategic surprise in both cases, but the aggressors also caused significant shock and panic, seeing entire units collapse before they could react to the rapidly advancing opponents.[39]

The effective use of the combination of deception, surprise, confusion, and shock has, just as Storr argues, outsized effects. The Egyptian assault across the Suez in 1973 caught the Israeli defensive force by complete surprise. After deceptive operations to help conceal significant preparations, the Egyptian attack was so successful at causing surprise and confusion that it placed the Israeli government in a state of shock, unable to respond effectively at either national or local tactical levels.[40] This was despite it being the expected location of a future front in case of war; then-defence minister Moshe Dayan would later recount that the attack 'came as a surprise, though it was not unexpected'.[41]

Lastly, the moral aspect of combat can be referred to as *moral cohesion*: the ties of familiarity, trust, and commitment among the members of a

---

[36] Jim Storr, *The Human Face of War* (London: Bloomsbury Publishing, 2009) 86.
[37] Friedman, *On Tactics*, 74.
[38] Ibid., 79.
[39] Storr, *The Human Face of War*, 92.
[40] George W. Gawrych, *The 1973 Arab-Israeli War: The Albatross of Decisive Victory* (Combat Studies Institute, US Army Command and General Staff College, 1996), 27–29.
[41] Gawrych, *The 1973 Arab-Israeli War*, 29.

unit that allow it to fight as a unit rather than a collection of individuals.[42] The importance of cohesion was recognized by Machiavelli and Clausewitz, both of whom found cohesion as important factors in war.[43] When the physical actions of the enemy produce sufficient mental effect to stress and traumatize the opponent, the moral cohesion of that unit falls apart. Retreats and headlong routes occur when that mental cohesion is broken. Units with a higher level of training and morale can withstand more stress than others. A unit that has better training, higher morale, and strong cohesion is then more effective in combat situations; cohesion is how a unit keeps its 'unity as a social group even in the intense environment of combat'.[44]

Perhaps the best example of all these tenets being applied in real life is the tactical action of the ambush. Ambushes are a combination of mass, manoeuvre, usually firepower, and tempo employed to produce surprise, shock, and confusion, which usually destroys moral cohesion for at least some time; oftentimes deception is used as well. The high number of tenets that underpin this tactic are why it has been so successful. The target unit, suddenly presented with a threat in a time and place where it was not expected, usually cannot fight back as a coherent unit, although those that can usually succeed in extricating themselves. Those that cannot might be eradicated.

Ambushes are famous throughout history, not the least being the Battle of Teutoburg Forest in 9 AD, when Roman legions under Publius Quinctilius Varus were wiped out by Germanic tribes in what is today Germany. That this ancient tactic still typifies combat in modern battles today should not be surprising, and through tactical theory we can understand why it remains effective. Other tactical actions will often use a combination of the tenets discussed above, albeit seldom with such coherent chain of effects as an ambush. Thus, the tenets are best thought of as useful in some combination, the composition of which varies depending on the situation. Combined, the tenets create systematic effects, which create conditions to be exploited.[45]

Relationship with Strategy

Tactics only really mean anything in connection with strategy. Strategy provides the all-important context for tactical action. Without strategy, tactics become merely actions to succeed in engagements, but lack purpose beyond

---

[42] Friedman, *On Tactics*, 21.
[43] Anthony King, *The Combat Soldier: Infantry Tactics and Cohesion in the Twentieth and Twenty-first Centuries* (Oxford: Oxford University Press, 2013), 13.
[44] King, *The Combat Soldier*, 13.
[45] Storr, *The Human Face of War*, 93.

the immediate needs. The connection between the two, and the subordinate nature of tactics has been generally accepted throughout the strategic studies and military communities.

There are many different conceptions of the relationship. In Clausewitz's version, tactics is everything that happens on the battlefield to produce, or fail to produce, a tactical victory. The outcome of tactical engagements is the 'currency' that is then 'cashed in' for strategy, which is the use of that currency to achieve the goal of the war.[46] Colin S. Gray makes this connection through the concept of strategic effect. Tactical engagement, no matter its scale, produces a strategic effect.[47] For this effect on the end of the war to mean anything, it must be inherently political. The idea that strategy is political begins with the works of Niccolo Machiavelli, the Renaissance-era political writer, and Christine de Pisan, a late medieval writer.[48] Thus, in Clausewitz's words, strategy can, 'never take its hand from the work for a moment'.[49] Because these effects are unpredictable, that is, the tactical input does not necessarily consistently produce a repeatable strategic output, the relationship is non-linear. There is no magic number of tactical victories that can predictably lead to strategic victory.

Despite the fame of Clausewitz, many major military organizations adhere to a far more Jominian, and thus linear, view of the relationship. The most common of which is the Lykke Model ends, ways, and means.[50] In this model, means (such as military forces) are used in such ways (plans and battles) to achieve an end (the goal or end state). The Lykke Model portrays tactics, operations (or campaigns), and strategy as building blocks. Tactics build up over time to produce operations, which build up over time to produce a strategy. Winning victories at every level will lead to winning the war. Although this is intuitive, and thus easy to teach, it is not reflected in military history. Some recent conflicts, notably the wars in Afghanistan, Iraq, and Vietnam, feature repeated American tactical successes that consistently fail to produce strategic victory; history is fraught with other examples. Frequent statements by American policymakers that there is no military solution whilst only military means are employed reveal a logical disconnect between tactics and strategy. This logical disconnect is a result of expecting a linear relationship between tactics and strategy when no such relationship exists. Indeed, therefore it is not checklists of principles being discussed here, rather toolboxes of tenets.

[46] See Clausewitz, *On War*, book 2, ch. 1.
[47] Colin S. Gray, *The Strategy Bridge: Theory for Practice* (Oxford: Oxford University Press, 2010), 31–33.
[48] Heuser, *Strategy Before Clausewitz*, 40–42.
[49] Clausewitz, *On War*, 127.
[50] Arthur F. Lykke, 'Defining Military Strategy', *Military Review* LXIX, 5 (May 1989): 2–8.

Tactical Theory in Practice

Tactical theory can and should be used to evaluate both the past through the analysis of historical case studies and potential futures through the planning process. It can also be used to evaluate the strengths and weaknesses of warfighting organizations. Take, for example, the differences between the United States Army and the United States Marine Corps. Despite serving the same country and residing in the same department of the US government, the Army and the Marine Corps are two vastly different organizations. Whilst the comparison is not strictly one to one as the Army is a ground force and the Marine Corps is a maritime force, the Marine Corps has participated in, organized for, and remains capable of enough ground combat—especially from the Pacific Theatre in the Second World War through Vietnam, Afghanistan, and Iraq—to bear the comparison.

David Kilcullen, the Australian counterinsurgency expert, advisor to the United States Department of Defense, and former Australian Army officer has described the US Army as 'campers' and the US Marine Corps as 'hikers'.[51] In other words, the Army essentially moves into an area, sets up strong points and infrastructure, then proceeds to fight from there. The Marines Corps, by contrast, never stops moving and instead fights from and within the context it finds on the ground already. Another way of saying this is that the Army is focused on mass and firepower along with the infrastructure required to support them and is more comfortable paying the price in terms of lesser mobility in exchange. A common maxim in the Army being 'firepower leads to maneuver'. The Marine Corps, however, is focused on manoeuvre and tempo, surely as a way to compensate for the lack of mass and firepower consequent to its lack of heavy equipment, infrastructure, and robust sustainment.

That certain tenets were prioritized over others was highlighted during the Vietnam War, where the Army conducted large-scale Search and Destroy operations against large enemy formations, real or not, where firepower was the primary tool to achieve their goals.[52] This reliance on firepower led to the overall US commander, General Abrams, bemoaning that Army commanders were unable to change their conception of war, as the use of heavy artillery and large-unit operations seemed to confirm their beliefs.[53]

---

[51] David Kilcullen, *Out of the Mountains: The Coming Age of the Urban Guerrilla* (Oxford: Oxford University Press, 2013), 35.

[52] Austin Long, *The Soul of Armies: Counterinsurgency Doctrine and Military Culture in the US and UK* (Ithaca, NY: Cornell University Press, 2016), 131–32.

[53] Long, *The Soul of Armies*, 131.

Abrams predecessor, General Westmoreland, even explicitly used the word 'firepower' as a one-word answer for how to win the war.[54] In contrast, the Marines refused to prioritize a firepower-centric approach, instead using small-unit operations through manoeuvre and tempo affecting the cohesion of the local population. This was despite the conventional-style fighting on the nearby border of North and South Vietnam.[55] Decades later, in Afghanistan, the Marines would continue to purposefully limit the use of mortars and airstrikes—firepower—to avoid civilian casualties, despite the tactical advantage it would provide.[56]

It not just equipment that influences these tactical preferences, but also mission and culture. The Army has a culture of mass because it has usually enjoyed the ability to generate mass; it has always been far larger than the Marine Corps. The Marine Corps has a culture of speed, aggression, and elite status because it has lacked the numbers and equipment for anything else and has always been able to be more exclusive when it comes to recruitment. Legal directives also influence this preference. By US law, the Marine Corps is responsible for amphibious operations and rapid crisis response. Both demands require Marine Corps forces to respond to situations with whatever and whomever it has on hand and then improvise from there. By contrast, the Army can be more deliberate with designing forces for a mission once it is received, as their mission entails fighting and sustaining large combat forces.

These tactical preferences do not just highlight strengths but weaknesses as well. The US Army has historically fared better later in wars once a robust logistics system has been developed. This tendency was seen in conflicts as diverse as the American Civil War, the First World War, and the Second World War. By contrast, the Marine Corps tends to be ready earlier in conflicts. During the Second World War, for example, the US Army was unable to rapidly deploy after Pearl Harbor so the initial troop contributions in both theatres fell to the Marine Corps.[57]

Lastly, theory plays a large role in producing these doctrinal differences. The US Army has, for over a century, been largely influenced by the theories of Jomini as his major American translator, Dennis Hart Mahan, father of the naval theorist Alfred Thayer Mahan, was Professor of Military Science at West Point and wrote many works largely influenced by Jomini. Jomini's focus

[54] John A. Nagl, *Learning to Eat Soup with a Knife: Counterinsurgency Lessons from Malaya and Vietnam* (Chicago: University of Chicago Press, 2005), 200.
[55] Long, *The Soul of Armies*, 126.
[56] Ibid., 211.
[57] Kenneth J. Clifford, *The United States Marines in Iceland, 1941–1942* (Washington, DC: Headquarters Marine Corps, 1970), 3.

on lines of operation, decisive points, and deliberate, predictable planning remains in evidence in Army doctrine today.[58] Jomini's vision of predictable, calculable, geometric tactics found fertile ground in the engineering-focused West Point of the nineteenth century.

By contrast, the US Marine Corps largely did not adopt a formal theory of war until quite recently. Prior to that, it principally developed amphibious doctrine and eschewed theory entirely. In 1989, however, it formally adopted FMFM-1 *Warfighting*, a fundamentally Clausewitzian conception of war heavily influenced by John Boyd. *Warfighting*, its 1997 update MCDP-1 *Warfighting*, and Marine Corps doctrine reflects a greater focus on uncertainty, probability, and complexity from Clausewitz, and whilst also taking manoeuvre and tempo from Boyd.[59] Whilst MCDP-1 *Warfighting* does not deny the importance of firepower, it does argue that to achieve victory in combat, manoeuvre and tempo should be used to achieve surprise and shock. It even argues that the aforementioned Marine small-unit operations in Vietnam were 'maneuver warfare'.[60]

Through approaching the comparison of the two services using tactical tenets, we are able to identify distinct differences in how they prioritize firepower contra manoeuvre in their doctrine, but also how they conceptualize the battlefield.

Conclusion

Military theory is of professional importance not just to practitioners, but also policymakers and academics. Theory's ability to foster communication between the three, to act as a lingua franca for debating the use of violent means for political ends, could improve a sometimes lopsided and ineffective discourse. Tactical theory is just as important as its more established cousin strategic theory, although for the simple reason that strategy can only be accomplished through tactics. The history of tactical theory through the centuries of military history bears out its importance. Not only have officers taken to writing it, but so have kings and emperors. War is of vital importance to the state and thus is of concern to all three points of Clausewitz's secondary trinity: the policymakers, the military professionals, and the people.

[58] Russell F. Weigley, *The American Way of War: A History of United Military Strategy and Policy* (Bloomington: Indiana University Press, 1997), 81–84.
[59] *MCDP-1 Warfighting*.
[60] Ibid., 36–38.

The tactical tenets depicted here offer a simple and teachable basis for tactical theory free of the usual jargon and acronyms that typify modern military writing found in doctrines. Nor is tactical theory only useful for analysing engagements in combat. As shown above, it can be used to evaluate warfighting organizations as institutions. Policymakers especially must understand the strengths and weaknesses of the forces that will execute policy. This is important when deciding whether or not to pursue war, but also how such organizations should be funded, designed, and administered.

7

Urban Warfare

Challenges of Military Operations on Tomorrow's Battlefield

Mikael Weissmann

Introduction

> The future of warfare lies in the streets, sewers, high-rise buildings, industrial parks, and the sprawl of houses, shacks, and shelters that form the broken cities of our world. We will fight elsewhere, but not so often, rarely as reluctantly, and never so brutally. Our recent military history is punctuated with city names—Tuzla, Mogadishu, Los Angeles, Beirut, Panama City, Hue, Saigon, Santo Domingo—but these encounters have been but a prologue, with the real drama still to come.[1]

It is often said that future combat will take place in dense urban areas, including in megacities, and the importance of urban warfare has been widely recognized. Today, it is agreed upon and accepted that the battlefields of tomorrow will include battles in urban terrain. This is a fact that could be observed in practice after the Russian invasion of Ukraine in February 2022. In short, to prepare for urban warfare has become a necessity.[2] This necessity is the result of a number of reinforcing trends, urbanization and technology being driving forces, the former makes it clear that cities are the centre of gravity and the latter forcing insurgency into the urban areas as it is providing the defensive advantage needed for irregular forces to survive.

[1] Ralph Peters, 'Our Soldiers, Their Cities', *Parameters* 26, 1 (1996).
[2] A number of labels are used for operations and combat in urban environments, including urban operations, military operations in urban terrain (MOUT), operations in built-up areas (OBUA), fighting in built-up areas (FIBUA), and Close Quarter Battle (CQB). The labels often have specific definitions in doctrine and handbooks. For the purposes of this chapter, the term urban warfare is used as a blanket term for different forms of operations and combat in urban terrain.

Mikael Weissmann, *Urban Warfare*. In: *Advanced Land Warfare*. Edited by Mikael Weissmann and Niklas Nilsson, Oxford University Press. © Mikael Weissmann (2023). DOI: 10.1093/oso/9780192857422.003.0007

The changing character of war, with a compression of time ('the death of distance'), with the information domain being the centre of gravity, with space and cyber domains in their own right, with AI coming to the fore-front of military thinking, can be added to the above.[3] In short, fighting asymmetrical warfare, where the weaker force must seek defence in urban areas, has become a necessity, in particular in the Global South where mega- and feral cities will become the new normal, sometimes even in the form of cross-border megaregions, creating previously unheard of complexity.[4]

Furthermore, future urban operations will need to meet challenges from both cross-domain and cross-conflict-spectrum fighting, since the grey zone between peace and war has grown. The former calls for multi-domain oper-ations, whilst at the same time handling urban warfare in an operating environment that is often situated in the grey zone between peace and war.

A future that includes urban warfare is widely recognized among practi-tioners. It is a case in point that General Mark Milley, then Chief of Staff of the US Army, now Chairman of the Joint Chiefs of Staff and the highest ranking officer of the US Armed Forces in 2016 stated '[I]n the future, I can say with very high degrees of confidence, the American Army is probably going to be fighting in urban areas', adding, 'We need to man, organize, train and equip the force for operations in urban areas, highly dense urban areas.'[5] A similar idea can be seen with regards to NATO, where a general consensus exists that NATO forces will be engaged in urban operations in the future, and the need for NATO Allies to strengthen their capabilities in the area is recognized.[6] In short, Lt. Col. Leonhard seems to have been correct when he argued in 2003 that, 'Urban areas should become our preferred medium for fighting. We should optimize our force structure for it, rather than relegating

[3] Zachery T. Brown, 'Unmasking War's Changing Character', *Modern War Institute*, 12 March 2019, https://mwi.usma.edu/unmasking-wars-changing-character/. Also see T. X. Hammes, 'The Chang-ing Character of War', 15 May 2022, https://keystone.ndu.edu/Portals/86/Future%20of%20Conflict.pdf; T. X. Hammes, 'Technologies Converge and Power Diffuses: The Evolution of Small, Smart, and Cheap Weapons', *Policy Analysis* no. 786, Cato Institute, 22 January 2021.

[4] Jeremiah Rozman, 'Urbanization and Megacities: Implications for the U.S. Army', The Institute of Land Warfare, the Association of the United States Army, ILW SPOTLIGHT 19–3, August 2019, https://www.ausa.org/sites/default/files/publications/SL-19-3-Urbanization-and-Megacities-Implications-for-the-US-Army.pdf; Margarita Konaev, 'The Future of Urban Warfare in the Age of Megacities', *Focus stratégique* 88 (March 2019); Joel Lawton and Lori Shields, 'Mad Scientist: Megacities and Dense Urban Areas in 2025 and Beyond', United States Army, Training and Doctrine Command (TRADOC) G-2, Fort Eustis, VA, 18 August 2016, https://community.apan.org/wg/tradoc-g2/mad-scientist/m/mdua/170637.

[5] Michelle Tan, 'Army Chief: Soldiers Must Be Ready To Fight in "Megacities"', *Defense News*, 5 October 2016.

[6] Philippe Michel-Kleisbauer, 'URBAN WARFARE', NATO Parliamentary Assembly, SCIENCE AND TECHNOLOGY COMMITTEE (STC), Sub-Committee on Technological Trends and Security, 20 November 2020, 12.

it to Appendix Q in our fighting doctrine, treating it as the exception rather than the norm. … Instead of fearing it, we must own the city [sic].'[7]

The need to plan for urban warfare has also been observed given the increasing frequency of operations in cities in the last two decades. After the September 11 attacks, the US military became entangled in war in Iraq and Afghanistan. At the same time as the US Army and the United States Marine Corps (USMC) fought al Qaeda supporters and the Taliban mainly in the rural farm areas and eastern mountains of Iraq, US forces also found themselves fighting in Baghdad, Fallujah, Tal Afar, Ramadi, Najaf, and many more urban areas.[8] This trend has continued, with major urban battles involving city attacks identified in the ongoing civil war in Syria, the war against the Islamic State in Iraq, Syria, and the Philippines, and in Ukraine.[9]

This chapter will address the daunting challenge of urban warfare on tomorrow's battlefield. In the first section, it will provide a brief background of the urban warfare phenomenon. It approaches urban warfare by asking why the field has now emerged after a long period of relative neglect. Thereafter, the chapter outlines the different challenges to and expectations for urban operations on today's and tomorrow's battlefields. Here, a number of key challenges will be addressed: the impact of rapid urbanization, multi-domain operations, the grey zone problems, and the impact of technology on urban operations, and the urbanization of insurgency. Finally, several conclusions will be drawn.

One problem in most urban warfare research, as well as in doctrine and handbooks, is a focus on superior and more technologically advanced Western regular forces, often the USA, conducting offensive operations against weaker, less technologically advanced irregular forces. Whilst this focus is of course not unjustified, given the short-term needs of the field, this chapter will take a broader perspective and engage throughout with the impact of the offensive/defensive dimension, types of force, power symmetry, and level of

---

[7] Lt. Col. Leonhard, U.S. Army cited in Stephen Graham, 'Imagining Urban Warfare: Urbanization and US Military Technoscience', in *War, Citizenship, Territory*, edited by Deborah Cowen and Emily Gilbert (New York, London: Routledge 2008), 41.

[8] Gian Gentile, David Johnson, Lisa Saum-Manning, Raphael Cohen, Shara Williams, Carrie Lee, Michael Shurkin, Brenna Allen, Sarah Soliman, and James Doty, *Reimagining the Character of Urban Operations for the U.S. Army: How the Past Can Inform the Present and Future* (Santa Monica, CA: RAND Corporation, 2017), 1.

[9] Recent examples include Aleppo, Syria, 2016; Ghouta, Syria, 2018; Deir ez-Zor, Syria, 2017; Ilovaisk, Ukraine, 2014; Kobani, Syria, 2014/2015; Debal'tseve, Ukraine, 2015; Ramadi, Iraq, 2015/2016; Fallujah, Iraq, 2016; Mosul, Iraq, 2016/2017; Raqqa, Syria, 2016/2017; Marawi, Philippines, 2017; Tal Afar, Iraq, 2017.

Other historical examples of city attacks in limited warfare where the attacking force attempted to kill the defenders or seize the city include Hue, Vietnam, 1968; Vukovar, Croatia, 1991; Sarajevo, Bosnia and Herzegovina, 1992–1996; Grozny, Chechnya, 1994/1995; Grozny, Chechnya, 1999/2000; Fallujah, Iraq, 2004. (John Spencer, 'The Eight Rules of Urban Warfare and Why We Must Work to Change Them', *Modern War Institute*, 12 January 2021, https://mwi.usma.edu/the-eight-rules-of-urban-warfare-and-why-we-must-work-to-change-them/).

Table 7.1 Dimensions of warfare

Dimensions	Us	Them
Offensive/defensive	Attacker	Defender
Type of force	Regular	Irregular
Power symmetry	Asymmetric/STRONG	Peer or near-peer adversaries
Technology	HIGH TECH	LOW TECH

technology (see Table 7.1). For example, how do we conduct urban warfare against peer or near-peer adversaries? How does the proliferation of civilian technology impact urban warfare?

Approaching Urban Warfare

> ... the worst policy of all is to besiege walled cities.
> **Sun Tzu, The Art of War**

Whilst urban warfare itself is nothing new, there are trends inexorably forcing battles to move to urban areas to a greater extent than ever. Rapid urbanization and new technologies are two forces moving warfare toward urban areas, whilst also impacting the manifestation of the urban battlefield and how urban battles are fought. The strategic environment is changing with population growth and inexorable urbanization, as global populations move to cities, often megacities with populations of over 10 million. Today, more than half of the world population lives in urban areas.

Furthermore, technological development not only forces battles into the city, for example when sensors eliminate the cover traditionally gained from darkness or forests, or so that irregular fighters can resist technologically superior forces, but also transforms the battlefield along the digital/cyber dimension, breaking down the border between kinetic and non-kinetic warfare. Technology also throws into question what is (identifiable) warfare, further increasing the need to account for non-conventional warfare, much of which can be expected to occur in the urban areas where half the world's population lives.

As wars tend to ultimately be decided where people live, armies need to organize, equip, and train to win fights in urban areas, including in megacities.[10] This is a daunting challenge, as military leaders have steered away from conducting operations in cities for 2,700 years. In 500 BC, Sun Tzu advised

---

[10] David Kilcullen, *Out of the Mountains: The Coming Age of the Urban Guerrilla* (Oxford, New York: Oxford University Press, 2013), 28.

against attacking walled cities, calling it the worst military policy of all, and doctrine as recent as the post-Second World War era advised avoiding, isolating, or bypassing cities altogether.[11] This has clearly changed, as military leaders recognize and prepare for a future of urban warfare.

The significant advantages of dense modern urban terrain to the defender, together with urban canyons—that is, streets flanked by buildings on both sides—and underground warfare, also explain why experience and doctrine advise avoiding cities. This is also why past US doctrinal manuals emphasized that urban areas should be avoided insofar as possible, since historical experiences, for example at Aachen, Metz, and Manila in the Second World War, Seoul during the Korean War, and Hue during the Vietnam War, show that urban combat can be extremely costly for both combatants and civilians.[12]

In fact, as argued by Ian Rigden, '[t]he urban environment is perhaps arguably the most difficult because it is among the people and it is a man-made environment with all the intentional and unintentional challenges that entails. ... There are rarely clear winners in urban warfare which, in the context of warfare in the twenty-first century, challenges the very concepts of winning and victory.'[13]

It should be noted that the city-avoidance doctrine can at least in part be traced to Cold War thinking regarding the eventuality of US ground forces confronting the Soviet Union in Western Europe, where fighting would take place not in large cities or urban areas but out in the open.[14] Not until the late 1990s, nearly a decade after the end of the Cold War, did US planners begin to realize that large urban areas could not be avoided, since they were the hubs of political, economic, and cultural significance.[15]

Looking further back, cities have always been centres of gravity, thus fighting has often been drawn toward cities. Perhaps a force needed to attack an urban area to destroy the enemy, achieve a strategic location, or access a capability needed for future operations. Often, an inferior defender sought shelter in urban terrain, which provides an inherently defensive advantage.[16] This

---

[11] Kenneth K. Goedecke and William H. Putnam, *Urban Blind Spots: Gaps in Joint Force Combat Readiness*, National Security Fellows Program, Paper, November 2019, Belfer Center for Science and International Affairs, Harvard Kennedy School, 6.

[12] David Johnson, 'Urban Legend: Is Combat in Cities Really Inevitable?', *War on the Rocks*, 6 May 2019, https://warontherocks.com/2019/05/urban-legend-is-combat-in-cities-really-inevitable/.

[13] Ian Rigden, 'The Poisoned Chalice: Urban Warfare in the Twenty-First Century and Beyond', in *A History of Modern Urban Operations*, edited by Gregory Fremont-Barnes (Cham: Palgrave Macmillan, 2020), 346.

[14] Gentile et al., *Reimagining the Character of Urban Operations for the U.S. Army*.

[15] Ibid.

[16] Louis A. DiMarco, *Concrete Hell: Urban Warfare from Stalingrad to Iraq* (Osprey Publishing, 2012), 15.

can also be seen today in, for example, Afghanistan, Iraq, and Syria, as well as historically.

However, there is one key difference between historical and present-day battles over cities. Historically, battles were fought *about* the city, but seldom *in* the city. Siege warfare entailed breaking through the outer walls thereby having conquered the city, in contrast to modern day house-to-house fighting which is a very different beast. Historically, siege warfare was common and can be traced back to antiquity. It was also common during the Middle Ages. In fact, not until the Second World War did extensive fighting within cities become a more common occurrence.

The historical fact of urban warfare does not, as we will see, mean that it has not changed. The character of warfare has changed, and the size and complexity of the urban terrain has grown exponentially. Furthermore, the international security environment has become more complex, the world more interconnected, and there is increasingly no clear distinction between war and peace, as we live in a grey zone where conflict is always ongoing, and where non-kinetic effects also play an important role.

This complexity has been recognized by military forces and scholars alike. To cite the UK Ministry of Defence, 'the urban environment will be one of the most challenging areas to operate in. The city, and its surrounds, will become an increasingly complex and ambiguous tapestry of multiple actors with shifting allegiances, in which we may be required to operate in a variety of ways, from major conflict at range to peace support and humanitarian operations'.[17] Professor Anthony King of Warwick University even argues for treating urban warfare as its own domain together with land, sea, air, space, and cyber: '[T]oday, urban warfare has coalesced into gruelling micro-sieges, which extend from street level—and below—to the airspace high above the city—as combatants fight for individual buildings, streets, and districts. At the same time, digitalized social media and information networks have communicated these battles to global audiences across the urban archipelago, with these spectators often becoming active participants in the fight'.[18]

Having clearly demonstrated the level of complexity of future urban warfare, it is now time to look closer at the future challenges, their impact, and the means of managing them.

---

[17] UK MOD Developments, Concepts and Doctrine Centre, *Future Operating Environment 2035* (14 December 2015), 55.

[18] Anthony King, *Urban Warfare in the Twenty-first Century* (Cambridge UK, Medford MA: Polity Press, 2021).

Future Challenges for Urban Warfare

> We talk about the three-block war, but we are moving quickly to the four-floor war. ... We are going to be on the top floor of a skyscraper... evacuating civilians and helping people. The middle floor, we might be detaining really bad people that we've caught. On the first floor we will be down there killing them. ... At the same time, they will be getting away through the subway or subterrain. How do we train to fight that? Because it is coming, that fight right there is coming I do believe with all my heart.
>
> **Brig. Gen. Julian Alford, the Marine Corps Warfighting Laboratory commander[19]**

As outlined above, at least four key areas pose fundamental challenges to expectations about fighting tomorrow's wars. This section addresses those areas, focusing first on urbanization, as the cause of increasingly urbanized warfare and the defining feature of the battlefield of the future. Thereafter, the focus moves to discussing multi-domain operations and the handling of grey zone problems. Thirdly, emerging, novel, and disrupting technologies are addressed as forces move battles into the city and alter how urban battles are fought. Finally, the fourth section analyses the irregular turn in urban warfare and the urbanization of insurgency, given the increasingly critical importance of urban areas for irregular and weaker actors seeking to challenge a superior or stronger opponent.

Urbanization

The rapid urbanization trend is one of the main reasons why urban warfare has been identified as a key area for the battles of the future. The most recent National Intelligence Council report, *Global Trends 2040*, sees the urbanization trend continuing, and expects the share of urban population to rise from 56 per cent, in 2020, to nearly two-thirds by 2040. Nearly all this growth is predicted to occur in the developing world, with urban residents of poor countries projected to increase by 1 billion, to more than 2.5 billion by 2040.[20] Furthermore, and of foremost importance for the future urban battlefield, both large and mega cities are increasing. It is estimated

---

[19] Cited in Jen Judson, 'US Troops Need Training to Battle in Future Megacities, Marine General Warns', *Defense News*, 25 May 2017, 3.
[20] National Intelligence Council, *Global Trends 2040: A More Contested World* (The National Intelligence Council 2021). P 20.

that more than 600 million people will live in almost 40 megacities by as soon as 2025–2030. Another approximately 400 million people will live in cities of 5–10 million people, and just over 1 billion will live in cities in the 1–5 million range.[21]

The urbanization trend does not stop here. In fact the 'peri-urban' or 'rur-ban' areas—the space between the city and the countryside—is growing faster than city centres. There is also an increase in the number of megaregions, metropolitan regions that spill over multiple jurisdictions, with at least 40 large bi- or tri-national metro-regions expected by 2030.[22] To this, add lit-toral cities. To cite David Kilcullen, '[a]lready in 2012, 80% of people on the planet lived within sixty miles of the sea, while 75% of large cities were on a coast. Of twenty-five megacities … at the turn of the twenty-first century, twenty-one were on a coast or a major river delta, while only four (Moscow, Beijing, Delhi, and Tehran) lay inland.'[23]

In short, the battlefield of the future is, if not a nightmare, at least a great challenge. Not only is the size of the urban terrain daunting,[24] but as strate-gists have long preferred avoiding the complex and messy environments of coastal cities, the fact that cities tend to develop on coasts complicates the task further. Coastal cities also often include waterways, like canals, river, inlets, and harbours, creating an overlapping need for sea and land capabilities.[25]

Challenges and Problems

Urban warfare is the most difficult form of warfare, being a high-cost, high-risk operation. With rapid urbanization, not only will the rate of urban warfare increase, but it will increase in complexity and scope as the scale of urban areas grows. For example, Fallujah was a densely populated city occupying an area of approximately 25 square kilometres, including its imme-diate surroundings, and with a population of between 250,000 and 350,000 people and 50,000 structures.[26] In contrast, Jakarta, the capital of Indonesia,

---

[21] European Strategy and Policy Analysis System, *Global Trends to 2030: The Future of Urbanization and Megacities*, 1, https://espas.secure.europarl.europa.eu/orbis/sites/default/files/generated/document/en/Think%20piece%20global%20trends%202030%20Future%20of%20urbanisation.pdf.

[22] National Intelligence Council, *Global Trends 2030: Alternative Worlds a Publication of the National Intelligence Council* (Washington, DC: National Intelligence Council, 2012).

[23] Kilcullen, *Out of the Mountains*, 30.

[24] See e.g. Lawton and Shields, 'Mad Scientist'; Mad Scientist Laboratory, '44. Megacities: Future Challenges and Responses', 12 April 2018, https://madsciblog.tradoc.army.mil/44-megacities-future-challenges-and-responses/; Dave Dilegge, Robert J. Bunker, John P. Sullivan, and Alma Keshavarz (eds), *Blood and Concrete: 21st Century Conflict in Urban Centers and Megacities* (Bethesda, MD: Small Wars Foundation, 2019); Konaev, 'The Future of Urban Warfare in the Age of Megacities'.

[25] Kilcullen, *Out of the Mountains*, esp. 263–94.

[26] Timothy S. McWilliams and Nicholas J. Schlosser, *U.S. Marines in Battle: Fallujah November–December 2004*, United States Marine Corps, 15 May 2022, https://www.usmcu.edu/Portals/218/FALLUJAH.pdf.

is an urban area of almost 35 million people covering an area of 16,262 square kilometres. Furthermore, at the time of the Second Battle of Fallujah in November–December 2004, only an estimated 500 civilians remained together with 3,000 to 4,500 insurgents.[27] Even Mosul, about 180 square kilometres with a population of 1.5 million, is dwarfed by a megacity like Jakarta.

The vertical dimension must also be considered. As JP 3-06 notes, '[v]olume, not area, is the more pertinent spatial measure of the urban environment' since a '10-story building may take up the same linear space on a two-dimensional map as a small field, but the building has eleven times the actual defensible space—10 floors plus the roof and any associated subterranean structures.'[28] Admittedly an extreme case, Hong Kong in 2018 had 8,733 high-rise buildings and 300 buildings surpassing 150 metres in height.[29]

Drawing on John Spencer's eight rules of urban warfare,[30] the defenders' advantage grows exponentially with the size and complexity of the city, as does how 'urban terrain reduces the attacker's advantages in intelligence, surveillance, and reconnaissance, the utility of aerial assets, and the attacker's ability to engage at distance'. The problem buildings pose 'as fortified bunkers that must be negotiated' increases in a large city, as does the defenders' ability to maintain 'relative freedom of maneuver within the urban terrain', and as do problems with the underground serving 'as the defender's refuge'. To give an example, the proceedings of the 2018 Multi-Domain Battle in Megacities Conference indicate that the army today does not have sufficient divisions to isolate and control one megacity, and that it would not be feasible for a coalition military force to conduct extensive combat operations across the whole expanse of a megacity.[31]

A challenge is also posed by complex, adaptive, and interconnected systems characterizing megacities. As observed by Spencer, 'Cities are complex adaptive systems—or more accurately, many systems of systems. … Like other complex systems, when it is touched, it changes, and the system's complexity makes it nearly impossible to truly know the second- or third-order effects

---

[27] Ibid., 6.

[28] Joint Chiefs of Staff, *JP 3-06, Joint Urban Operations* (2013), I-3.

[29] Hana Davis, 'How Hong Kong Rose to Become Tallest City in the World', *South China Morning Post*, 30 June 2018, https://www.scmp.com/news/hong-kong/community/article/2152952/how-hong-kong-rose-become-tallest-city-world.

[30] Spencer, 'The Eight Rules of Urban Warfare and Why We Must Work to Change Them'.

[31] Russell W. Glenn, Eric L. Berry, Colin C. Christopher, Thomas A. Kruegler, and Nicholas R. Marsella, eds, *Where None Have Gone Before: Operational and Strategic Perspectives on Multi-Domain Operations in Megacities*, Proceedings of the 'Multi-Domain Battle in Megacities' Conference, 3–4 April, 2018, Fort Hamilton, New York, 11–13; Konaev, 'The Future of Urban Warfare in the Age of Megacities'.

of those changes.'[32] In short, assessing the full effect of one's actions in an urban setting, both within the area itself and effects in other interconnected cities across the globe, is arguably an impossible task (see also the section on Technology below).

With size come new tactical challenges that place new demands on doctrine, training, and partnerships. The combined effect of skyscrapers and high-rise buildings, tunnels, and the sheer density of today's cities challenges such basic elements of warfare such as fires, manoeuvre, communication, and situational awareness. Large cities also challenge electronic and cyber capabilities, given difficulties communicating between floors in high-rise buildings and at subterranean levels, for example (not to mention the challenge of fighting in subterranean environments and in high-rise buildings). Buildings and other urban features also hamper the efficiency of weaponry, often acting as fortifications. For example, a study conducted by the Bundeswehr in the late 1990s found that munitions were unfit for modern combat conditions; the 20-mm gun arming their Marder infantry-fighting vehicle lacked penetration power and the Leopard tank's multipurpose (MZ)25 12-cm hollow-charge shell was unable to blast a hole big enough to penetrate a building.[33] The complexity of urban areas also often provides the defender with distinct advantages and the ability to maintain the initiative.[34]

Given the added layers of complexity in urban warfare, not found in operations in unpopulated, rural terrain, the demand for intelligence is paramount. This is particularly so given that cities are centres of human activity, where the civilian population often outnumbers enemy combatants. Thus, there is a need to understand the civilian population as well as the enemy. It is essential to find a good mix of different intelligence sources, including Human Intelligence (HUMINT), Signals Intelligence (SIGINT), and Open-Source Intelligence (OSINT) (but also Communications Intelligence (COMINT)), Imagery Intelligence (IMINT), Geospatial Intelligence (GEOINT), and Measurement and Signatures Intelligence (MASINT)). It is important to develop an advanced system for operational assessments, analysis, and planning, including everything from skilled analysts to AI- and machine-learning capabilities. Future urban warfare is very much a big data affair, where at issue might be whether a given analysis asks the correct question of a

---

[32] Graham, 'Imagining Urban Warfare'; Stephen Graham, *Cities under Siege: The New Military Urbanism* (London, New York: Verso, 2011); Stephen Graham, *Vertical: The City from Satellites to Bunkers* (London: Verso, 2018); John Spencer, 'The City Is Not Neutral: Why Urban Warfare Is So Hard', *Modern War Institute*, 22 March 2020, https://mwi.usma.edu/city-not-neutral-urban-warfare-hard/.

[33] Alexandre Vautravers, 'Military Operations in Urban Areas' (en), *International Review of the Red Cross* 92, 878 (2010).

[34] Gentile et al., *Reimagining the Character of Urban Operations for the U.S. Army*, 119.

system, rather than answering it itself. If this is not done, one will inexorably lag behind in the OODA-loop. The main challenges to tackle here are (1) the collection, processing, and dissemination of information (so-called 'fog of information' problems), (2) intelligence and the role of the security function in the planning process (information dissemination between and within levels), and (3) continuous assessment and operational adaptation (flexibility).

Achieving Success

The key for success in operations and combat on the future battlefield is as simple as it is difficult to achieve: the daunting challenges and problems of urban warfare must not be avoided or downplayed. The difficulty of this task makes it even more important to be as well prepared and trained as possible. Because urban warfare will arise. Despite preferences for avoiding urban terrain, you will simply not be able to (and be victorious). Preparation requires building intelligence capabilities suitable to the urban environment. Good leaders and fit, well-trained soldiers are also, as always, essential. Soldiers must be well educated and trained in urban warfare tactics.

It is also important, particularly in a European context, to plan for contingencies beyond offence. The defence of urban areas should be planned for. Similarly, most urban warfare writings assume that the opponent is irregular fighters, not a regular army. This may also change in a European context, where armies must also train for contingencies where the adversary fields regular forces. Learning to fight against regular forces may also be useful elsewhere. Often, as in Afghanistan and elsewhere, the opponent—or their units—have been professionally trained and are furthermore battle tested (and reasonably equipped). With the proliferation of the private military industry, one must also be prepared to meet highly trained private soldiers, who are often former regular soldiers.

There is also a problem related to power symmetry, we are not well equipped for fighting peer- or near peer adversaries in urban terrain, nor for the idea that we are the weak part of an asymmetric power capability. What if we cannot compartmentalize and separate the opponent? What if we must fight outnumbered? These contingencies must be addressed. Part of the problem here is that much of the research is done by the USA who wield incomparable military power, and Israel, whose situation is unique. Much can be learned from the USA and Israel, but it is also important to remember one's own situation and needs, as well as capacities.

A similar situation applies with regard to technology (see the section on Technology below), although here the technological breakthroughs also create capabilities available beyond militaries, in the form of unmanned aerial vehicles, or using the internet for surveillance and control. Yes, these provide an edge, but there is a quantitative aspect.

In conclusion, thought must be given to future wars and those one is expected to participate in. Megacities do apply in some cases, particularly for actors with expeditionary capability and ambitions in the developing world. For others, megacities are less relevant. In Europe, fighting in megacities is not a key task. Fighting irregular opponents in dense, confined urban terrain is central in Israel, yet may be less so in Estonia. Lessons can and should be learned, but equally important is understanding one's own situation and probable future fights.

As we will see in the next sub-section, there is also a need to be able to master multi-domain operations in a grey zone setting, utilize existing technology to get an edge, when fighting opponents with a natural defensive advantage in urban terrain.

Multi-domain Operations and Grey Zone Problems

The next challenge is the need for multi-domain operations (MDO) and the impact of grey zone problems.

As the volatility and intensity of the international security environment have grown in recent years, the grey zone between peace and war has expanded considerably.[35] Cities, the interconnected hubs of population and power, are the nexus of this grey zone, where future conflicts and wars are largely expected to take place. The challenges related to hybrid threats and hybrid warfare, and the need to manage a range of hybrid measures, are today recognized globally among experts and practitioners as well as key international organizations such as NATO and the European Union (EU). The battlefield of the future clearly exists in the grey zone between war and peace. In this grey zone, non-kinetic effects replace, or mix with, kinetic effects. A synergistic assortment of military and non-military activities will be carried out, ranging from different forms of strategic communication, through active measures such as intrusions, special operations, sanctions, and subversions, and even the use of masked soldiers, like the so-called green men in Crimea,

---

[35] Niklas Nilsson, Mikael Weissmann, Björn Palmertz, Per Thunholm, and Henrik Häggström, 'Security Challenges in the Grey Zone: Hybrid Threats and Hybrid Warfare', in Mikael Weissmann, Niklas Nilsson, Björn Palmertz, and Per Thunholm, eds, *Hybrid Warfare: Security and Asymmetric Conflict in International Relations* (London: I.B. Tauris, 2021).

cyberattacks, sabotage, and terror or proxy warfare, all without constituting actual war.[36]

The ability to conduct MDO operations is crucial to success here, as the five domains and the information dimensions all come together in the grey zone, with the cities as the centre of gravity. In future warfare, not only will the cyber and information domains be of upmost importance, but warfare itself will occur across the five domains as well as in the information environment. The battlefield will not be geographically limited, but in an interconnected world will have an impact on a global level. This all comes together in the cities. Thus, the urban environment is a key context where different countries must be prepared to defend against and counter a wide range of hybrid attacks, threats, and influence operations, be they 'little green men', disinformation campaigns, sabotage, intelligence operations, election-influence operations, or cyberattacks, to mention but a few possibilities.

The complexity and the importance of cities are both widely recognized. To give an example, the US Army notes that the emerging operational environment is multidimensional with

[f]our interrelated trends ... shaping competition and conflict: adversaries are contesting all domains, the electromagnetic spectrum (EMS), and the information environment ... smaller armies fight on an expanded battlefield that is increasingly lethal and hyperactive; nation-states have more difficulty in imposing their will within a politically, culturally, technologically, and strategically complex environment; and near-peer states more readily compete below armed conflict making deterrence more challenging.[37]

They also recognize the importance of cities.

Dramatically increasing rates of urbanization and the strategic importance of cities also ensure that operations will take place within dense urban terrain. Adversaries, such as China and Russia, have leveraged these trends to expand the battlefield in time (a blurred distinction between peace and war), in domains (space and

---

[36] Mikael Weissmann, 'Conceptualizing and Countering Hybrid Threats and Hybrid Warfare: The Role of the Military in the Grey Zone', in *Hybrid Warfare: Security and Asymmetric Conflict in International Relations*, edited by Mikael Weissmann, Niklas Nilsson, Björn Palmertz, and Per Thunholm (London: I.B. Tauris 2021). See also US Army, 'The U.S. Army in Multi-Domain Operations 2028', TRADOC Pamphlet 525-3-1, 6 December 2018, https://api.army.mil/e2/c/downloads/2021/02/26/b45372c1/20181206-tp525-3-1-the-us-army-in-mdo-2028-final.pdf; Ministry of Defence, 'Joint Concept Note 1/20, Multi-Domain Integration', November 2020, https://assets.publishing.service.gov.uk/government/uploads/system/uploads/attachment_data/file/950789/20201112-JCN_1_20_MDI.PDF.
[37] US Army, 'The U.S. Army in Multi-Domain Operations 2028', vi.

cyberspace), and in geography (now extended into the Strategic Support Area, including the homeland) to create tactical, operational, and strategic stand-off.[38]

It should be noted here that it is not only great powers or states that wield such leverage, but all types of actors do so to some degree.

There is also a need to prepare for hybrid urban combat, as we can expect not only conventional urban combat but also the need to engage in an internal security role, fighting adversaries such as terrorists and revolutionaries as well as carrying out urban operations and combat that is more similar to traditional police work than traditional military combat. The UK operations in Belfast and Londonderry, and the French experience in Algiers, are examples of the latter situation. Hybrid urban combat requires a more sophisticated military capability than traditional combat, as military forces must be able to operate simultaneously across the entire spectrum of urban combat intensity. This includes not only special operations capability but also civil affairs expertise, sophisticated methods for intelligence gathering, and close policy coordination between the military and politicians.[39]

Achieving Success

Success on tomorrow's urban battlefield requires not only the ability to conduct MDOs, but also developing capabilities to engage in the information environment. Success in the land, maritime, air, space, and cyber domains is insufficient to win a city; one must also win the battle of narratives in the information sphere that, together with the cyber domain, is predicted to be the centre of gravity in future conflicts. Furthermore, this must be done across the spectrum of conflict, from peace through the grey zone, as well as in war.[40] One must also prepare for all levels of combat intensity, from conventional warfare to what would normally fall within policing and humanitarian relief operations.[41] As observed by Stephen Graham, '[n]othing lies outside the battlespace, temporally or geographically. Battlespace has no front and no back, no start nor end.'[42]

---

[38] Ibid.

[39] DiMarco, Concrete Hell, 212. Also see Alice Hills, Making Mogadishu Safe: Localisation, Policing and Sustainable Security: Localisation, Policing and Sustainable Security (London: Routledge, 2019); Alice Hills, Future War in Cities: Rethinking a Liberal Dilemma (London: Frank Cass, 2004); Alice Hills, 'Making Mogadishu Safe', The RUSI Journal 161, 6 (2016).

[40] Frank G. Hoffman, The Contemporary Spectrum of Conflict: Protracted, Gray Zone, Ambiguous, and Hybrid Modes of War, 5 October 2015, https://www.heritage.org/military-strength-topical-essays/2016-essays/the-contemporary-spectrum-conflict-protracted-gray; Mikael Weissmann, 'Hybrid Warfare and hybrid Threats Today and Tomorrow: Towards an Analytical Framework', Journal on Baltic Security 5, 1 (2019); Weissmann, 'Conceptualizing and Countering Hybrid Threats and Hybrid Warfare'.

[41] Hills, Making Mogadishu Safe; Hills, Future War in Cities.

[42] Graham, Cities under Siege, 31.

To be able to handle the outlined challenges, doctrines and handbooks must be developed that pay attention to the increasing importance of urban warfare. It is also essential to train for multi-domain operations in urban settings. Cross-domain integration and the information sphere are therefore crucial. The information sphere does not only include technology, although that is admittedly important, but also the battle of narratives on the local, regional, and global level. Everything is connected, and the public view of the population—among adversaries, adversary population, at home and elsewhere—is crucial and cannot be taken for granted. This is not only a result of what you say, but also very much what you do (or do not do). Thus, urban warfare is about more than combat and 'winning battles'. It requires collaboration not only across domains, but also between the military and civilian spheres.

It is also important to think outside the asymmetrical warfare box, preparing for contingencies other than taking the offensive in an asymmetric conflict against a non-peer adversary, which tends to be the focus of most current research, particularly in the US literature. However, the idea of defensive urban operations is relevant in a European context, in particular in the Baltics, where the main focus is the deterrence of potential Russian aggression. Here 'U.S. and NATO forces could create conditions in urban areas in the Baltics that make it impossible for the Russians to overrun them rapidly, thus removing the possibility of a fait accompli and thereby changing their risk calculation to preclude assumptions of an early, cheap success.'[43]

It is also important to consider the technological balance. Besides the obvious case of peer or near-peer adversaries, the less obvious situation of opposing irregular forces becomes more and more likely with increases in the availability of technology. This is so regarding, for example, the increased availability of UAVs, and the equalizing capability of irregular forces to utilize the cyber domain despite the technological superiority of regular forces. Non-state armed groups are capable of utilizing social media not only to fight the 'battle of narratives', but also for recruitment, propaganda, and even the coordination and organization of combat operations.[44] This leads us to the next challenge, namely technology.

[43] Gentile et al., *Reimagining the Character of Urban Operations for the U.S. Army*, 60.
[44] For examples, see David Kilkullen's presentation on 'Emerging Patterns of Adversary Urban Ops: Insights from the NATO Urbanisation Program', RUSI Urban Warfare Conference 2018, available at https://www.youtube.com/watch?v=mbxknQrNEgY&t=4075s (starts at 6:17).

Technology

The breakthroughs in technology have not only forced the battle to the city, but emerging and novel technologies also have a great impact on battles and combat itself.[45] The physical terrain, infrastructure, and civilian presence in urban areas are major operational challenges, to which the adoption and development of new technology is a potential solution. The availability and quality of UAVs and sensor technology have increased greatly, whilst battlefield information at the tactical, operational, and strategic levels has also become available at greater scale. This is very important in the rapidly changing and chaotic urban environment, since these and other technologies enhance intelligence, surveillance, and reconnaissance (ISR) and for command and control, which is particularly important in the type of joint multi-domain operations that need to be the focus in urban operations. These technologies also assist in force protection and the limitation of collateral damage, as well as protecting and controlling the civilian population.

The use of UAVs is not new; they have been used by military forces for many years in a broad range of tasks. In the context of urban warfare, their reconnaissance role has been the most important one. They also play an important role in target identification and precision targeting, enhancing fighting power, and helping to reduce collateral damage. Both small and large drones may be used to enhance battlespace awareness, although at least against peer or near-peer adversaries the latter are limited by being observable by radar. UAVs are also part and parcel of the US Defence Advanced Research Projects Agency's Urban Reconnaissance through Supervised Autonomy (URSA) project, where the aim is to find ways to use autonomous systems to help the military detect hostile forces in urban environments and positively distinguish combatants from civilians before own forces come in contact.[46] Drones can also deliver warning signals to any humans they encounter and forward information on the response, together with video and location data, to military personnel who can in turn decide how to respond to a situation.[47]

---

[45] Michael Raska, 'The Sixth RMA Wave: Disruption in Military Affairs?', *Journal of Strategic Studies* 44, 4 (2021); Kelley M. Sayler, *Emerging Military Technologies: Background and Issues for Congress*, CRS Report R46458, updated 10 November, Congressional Research Service 2020. Also see the special issue on Defence Innovation and the 4th Industrial Revolution: Security Challenges, Emerging Technologies, and Military Implications, edited by Michael Raska, Katarzyna Zysk, and Ian Bowers, of which this article is a part (*Journal of Strategic Studies*, 44, Issue 4 (2021)).

[46] Lauren C. Williams, 'Can AI and Autonomous Systems Detect Hostile Intent?', *Defense Systems* 4 October 2021.

[47] Paulina Glass, 'Here's the Key Innovation in DARPA AI Project: Ethics from the Start', *Defense One* 15 March 2019.

One important development in drone technology is the emerging proliferation of what are called 'swarms', that is 'large numbers of simple, low cost, expendable systems that are interconnected'.[48] Swarms are argued to have the potential to change how we fight, with large autonomous swarms of drones flying and operating together as a single unit, with the capability to autonomously alter their behaviour and action based on intercommunication.[49] Such drones will also have great potential as sensors, able to identify threats and targets and relay relevant information both to each other and back to base for further assessment and action.

Moving on, sensors are one of the key technologies for the future of urban warfare. Sensors encompass a wide range of technologies and devices, including radar, acoustic, thermal, optics, seismic, magnetic, active sensors, smart sensors, nano sensors, and wearable sensors. For example, sensors today can enable soldiers to see through walls and detect fired projectiles. The use of unattended ground sensors has increased among high-tech forces such as the US and NATO to enhance their intelligence, surveillance, and reconnaissance abilities to a degree limiting adversaries' possibilities for cover and concealment. This is also why huge R&D investment has been made in developing new forms of concealment. Cheap and manoeuvrable micro- and nano-drones have also been developed for use in reconnaissance and surveillance, as has wearable sensor technology providing location and navigation data and uninterrupted communication between troops and UAVs in areas where GPS signals are weak or absent.[50] The importance of the need for uninterrupted communication should not be underestimated, since communication in urban terrain often creates particular difficulties.

Another important area is artificial intelligence (AI), used increasingly on all levels. For example, Israel has developed the Fire Weaver, 'a networked sensor-to-shooter system' that 'connects forces on the battlefield to a network that works with advanced computer vision technology and artificial intelligence algorithms to aid in targeting for commanders and soldiers. ... The new system allows leaders to use a host of resources at the tactical level, from drones to forward observers who are networked so that military leaders can see the same battlefield and targets from different angles. An increasingly

---

[48] Michel-Kleisbauer, 'URBAN WARFARE', 6. More formally defined: 'multiple unmanned systems capable of coordinating their actions to accomplish shared objectives' (Zachary Kallenborn and Philipp C. Bleek, 'Swarming Destruction: Drone Swarms and Chemical, Biological, Radiological, and Nuclear Weapons', *The Nonproliferation Review* 25, 5–6 (2018)).

[49] Zachary Kallenborn and Philipp C. Bleek, 'Drones of Mass Destruction: Drone Swarms and the Future of Nuclear, Chemical, and Biological Weapons', *War on the Rocks*, 20 February 2019; Kallenborn and Bleek, 'Swarming destruction. See also T. X. Hammes, 'The Future of Warfare: Small, Many, Smart vs. Few & Exquisite?', *War on the Rocks*, 7 August 2015; Shmuel Shmuel, 'The Coming Swarm Might Be Dead on Arrival', *War on the Rocks*, 10 September 2018.

[50] Konaev, 'The Future of Urban Warfare in the Age of Megacities'.

digitized battlefield requires a system to digest all the data coming in from various sensors and potential shooters.'[51]

So far, the application of autonomous systems has been limited by their dependence, on some level, on direct human control. With the proliferation of data provided by sensors, and the advances in AI, the need for human control will diminish over time. Autonomous ground vehicles will also improve the survivability and resilience of ground troops in an urban environment. Several countries are already researching robotic vehicles for use in ground supply and medical evacuation, two dangerous and resource-intensive tasks. Systems have also been developed to improve force protection, and are already in use investigating tunnels, caves, and buildings before sending in soldiers. Unmanned Ground Vehicles (UGVs) have also been developed.[52] Both Israel and Russia have fielded UGVs in battles. Russia has mainly used UGVs in Syria.[53] In contrast, Israel's Carmel Armoured Combat Vehicle is particularly suited for urban combat; the system integrates advanced artificial intelligence and autonomous capabilities to enhance mission effectiveness for the Israel Defence Forces (IDF).[54] The importance of unmanned vehicles cannot be underestimated, as recent experience, such as in Fallujah, Baghdad, or Mogadishu, has shown a high casualty rate among soldiers in urban operations particularly due to IEDs, mines, and sniper fire.

Two other areas where technology will have an impact on urban warfare are Augmented Reality (AR) and biometrics. The former has great potential, as it allows for moving beyond the traditional 2D map, which is inadequate for the three-dimensional urban battlefield where the vertical dimension is essential.[55] Not least, benefits may be drawn from tactical augmented reality (TAR), helping improve soldiers' ability to locate themselves, friendly

[51] Seth J. Frantzman, 'Israel Finds an AI System to Help Fight in Cities', *C4ISRNET*, 5 February 2020, https://www.c4isrnet.com/battlefield-tech/2020/02/05/israel-finds-an-ai-system-to-help-fight-in-cities/.

[52] Michel-Kleisbauer, 'URBAN WARFARE'.

[53] Sten Allik, Sean Fahey, Tomas Jermalavičius, Roger McDermott, and Konrad Muzyka, 'The Rise of Russia's Military Robots: Theory, Practice and Implications', International Centre for Defence and Security, Estonia, February 2021, https://icds.ee/wp-content/uploads/2021/02/ICDS-Analysis_The-Rise-of-Russias-Military-Robots_Sten-Allik-et-al_February-2021.pdf; Sebastien Roblin, 'What Happened When Russia Tested Its Uran-9 Robot Tank in Syria?', *The National Interest*, 7 April 2021, https://nationalinterest.org/blog/reboot/what-happened-when-russia-tested-its-uran-9-robot-tank-syria-182143; David Hambling, 'Russia's Autonomous Robot Tank Passes New Milestone (and Launches Drone Swarm)', *Forbes*, 2 September 2021, https://www.forbes.com/sites/davidhambling/2021/09/02/russias-autonomous-robot-tank-passes-new-milestone-and-launches-drone-swarm/.

[54] ESD Team, 'Israel's Carmel Programme Charting Future Concepts for Mounted Combat', *European Security & Defence*, 7 February 2020, https://euro-sd.com/2020/02/articles/16078/israels-carmel-programme-charting-future-concepts-for-mounted-combat/; Michael Peck, 'Carmel: Israel Unveils New Stealth Street-Fighting Tank', *The National Interest*, 28 September 2019, https://nationalinterest.org/blog/buzz/carmel-israel-unveils-new-stealth-street-fighting-tank-72491.

[55] Xiong You, Weiwei Zhang, Meng Ma, Chen Deng, and Jian Yang, 'Survey on Urban Warfare Augmented Reality', *International Journal of Geo-Information* 7, 2 (2018); Yaakov Lappin, 'Israel's Rafael Reshapes Urban-warfare with AI, Augmented Reality', Israel Hayom, 2 February 2020, https://www.israelhayom.com/2020/02/02/israels-rafael-revolutionizes-urban-warfare-with-ai-augmented-reality/.

soldiers, and adversaries compared to using traditional night vision googles and GPS.[56] Biometrics is also useful in the urban setting, where the mixture of foes and civilians creates a need for an ability to identify hostile individuals and non-state actors. Automated identification and the analysis of different behaviours and biological characteristics is one way to do this.[57] Biometric technologies, which use unique attributes like fingerprints, facial or ocular measurements, DNA, cardiac signatures, and voice or gait patterns to identify individuals, have been used for decades, but the possibility to combine such identifiers with advances in artificial intelligence (AI) and Big Data analytics expands their applicability tremendously.[58]

Loitering munition will become increasingly important in urban warfare, as they can be used by soldiers on the ground to reduce radar, visual, and thermal signatures, making them more difficult to find, track, and defeat. This is important as a countermeasure to the proliferation of sensor technology and UAVs.

Social media also poses challenges. Traditionally, technological superiority has enabled information superiority, in the form of influence and control over the flow of information in and out of the area of operations.[59] As argued by Margarita Konaev, 'information superiority and asserting control over the information environment is all the more critical in urban warfare, as it allows the state's force to cut off local hostile forces from their strategic leadership, prevent them from disseminating their message and from communicating with the city's civilian population and the outside world, shape public opinion in their favour and win the "battle of narratives".[60] States' superiority in the information sphere has been challenged by platforms like Facebook, Twitter, and YouTube.[61] In fact, not only do all conflicting parties use social media platforms to spread their version of reality, non-state groups have also proven very capable of doing so.[62]

[56] E.g. David Vergun, 'Heads-up Display to Give Soldiers Improved Situational Awareness', US Army, 20 September 2021, https://www.army.mil/article/188088/heads_up_display_to_give_soldiers_improved_situational_awareness.
[57] Mark Lunan, 'Biometrics', *The Three Swords Magazine* 33 (2018); Kelley M. Sayler, *Biometric Technologies and Global Security*, CRS IF11783, updated March 30, Congressional Research Service 2021.
[58] Sayler, 'Biometric Technologies and Global Security'.
[59] Konaev, 'The Future of Urban Warfare in the Age of Megacities'.
[60] Ibid., 39.
[61] E.g. ibid.; P. W. Singer and Emerson T. Brooking, *LikeWar: The Weaponization of Social Media* (Boston, MA: Houghton Mifflin Harcourt, Mariner Books, 2019[2018]); David Patrikarakos, *War in 140 Characters: How Social Media is Reshaping Conflict in the Twenty-first Century* (New York: Basic Books, 2017).
[62] E.g. Anna Leander, 'Digital/commercial (in)visibility', *European Journal of Social Theory* 20, 3 (2017); Bozorgmehri Majid, 'Recruitment of Foreign Members by Islamic State (Daesh): Tools and Methods', *Journal of Politics and Law* 11, 4 (2018).

Achieving Success

The above outline of new technology's impact on urban warfare paints an apparently promising picture, in which technology can be key for success urban warfare. This is all very well, but experience has also shown that the underlying principles of technology, as well as the technologies themselves, tend to break down in cities.

It is clear that breakthroughs in technology are crucial for the future of urban warfare. It might seem like technology, especially sensors and unmanned systemic combined with AI, is a panacea. This may be so, but it is also important to be cautious. Throughout history, revolutions in military technology have often been expected to change everything. The reality never turns out to be that simple. In the case of urban warfare, we can expect the fights of the future to be at least as dirty as those of the past. No other environment is as complex—in physical and human terms—as cities, and cities have never been so complex or interconnected as today. Yes, technology will help. But penetrating walls, and clearing house to house, and room to room, are hardly tidy tasks, even with improved technology. David Bellavia's memoirs of his experiences from Fallujah, *House to House: A Soldier's Memoir*, here offers a telling tale.[63] Whilst not being an operation and combat with all the tools of the future, it shows the difficulty of fighting a non-peer irregular opponent despite superiority in force and technology. Unless you want tomorrow's war to be fought only with unmanned vehicles and robots, or by flattening enemy cities to the ground, urban warfare will remain a dirty business. Furthermore, even if you chose unmanned combat or total destruction you might win the fight, but still lose the war, which is not contained to the battle zone, but is interconnected and ultimately embedded in the information sphere and the battle of narratives.

Dense concrete environments drastically reduce the advantages of superior technology, since buildings and other infrastructure mask targets and create urban and suburban canyons in which to hide and manoeuvre. There is a reason why so much emphasis has been put on developing doctrine, training, and equipment to fight underground.[64] To give a specific example of the scale of this investment: in 2017, the US Army launched a $572 million effort to train and equip twenty-six of thirty-one active combat brigades for fighting in

---

[63] David Bellavia and John R. Bruning, *House to House: An Epic Memoir of War* (London: Simon & Schuster, 2007).

[64] See Jeremiah Rozman, 'The Army Is Preparing to Go Underground', *RealClearDefense* 3 July 2019, for an overview of efforts. See also Todd South, 'The Subterranean Battlefield: Warfare is Going Underground, into Dark, Tight Spaces', *Military Times* 25 February 2019; Modern War Institute, 'The Elephant in the Tunnel: Preparing to Fight and Win Underground', 18 March 2019, https://mwi.usma.edu/elephant-tunnel-preparing-fight-win-underground/.

large-scale subterranean facilities under dense urban areas.[65] There are also initatives addressing areas such as multi-domain battle (MDB) in megacities, bio-convergence, and the soldier of 2050, addressing the 'Gen Z' perspective in relation to the operational environment and national security challenges.[66]

Also, the existing warfare literature is biased toward the stronger and technologically superior force fighting against a non-peer, irregular, and less technological adversary. It is worth considering the implication of urban warfare against a peer or near-peer opponent from the perspective of their mutual possession of advanced technology. Furthermore, contingences should be considered in which one does not have control of the area of operation, or superiority in force, or the offensive advantage of choosing the time and place of fighting. Lastly, not only has enabling technology been developed, but also counter-measures.

When fighting an equally high-tech opponent, concrete and tunnels may interfere with sensors, but so also may electronic warfare counter measures, creating a contested communications environment. This must be taken into consideration, as well as the opponent using offensive cyber capabilities. Nor can you expect that you have intelligence superiority, as it may be both challenged and a target for deception. In fact, if history is correct, urban warfare between peers might be the most recognizable contingency, harkening back to Stalingrad 1942–43, Manilla 1945, or Hue 1968.

The Urbanization of Insurgency

After the Cold War, the urbanization of insurgency has become a factor. Urban battle spaces have always been to the defenders' advantage, as 'the physical environment tends to mitigate many technological advantages held by the attacker; the presence of civilians can greatly complicate the operations of attacking forces, while sometimes also providing cover and concealment to the defender; and it opens the battle to modern media scrutiny.'[67] With the urbanization and technology megatrends, moving the fight to urban areas is arguably the only way for irregulars to win future battles against high-tech regular forces. Not only is it easier to defend an urban environment, but one

[65] Matthew Cox, 'Army Is Spending Half a Billion to Train Soldiers to Fight Underground', *Military.com*, 24 June 2018, https://www.military.com/daily-news/2018/06/24/army-spending-half-billion-train-troops-fight-underground.html.

[66] In many cases such innovations are being conducted as collaborative partnerships and dialogues between academia, industry, and government. A good example here is the US Army Mad Scientist Laboratory initiative.

[67] DiMarco, *Concrete Hell*, 24–5.

cannot win today holding fields and forests, since urban areas have people and power.

Today's sensors and high-precision weapons limit operational and tactical manoeuvres in open terrain (including forests). Commanders who lack technological capacities will simply find cities appealing terrain, especially since they often know the city better and have a superior ability to mobilize their resources and population compared to their opponent. To this can be added the tendency of insurgencies to have more flexible rules of engagement, as well as interpretation of laws of war. It is also in the city, at close range, that the relative inefficiencies of the weapons used by insurgents are negated. The city also works as protection, as the effect on the urban terrain of military actions, or one's own fortification work, makes it easier to defend and harder to attack.[68]

Here, the cyber and information dimensions should be considered, which not only add a social media dimension to warfare, but also an array of open-source material, access to services like Google Maps, photo sharing, coded communication, different connected sensors, and increasingly cheap and capable UAVs. For example, a connected surveillance camera today costs £30 at a local hardware store (or online). As cities are interconnected, physical presence is not always needed on-site—for either side—since forces can be commanded, controlled, and launched from anywhere, as long as they are connected. The cyber dimension goes beyond the information sphere and the battle of narratives, as not only states can use different forms of cyberattack. In interconnected cities, it is also possible for defenders in the Global South to move the battle to the homes of the adversary, conducting counter-attacks in Brussels, London, Tokyo, or Washington.

It should be noted here that the main drivers of technological developments are no longer the military, but the civilian sector . Thus, commercially available technological advances today also benefit non-state actors, who can incorporate cheap, off-the-shelf products in their operations. One good example is the availability of cheap, commercial drones providing non-state actors with at least a limited air force capability that may least interfere with, if not challenge, the dominance of conventional forces. Non-state actors like ISIS, Hezbollah, Houthi rebels in Yemen, and the Russia-backed militants in eastern Ukraine demonstrate the potential use of commercially available drones, as well as military-grade UAVs, for reconnaissance, surveillance, and even combat in Syria, Iraq, and eastern Ukraine.[69]

---

[68] E.g. Spencer, 'The City Is Not Neutral'.
[69] Konaev, 'The Future of Urban Warfare in the Age of Megacities'.

A similar case can be made regarding the cyber domain and the information sphere, where non-state actors have shown increasing adaptability in using and combining expertise to spread propaganda globally and contest the battle of narratives, recruit supporters internationally, and draft recruits. These actors have also demonstrated an ability to utilize the interconnected world, both moving the fight out of the city and home to their opponents, and enabling supporting to get involved in the battle from afar.

Achieving Success

There will be fighting on the ground in cities. Unless one wishes to raze cities, house to house fighting will be necessary. Technology may help, but it would be overoptimistic to expect technology to replace the need for the human soldier. Thus, the role of western forces against irregular forces in urban combat must be considered: whether and to what extent we engage with our own ground forces; whether they cooperate with indigenous forces; what role do they play, as advisors, reserves, enablers? executing close combat? or rather focusing on intelligence, surveillance, and reconnaissance systems and precision strikes?[70]

Population control must also be considered. Fighting insurgencies in a city, by definition, complicates distinguishing civilians from foes. Here, one also needs to ask whether the civilian populations should be evacuated to enable operations, and whether this is possible. However, historically, populations have remained even after evacuation. Furthermore, it is not realistic to evacuate megacities. Where should the 35 million inhabitants of Jakarta be moved to?

In short, civilians will be at hand during urban warfare. They will impact the battle space, as they can both constrain and enable operations. This is particularly so as any city has an abundance of cell phones, and ways to relay messages both within and beyond the city.

Urban defenders will also be able to maintain their freedom of movement within their defences. Here, they 'can prepare the terrain to facilitate their movement to wherever the battle requires. They can connect battle positions with routes through and under buildings. They can construct obstacles to lure attackers unknowingly into elaborate ambushes because of the limited main avenues of approach in many dense urban environments.'[71] This creates a situation where the use of available technology for ISR will be crucial, and where the benefits of multi-domain operations must be utilized, since the

---

[70] See e.g. Johnson, 'Urban Legend: Is Combat in Cities Really Inevitable?'.
[71] Spencer, 'The Eight Rules of Urban Warfare and Why We Must Work to Change Them'.

synergies to be gained are necessary to win in a battlefield that favours the defender.

Conclusion: Eleven Takeaways about Urban Warfare

It should by now be clear not only that the introductory statement that '[t]he future of warfare lies in the streets, sewers, high-rise buildings, industrial parks, and the sprawl of houses, shacks, and shelters that form the broken cities of our world'[72] was correct, but also that this is just the beginning as the urban battlefield reaches far beyond the city limits. As we have seen, the character of war is changing, cities are interconnected, the grey zone between war and peace is increasing, and the information sphere has become a centre of gravity, consequently the urban battlefield knows no borders but reaches across the physical and temporal domains.

Having outlined the challenges of urban warfare on tomorrow's battlefield—urbanization, multi-domain operations, the grey zone problems, technology, and the urbanization of insurgency, eleven lessons about urban warfare can now be outlined.

> **Takeaway 1: Urbanization turns the future urban battlefield into a nightmare.** First, but possibly most important, urbanization turns the future urban battlefield into a possible nightmare. This is a fact where resistance is futile and should not be attempted, instead it needs to be accepted. The focus should simply be on accommodating and adopting to the new reality of urban operations and warfare, rather than trying to develop ways to avoid urban areas. Avoidance is like asking for failure, as it is not always possible to choose the battlefield and it is therefore better to prepare thoroughly for the eventuality or urban warfare.

> **Takeaway 2: Multi-domain operations are crucial for success.** The ability to conduct multi-domain operations is crucial for success. Future urban operations will need to meet the challenges from cross-domain and cross-conflict-spectrum fighting. In future warfare, not only will the cyber and the information domains be of outmost importance, but warfare itself will occur across the five domains as well as in the information environment.

---

[72] Peters, 'Our Soldiers, Their Cities'.

Takeaway 3: Urban battles will take place in the grey zone. You need to prepare for urban battles that will take place in the grey zone between peace and war, where the five domains and the information dimensions all come together, with the cities being the centre of gravity. You need to be prepared to conduct urban warfare in a legal state of non-war as well as war, alone as well as in collaboration with civilian actors.

Takeaway 4: The urban battlefield knows no physically borders. Do not expect the urban battlefield to be geographically limited to a physically defined area. The world is interconnected, nowhere more so than in cities. What happens in one place will have an impact on a global level. There is simply no such thing as 'outside the battle space'. You need to be prepared to defend against and counter a wide range of hybrid attacks, kinetic as well as non-kinetic, hybrid threats, and influence operations everywhere, including in yours and your partners' home country.

Takeaway 5: The importance of the information environment cannot be underestimated. The importance of the information environment cannot be underestimated. If you cannot win the 'battle of narratives' you will not be able to achieve victory. This battle of narratives happen on the local, regional, as well as the global level. Everything is connected, and the perception of the public—among adversaries, adversary population, at home and elsewhere—is crucial and cannot be taken for granted. It should here be stressed that perception is not only a result of what you say, but also what you do (or do not do, or do not say). Thus, urban warfare is about more than 'combat' and 'winning battles'. It requires collaboration not only across domains, but also between the military and civilian spheres.

Takeaway 6: Breakthroughs in technology are crucial for the future of urban warfare. Novel, emerging, and breakthrough technologies will be crucial for the future of urban warfare. Whilst technologies might appear to resolve the problems of urban warfare, especially with the use of sensors and unmanned systemics combined with AI, it should be stressed that technology should not be perceived as a panacea, and some caution is advised. If history has taught us anything, it is that whilst revolutions in military technology have often been expected to change everything, reality has frequently turned out to be less straightforward. More concretely, we should not expect future fights in cities to be any less dirty than those of the past. There are no other environments as complex in physical and human terms as cities, and the cities themselves have never been as complex and interconnected as they are today.

Takeaway 7: The demand for intelligence is paramount. The demand for intelligence is paramount given the added layers of complexity in urban warfare compared with operations in rural areas. Cities are not only interconnected and complex centres of human activity, but also an environment where the civilian population regularly outnumbers enemy combatants. Thus, it is essential with good intelligence, of all types, to understand the civilian population as well as the enemy. Here future urban warfare is expected to be very much a big data affair, where at issue might be whether a given analysis asks the correct question of a system, rather than answering it itself.

Takeaway 8: Think beyond the asymmetrical warfare. There is a need to think beyond the asymmetrical warfare box, where offensive operations against irregular, often low-tech, non-peer adversaries are in focus. There is a need to prepare for contingencies against high-tech, peer- or near peer adversaries (and in some cases superior adversaries). This is of particular importance not least in a European context, where there is a need to plan for defensive contingencies against high-tech adversaries with regular forces.

Takeaway 9: Plan for your own, not others' urban wars. Your own needs and operating environments should be in focus. Each country needs to ensure sufficient focus is put on safeguarding its own needs and preparing for the kinds of urban warfare it expects in its own operating environment. In short, plan for the wars you expect to fight. For example, megacities will not be a concern for all land forces, but is something of major interest for actors with expeditionary capability and ambitions in the developing world. There are of course lessons to be learned from other environments, including combat in megacities, but one should select and adopt according to one's own needs, capabilities, and resources.

Takeaway 10: All urban warfare will have a civilian dimension. All urban warfare will have a civilian dimension. The presence of civilians will impact the battle, both as a constraining and enabling force. It is essential that their presence is acknowledged and included in the operational planning, ranging from adapting behaviour and fire from own forces to avoid unwanted secondary effects, how the information sphere is utilized, to policing and population control.

Takeaway 11: The urbanization of insurgency is a matter of fact, not a possibility. Finally, it should be recognized that the urbanization of insurgency is a matter of fact, not a possibility. With the urbanization and technology megatrends, moving the fight to urban areas is simply the only way for irregular forces to have chance to win future battles against stronger high-tech opponents. Not only is the urban environment to the defenders' advantage, in addition one can neither hide in, nor win by holding, fields and forests, since urban areas hold the centre of people and power.

To sum up, urban areas will be an increasingly important arena for future land warfare. Urban operations and warfare should therefore acquire a greater significance in our understanding of the operational environment. With large cities being the centre of gravity for political and economic interaction and although urban warfare is a nightmare that one reasonably hopes to avoid, it is not always possible to choose the battlefield and it is therefore better to prepare thoroughly for this eventuality.

8

Emerging Technologies

From Concept to Capability

Jack Watling

Introduction

There is a remarkable consistency in how the future battlefield is portrayed. From science fiction to computer science,[1] from the big screen to the small,[2] and from the US Army to the Armed Forces of the Russian Federation,[3] there are pervasive and persistent themes. In this imagined future swarms of autonomous drones scour the battlefield. The command team, surrounded by touch screens and wearing smart glasses, interact through artificial intelligence with a graphically complex and yet seamlessly relevant representation of the battlefield in real time, making decisions that unleash precise effects that simultaneously conform to human intent, yet do not require sustained oversight. The date of this imagined future varies. Overly optimistic portrayals had anticipated it to have arrived by now. Within militaries it was hoped for by the 2030s. This has slid to 2040 as technological advances have refused to track with the narrative, and in the USA it is now scheduled for 2050.

There are several factors that seem to give this vision a gravitational hold on the imagination. The vision of a command team with access to all relevant battlefield information at their fingertips is simultaneously exactly what military commanders wished they had, and is extremely convenient for story-

---

[1] Compare the use of autonomous UAVs in Peter Singer and August Cole, *Ghost Fleet: A Novel of the Next World War* (New York: Houghton Mifflin Harcourt, 2015), and Professor of Computer Science Stuart Russel's depiction of assassination by UAV in, Stewart Sugg, *Slaughterbots* (Space Digital, 2017).

[2] Consider for example the centrally coordinated UAV attack in the dystopian future of Steven Spielberg, *Ready Player One* (Warner Bros, 2018) or the C2 architecture shown in Gavin Hood, *Ender's Game* (Summit Entertainment, 2013), and the portrayal of future technologies in Infinity Ward, *Call of Duty: Infinite Warfare* (Activision, 2016).

[3] Compare US Futures Command concepts and those put out by the Russian Federation, see LTG Sean MacFarland, 'TRADOC Mad Scientist 2017 Georgetown: Welcome to Day 2 w/ LTG Sean MacFarland' TRADOC G-2 OE Enterprise (9 August 2017): https://www.youtube.com/watch?v=Cp3NqSzSnTg, accessed 8 January 2020; also note the consistent themes in how the US perceived its adversaries' future capabilities, see John Allen and Amir Hussain, 'On Hyperwar', *Proceedings Magazine* 143, 7 (2017), 1373.

Jack Watling, *Emerging Technologies*. In: *Advanced Land Warfare*. Edited by Mikael Weissmann and Niklas Nilsson, Oxford University Press. © Jack Watling (2023). DOI: 10.1093/oso/9780192857422.003.0008

tellers, who through this plot device can give their audience an understanding of widely dispersed events, whilst only needing to track a limited number of characters. Similarly, the swarm of armed quadcopters are close enough to real capabilities to be plausible, whilst futuristic enough to suggest progress. How the AI determines what information is relevant to the command team, or how such small quadcopters have the fuel to cover so much ground, are the kinds of questions that can be wished away when systems are powered by CGI.

Such questions are not mere technical details to be worked out, however. They are major technological hurdles standing between the present and this envisaged future. Some will be overcome. Some will not. And the second order effects of some of the solutions found will likely change how these capabilities are ultimately employed, either reflecting the constraints imposed by technology, or because better methods of employment become possible. It is the journey from concept to capability, and the deviations in course this may cause, that this chapter seeks to chart. The chapter examines several emerging technologies, widely anticipated to transform land warfare, unpacks the practicalities of their employment, and how this is likely to shape their eventual use. The four technologies to be considered in sequence are autonomous systems, layered precision fires, high fidelity sensors, and artificial intelligence. The chapter concludes by considering these capabilities in combination, and their collective impact on established principles in land warfare.

Autonomous Systems

There is a structural problem in the discourse surrounding autonomous weapons because it is dominated by an imaginary end state without reference to the process that will see autonomy become an increasing component of military systems. The discourse centres on whether the end state should be pre-emptively banned,[4] with the Secretary General of the United Nations declaring that 'machines that have the power and the discretion to take human lives are politically unacceptable, are morally repugnant, and should be banned by international law'.[5] He is likely to be proven wrong, and it is important to understand why. Article IV of the Outer Space Treaty, states that 'the establishment of military bases, installations and fortifications, the

---

[4] Stephen D. Goose and Mary Wareham, 'The Growing International Movement Against Killer Robots', *Harvard International Review* 37, 4 (2016), 29.

[5] António Guterres, 'Remarks at "Web Summit"', (5 November 2018): https://www.un.org/sg/en/content/sg/speeches/2018-11-05/remarks-web-summit, retrieved 12 May 2019.

testing of any type of weapons and the conduct of military manoeuvres on celestial bodies shall be forbidden'.[6] In spite of this agreement many military capabilities controlled by the treaty's signatories depend on infrastructure in space,[7] whilst a number of signatories have developed platforms for doing damage to one another's infrastructure in space,[8] without being found to have breached the treaty. The US Army's new operating concept—multi-domain operations—defines space as a contested domain of warfare,[9] whilst NATO considers space to be an operational domain.[10] This is legally possible because the definition of a 'weapon' in space was not clearly defined. Far from being a problem with the treaty's language, however, it seems more sensible to conclude that it is a problem with the notion of trying to ban something that does not exist, since it seems unreasonable to expect diplomats to create technically precise definitions to regulate non-existent technologies. In the context of lethal autonomy, as Nehal Bhuta and Stavros-Evdokimos Pantazopopulos have observed:

> As functions and tasks are delegated piecemeal, exactly what constitutes 'human control' over an existing technology integrated into a new technological system may be very difficult to know ex ante. It is only as such complex human–machine systems are assembled, tested and used that we may fully and concretely appreciate whether, and to what extent, human judgement and human decision making remain significant variables in the functioning of the system.[11]

Evaluating the impact of autonomous technology on the future battlefield should therefore focus on how and why autonomy is adopted in military systems as a piecemeal process. In this light, the futility of trying to preemptively determine thresholds of 'meaningful human control' becomes all too apparent. Anti-tank mines have no meaningful human control, whilst

---

[6] Treaty on Principles Governing the Activities of States in the Exploration and Use of Outer Space, including the Moon and Other Celestial Bodies (1967), Article IV: https://www.ifrc.org/docs/idrl/I515EN.pdf, retrieved 12 May 2019.

[7] GPS being a prime example.

[8] China has been developing a wide range of anti-satellite capabilities, first demonstrated in 2007, see Carin Zissis, 'China's Anti-Satellite Test', The Council on Foreign Relations (22 February 2007): <https://www.cfr.org/backgrounder/chinas-anti-satellite-test>, accessed 17 February 2019. Other states have similar capabilities.

[9] 'The US Army in Multi-Domain Operations', TRADOC Pamphlet 525-3-1 (6 December 2018).

[10] Alexandra Stickings, 'Space as an Operational Domain: What Next for NATO?', RUSI Newsbrief, 15 October 2020.

[11] Nehal Bhuta and Stavros-Evdokimos Pantazopopulos, 'Autonomy and Uncertainty: Increasingly Autonomous Weapons Systems and the International Legal Regulation of Risk', Nehal Bhuta, Susanne Beck, Robin Geiß, Hin-Yan Liu, and Claus Kreß, *Autonomous Weapons Systems: Law, Ethics, Policy* (Cambridge: Cambridge University Press, 2016), 286.

smart-sea mines are discerning in their targets.[12] Searching for a definition that does not outlaw legitimate military technologies tends toward definitions that are unlikely to ever exist because they would lack any military utility. Whilst activists seek to define a hypothetical future, states are building a wide range of capabilities that incorporate increasing levels of autonomy.[13] To understand the impact of these developments, we must consider where these capabilities are being pursued, why, and what this means for their employment on the battlefield.

Militaries are developing autonomous systems for three reasons: to reduce the crew commitment and thereby enable fewer soldiers to wield more capability; to increase the resilience of systems by reducing protection requirements for crew and vulnerable command links for un-crewed systems; and to improve the dependability of functions that humans struggle to perform under stress. Finally, there are some new functions that would not have been possible with a crewed or remotely-crewed system that an autonomous system might perform. We can therefore begin to extrapolate what autonomous systems are likely to be tasked with doing.

The first area where we can envisage the increasing use of autonomous systems is logistical support between the brigade support area and battlegroups, and in last mile resupply of sub-units.[14] There are multiple reasons for this. The increasing range and accuracy of tactical munitions,[15] combined with the limited protection on resupply vehicles makes these functions dangerous. They also pose a risk of dispersed medical commitments in depth.[16]

---

[12] The MK60 Captor Mine for instance can distinguish between surface vessels and friendly and hostile submarines, launching a homing torpedo without human oversight: see http://www.vp4association.com/aircraft-information-2/32-2/mk-60-captor-mine/, accessed 12 May 2019.

[13] UK and US joint exercises have begun to include military robots, see Andrew Tunnicliffe, 'Robotic warfare: training exercise breaches the future of conflict', Army Technology (19 December 2018): <https://www.army-technology.com/features/military-robotics-warfare/>, accessed 12 May 2019. Russia has deployed UVGs to Syria, and is refining its platform based on the limited successes of its deployment. The platform's functions are also likely to become increasingly autonomous. See, 'Combat Tests in Syria Brought to Light Deficiencies of Russian Unmanned Mini-tank', Defence Blog (18 June 2018): <https://defence-blog.com/army/combat-tests-syria-brought-light-deficiencies-russian-unmanned-mini-tank.html>, accessed 12 May 2019. For China see Elsa Kania, 'Battlefield Singularity: Artificial Intelligence, Military Revolution, and China's Future Military Power', Center for a New American Security (28 November 2017): <https://www.cnas.org/publications/reports/battlefield-singularity-artificial-intelligence-military-revolution-and-chinas-future-military-power>, accessed 12 May 2019.

[14] Experimentation with these capabilities has been ongoing for several years, see https://www.gov.uk/government/publications/accelerator-competition-autonomous-last-mile-supply/accelerator-competition-autonomous-last-mile-resupply, accessed 5 April 2021; this has led to multinational collaboration, see https://www.army.mil/article/227647/us_uk_coordinate_autonomous_last_mile_resupply, accessed 5 April 2021; and to testing on operations.

[15] As demonstrated recently in Nagorno-Karabakh, see Jack Watling and Sidharth Kaushal, 'The Democratisation of Precision Strike in the Nagorno-Karabakh Conflict', RUSI Commentary, 22 October 2020.

[16] Medical support also has an EMS signature that is hard to suppress, author interviews with British Army medical teams, Tidworth, January 2021.

The reliability of autonomous systems in this area is likely to be able to leverage technological development in the civilian sector, enabling faster development.[17] In the close fight, the use of autonomously navigating mules should allow infantry to carry more and heavier equipment whilst simultaneously reducing the load on personnel.[18] The importance of autonomy as opposed to a remote-controlled system here lies in freeing up capacity among combat troops to fight.[19] Until autonomous navigation enables the vehicle to execute commands within the direct fire zone, such systems will likely stay two bounds back,[20] because replacing runners with someone staring into a screen is not an efficiency and comes with an increased training burden.

Once autonomous navigation is able to function reliably in the direct fire zone, it is likely that autonomous systems will begin to carry weapons systems. This will enable dismounted infantry to manoeuvre with heavier weapons and may liberate infantry from some of the weight of ammunition. It is unlikely that these systems will replace section level support weapons. The movements of an infantry sections are highly complex, context dependent, and rely upon teamwork. It is unlikely autonomous systems will be able to do this in the foreseeable future. However, base-of-fire teams, which must be set up and deliver sustained effects to suppress the enemy, require less complex movement. An emplaced autonomous system in this role could be tasked with suppressing an area, or engaging targets in a defined kill box. This would likely remain under close human supervision, not least because of the need to coordinate these fires with the manoeuvre of assaulting sections. The automation would be in flagging targets and, once ordered to engage, alignment of weapons.[21] We may also envisage this base-of-fire team drawing upon complex sensors that an autonomous platform—having a motor and therefore power—could employ, which infantry could not. This might include radar and electro-optical sensors or tethered UAS. Preliminary testing of tethered autonomous reconnaissance and base-of-fire teams, attached to a platoon, show that they can enable a successful assault of an enemy force

[17] Numerous private sector firm are investing heavily in autonomous logistics vehicles, see Sean O'Kane, 'Daimler Is Beating Tesla to Making Semi-Autonomous Big Rigs', The Verge, 11 January 2019.

[18] David Hambling, 'The Overloaded Soldier: Why US Infantry Now Carry More Weight Than Ever', Popular Mechanics, 26 December 2018.

[19] Author observation of experimentation with dismounted infantry in November 2019. Light Infantry found the cognitive burden of managing remote systems deleterious to their tempo of manoeuvre.

[20] Tactical teaming is being trialled in force protection roles, see Wyatt Olson 'Air Force Robot Dogs Patrol Where Airmen Would Rather Not Tread', Stars and Stripes, 22 November 2020; and in more permissive environments, see 'Milrem Robotics' THeMIS UGV Finishes Mission Deployment in Mali', Army Technology, 6 May 2020.

[21] Targeting methodology demonstrated in briefing to author, US Army Futures Command, February 2021.

at equal strength, eliminating the need to secure a 3:1 force ratio requirement to conduct offensive manoeuvre.[22] This combination of firepower and situational awareness may enable a very small number of personnel to deny a large area of ground, making such sections perfect for screening flanks with area effect and anti-tank weapons, or providing overwatch to guard against UAS. By automating the firing posts, this would free up more personnel to be dedicated to the assault sections, increasing firepower and available combat mass without expanding the size of the platoon. The denial of ground by such means would also reduce the time taken to emplace and remove mines or other obstacles.

There are several limiting factors in the employment of autonomous ground vehicles that are not likely to be resolvable without as yet unforeseen technological advances. The first is power. Most autonomous platforms that are small enough to support light infantry rely on power packs with limited endurance.[23] This imposes a disjointed tempo of movement, and suggests capabilities are only likely to be available for a limited period. Frontline units cannot 'go-static' for prolonged periods, or hand over their organic lethality to a system that is intermittently available. The second issue is maintenance and repair. Mechanical vehicles can often be maintained by crews unless they have suffered serious damage or wear.[24] With digital systems, however, maintaining them can often requires expertise that is not widely held in military formations,[25] and it would be uneconomical to train combat arms to carry out this work. Fixing software issues is likely to rely on contractor support and be concentrated in specialist teams within a force's combat service support functions. The combination of these and other factors mean that autonomous systems designed to function in a close fight are likely to be assets held at higher echelons and then assigned to support lines of effort, rather than being held organically by these units. This also reflects the fact that whilst such systems may be getting cheaper, the sensors alone that enable them to function

[22] Author interview, Two Senior British Officers responsible for UGV experimentation, Rollestone Camp, November 2018; Author observation of light infantry platoon attack, Salisbury Plain, November 2018; Author interview, a UGV engineer overseeing development of the capability, Farmborough, July, 2019.

[23] Since acoustic signature is an issue during the approach to contact and sensor systems require electrical power, hybrid or electric motors have predominated. Although the cost of batteries is projected to decrease significantly, see file:///Users/user/Downloads/kjna29440enn.pdf, accessed 5 April 2021; and the bulk storage availability is expected to similarly increase exponentially, see https://www.irena.org/-/media/Files/IRENA/Agency/Publication/2019/Sep/IRENA_Utility-scale-batteries_2019.pdf, accessed 5 April 2021, actual energy stored within a given cell is projected to increase incrementally.

[24] Eric Peltz et al., *Diagnosing the Army's Equipment Readiness: The Equipment Downtime Analyzer* (Santa Monica, CA: RAND, 2002), xvi.

[25] Nina Kollars and Emma Moore, 'Every Marine a Blue-Haired Quasi-Rifleperson?', War on the Rocks, 21 August 2019.

are relatively expensive pieces of equipment, especially once hardened to survive battlefield conditions.[26] These platforms therefore are not likely to be ubiquitous on the future battlefield. They will be under high demand with a limited supply.

At higher echelons we may expect autonomous systems to play multiple supporting functions. Standoff ISR platforms are likely to fly with increasing levels of autonomy, reducing the footprint of the base stations that manage their orbits, and freeing human operators to focus on the returns from their sensors.[27] Deception platforms—whether dummy vehicles or electronic deception systems, are similarly likely to become increasingly autonomous, travelling along logical but irregular routes, and emitting signatures that generate realistic patterns for adversary standoff sensors. A further element of deception will be decoys, and here we may see a use for swarming technology, utilizing small teams of UAVs to collectively confuse and disrupt adversary precision strikes as a form of passive point defence.[28] Another plausible use for autonomous systems will be as communications relays. In an increasingly contested EMS environment, where overpowering jammers will require significant power output, which all but guarantees detection, units may increasingly rely upon line-of-sight relays that the enemy will struggle to detect and align electronic warfare (EW) assets to disrupt.[29] Autonomous aerial vehicles, able to loiter in orbits between units, and automatically align their antenna, would enable line of sight communications to be extended beyond the horizon. Again, here we may assume that a section attached to a signals formation would maintain these systems, but that the autonomous system would manage navigation, and maintain alignment without direct human control.

There are also a range of functions that are likely to remain under remote control rather than be assigned to autonomous systems. There is considerable interest in removing crews from engineering support equipment such as

---

[26] DARPA testing in denied EMS, for example, shows that survivable systems will not be cheap, see Brandon Knapp, 'These drone swarms survived without GPS', C4ISRNet, 4 November 2018.

[27] This is an extrapolation of the current trajectory from Predator UAVs initially being actively flown by an operator with a significant lag and limited responsiveness in controls leading to crashes, to the modern Protector, which will fly where directed, managing its own flight surfaces; see Justin Bronk, 'Swarming Munitions and the Myth of Cheap Combat Air Mass', in *Necessary Heresies: Confronting the Myths Distorting Conetmpoary Thought on Defence* edited by Justin Bronk and Jack Watling (London: Taylor and Francis, 2021), pp. 49-60.

[28] See the Gremlin programme for example, although it is currently geared around offensive mission sets, https://www.darpa.mil/program/gremlins, accessed 15 May 2022.

[29] Tyler Rogoway, 'The RQ-180 Drone Will Emerge from the Shadows as the Centerpiece of a Air Combat Revolution', The Drive, 1 April 2021.

assault bridges, diggers, and breaching vehicles.[30] This is to reduce their signature and remove personnel from highly exposed platforms. These are likely to be sub-optimal for autonomous systems, however. To begin with breaching, most of the effort in the development of autonomous navigation systems is in obstacle avoidance. A breaching vehicle, by contrast must intentionally collide with obstacles. How it approaches doing this must be informed by data concerning the density and construction of the obstacle, and how it can best be broken up using a range of tools. For an autonomous vehicle to be able to judge the density of terrain, as well as its shape, would require highly sophisticated and vulnerable sensors. Nor is there much ability to leverage development in the civilian sector for such activities. For bridging, there is similarly a need to judge both the strength of the banks of a gap, and to synchronize the location of the bridge with the timings of ground manoeuvre elements that must cross it. Given these constraints it is reasonable to assume that these systems are likely to become increasingly remote controlled—often by command line—but not autonomous in their operation.

The concepts outlined above are ones currently being developed or tested. Nevertheless, they are unlikely to be fielded until the 2030s. At present, for example, autonomous navigation systems struggle to function above 12 kmph when off road because of the amount of data processing required to accurately interpret their surroundings.[31] This is in the absence of dense fog, sleet, smoke, shrapnel, electronic attack, or any of the other myriad elements that would pose further challenges on an actual battlefield. The above is therefore realistic but ambitious. And yet it suggests that autonomous platforms will differ substantially in their employment to the vision that is usually presented. Autonomous platforms will not be ubiquitous. Nor will they be organic to every element, working in seamless human–machine teams. Instead, they will be available in limited numbers, maintained by specialists, and employed in support of lines of effort. They will not be adaptive and responsive but optimized to perform narrow and specific functions with greater reliability and effectiveness than human operators. They will likely accelerate the tactical tempo of operations where they are emplaced, but the need to emplace them will impose a constraint on operational tempo. Finally, vehicles that need less protection because they lack a human crew, may be smaller, lighter, and cheaper to maintain than the equivalent crewed platforms. That does not, however, make them cheap. Commanders will need

[30] Sydney Freedberg, 'US, UK Test Robot Breachers, Drones in Germany', Breaking Defense, 6 April 2018.
[31] Author interview, several officers and engineers overseeing autonomous vehicle experimentation, Salisbury Plain, November 2020.

to exercise judgement as to when and where these assets are committed. In the right circumstances they may drastically increase the lethality of a force. In the wrong circumstances—without the wide mesh of capabilities to allow them to effectively apply their narrow function for which they are optimized—they will likely be out manoeuvred and destroyed.

Layered Precision Fires

The idea of swarms of hunter-killer UAVs, autonomously scouring the battlefield for targets has become a persistent trope in portrayals of future war.[32] The components of such a system are viable but there are significant problems with applying them in combination.[33] Sensors are increasingly able to identify targets within a defined area with more munitions becoming active seekers. This is certainly achievable against vehicles.[34] Targeting infantry is much less assured. Targeting human signatures is possible, but in a dense urban environment—for example—the majority of human signatures would not be targets, whilst there are fewer unique identifiers to avoid false positives. We may envisage precision munitions being launched by tactical units to engage snipers or other defined groups.[35] These will need to be carried and will therefore be available in small numbers with a limited range. The moment a precision strike is attempted at scale then there must be a launch platform. The vulnerability of that platform quickly forces it back from the Forward Line of Own Troops (FLOT) and with the increase in range comes a corresponding increase in energy requirements for the munition, and therefore size and weight. This rapidly increases the cost, driving down the number of munitions that can be employed in this manner. A further factor shaping the use of such standoff capabilities is the latency between launch and effect and the time this affords to enact countermeasures. A large swarm—necessitating a launch platform—being cued from some distance, will have a significant radar and electronic signature, and, for small systems, will be highly vulnerable to electronic attack. As EW systems become organic to more formations such defences will require loitering munitions to be partially hardened, further driving the cost of each munition up and the size of a viable swarm down.

[32] Space Digital, 'Slaughterbots', 12 November 2017, https://www.youtube.com/watch?v=9CO6 M2HsoIA, accessed 5 April 2018.

[33] Jack Watling and Nicholas Waters, 'Achieving Lethal Effects with Small Unmanned Aerial Vehicles', The RUSI Journal 164, 1 (2019), 40–51.

[34] Author Briefing, MBDA, London, October 2020.

[35] Pursuing the Switchblade line of development, see https://www.avinc.com/tms/switchblade, accessed 8 January 2020.

We may therefore infer some conclusions as to the trajectory of land precision fires on the future battlefield.

First, it is clear that tactical precision fires will become organic to manoeuvre elements in limited numbers. This should increase the tempo with which they can dislodge a static enemy defence from complex terrain. It also means that combat units will need situational awareness above them during close combat for their own protection.[36] These systems will likely be automated sufficiently to climb from launch and present targets, then strike as directed, but are unlikely to actively hunt. Simplicity will be essential in preventing users being fixed crewing UAVs and will also drive down the cost. Such capabilities will be limited in both range and endurance.[37]

Secondly, longer range precision fires are destined to become significantly more widely available and to proliferate to sub-peer adversaries, in the form of loitering munitions.[38] Large salvos are less likely because of the vulnerability of these systems to dedicated countermeasures. Being relatively slow flying, they can be detected and engaged. With a limited area of regard, they could also miss targets unless successfully cued on. Nevertheless, salvos of twelve to eighteen loitering munitions seem eminently feasible out to ranges of 500 km.[39] If a proportion of these utilize EW capabilities to safeguard the salvo, and the munitions are cued onto the correct area, then they can deliver precise effects against high value targets at reach. The impact of such capabilities is significant because it gives brigades—at a small logistical footprint—the ability to deliver high impact cross-boundary fires. If coordinated by higher echelons, this means that if a concentration of high value targets are discovered in the enemy's second echelon, a large volume of munitions could rapidly converge from multiple directions to deliver operationally significant effects.[40] The persistent threat of such capabilities against forces far out of contact must reshape sustainment, command and control, and force protection engineering. This is likely to increase the threat to forces within

---

[36] The emphasis on CUAS is misguided as the sensors involved have much wider application. Author briefing, US Army Futures Command, February 2021.

[37] Israel, the USA, Australia, and the UK are all experimenting with such systems, including in operations, see Seth Frantzman, 'Israel acquires FireFly loitering munition for close combat', C4ISRNet, 5 May 2020.

[38] Loitering munitions are already manufactured by a number of states and have increasingly proliferated to proxies and non-state actors, see https://dronecenter.bard.edu/files/2017/02/CSD-Loitering-Munitions.pdf, accessed 15 May 2022.

[39] Tactical platforms are already carrying multi-munition canisters so that a battery could realistically generate 18 munitions, see https://www.defenseworld.net/news/28009/Israeli__Estonian_Firms_Develop_Unmanned_Vehicles_Mounted_Loitering_Munition_Launcher, accessed 5 April 2021.

[40] A phenomenon that is already being demonstrated using complex sensors and crewed systems like aviation but can be more widely replicated with un-crewed munitions working in conjunction with crewed platforms, see John Mead, 'Winning the Firefight on the "Road to Warfighter"', British Army Review 175 (Summer 2019), 70.

the zone where penetrating ISTAR is sufficiently dense to confirm returns from standoff ISR assets. In practice this will likely impose persistent attrition between the brigade and divisional support area, pushing many functions up an echelon and therefore increasing the need for organic capabilities at lower echelons. In this context the primary fight at the divisional level is likely to become sustainment.[41]

The proliferation of sensors makes the timing of strikes critical. With the greater range and fidelity of sensors comes greater potential for the interception of loitering munitions. Layered interceptors are becoming more reliable so that whereas sustainment assets remain difficult to protect as they move, fixed installations will likely be grouped in defended nodes. Striking those nodes will require munitions that are able to evade countermeasures. Such munitions—from missiles following quasi-ballistic trajectories to hypersonic glide vehicles—will be able to deliver devastating effects but, owing to the cost of such capabilities, they will be available in very limited numbers. Within the divisional and corps deep, therefore, the development of a long-range precision strike is likely to see a higher echelon struggle between sensors and deceivers that will play a critical role in shaping the endurance of units in the close fight. Commanders on the future battlefield will be able to strike what they want throughout operational depth, but they will not be able to do so repeatedly. Determining when and where to apply such effects will therefore be a critical judgement for higher level commanders. This may also reverse the traditional tendency to hold back more powerful capabilities in reserve. Because effects in the deep will likely enable victory in the close these higher echelon capabilities are liable to be applied early and to have a disproportionate impact on the success or failure of forces committed in the close.

High Fidelity Layered Sensors

The fidelity of modern sensors will have a disruptive effect on ground manoeuvre. Ground Moving Target Indication and Synthetic Aperture Radar have been around for some time.[42] However, the refinement of Active Electronically Scanned Arrays (AESA) has allowed high fidelity radar to be mounted on small mobile platforms. These systems are expensive. But they

[41] Jack Watling, 'Sustainment Is the Division's Hardest Responsibility', RUSI Defence Systems, 13 January 2021.
[42] John Richards, 'GMTI Radar Minimum Detectable Velocity', Sandia Report 1767 (Albuquerque: Sandia National Laboratories, 2011), April 2011, https://www.osti.gov/servlets/purl/1011708.

will nevertheless become available at all echelons.[43] Aviation mounting AESA radar can now orbit the divisional support area and monitor activity through the enemy's corps deep.[44] Those supporting brigades will need to be at a lower altitude to avoid being shot down but will still give a view into the enemy's divisional support area. Battlegroups meanwhile—with AESA radar mounted on recce vehicles—will reliably track dismounted infantry in any terrain out to 6 km.[45] Dedicated penetrating recce vehicles will struggle to apply these techniques because the emissions from such radar would give away their position. But as passive collection capabilities become more capable and widely available this will enable concealed listening posts to capture a vast quantity of data on movements around them in real time.

Beyond radar, there are a panoply of other sensors that are becoming increasingly capable and miniaturized. Infrared and thermal optics are already ubiquitous at all echelons.[46] Moreover, UAVs allow these sensors to be rapidly pushed forward. Passive EW collection is a major area of investment in the USA, Russia, and China.[47] Acoustic sensors can identify systems at considerable ranges.[48] Space-based observation is becoming widely accessible—even to non-state actors—because of commercially run constellations. Finally, whereas historical battlefields have been largely devoid of third-party sensors other than journalists, the future landscape is permeated by thousands of sophisticated cameras and active collectors as civilians video and share events unfolding around them.[49] Civilian infrastructure is now bristling with detection systems that can be hacked and exploited by militaries to provide even more data to find and monitor the enemy.

The distribution of highly capable sensors throughout the force has led many to a vision of future warfare in which commanders will stand at the heart of an information system providing them with total and immediate

---

[43] Justin Bronk, 'Technological Trends', in *The Future Conflict Operating Environment Out to 2035*, edited by Peter Roberts (London: RUSI, 2019), 61–68.

[44] Author observation, RNAS Yeovilton, August 2020.

[45] Author observation, Salisbury Plain, December 2018.

[46] Even among non-state actors such as the Taliben and Houthis, see https://reliefweb.int/sites/reliefweb.int/files/resources/%5BEN%5DLetter%20dated%2027%20January%202,020%20from%20the%20Panel%20of%20Experts%20on%20Yemen%20addressed%20to%20the%20President%20of%20the%20Security%20Council%20-%20Final%20report%20of%20the%20Panel%20of%20Experts%20on%20Yemen%20%28S-2020-70%29.pdf, accessed 5 April 2021.

[47] Bryan Clark, Whitney M McNamara, and Timothy A Walton, *Winning the Invisible War: Gaining an Enduring U.S. Advantage in the Electromagnetic Spectrum* (Washington, DC: Center for Strategic and Budgetary Assessments, 2019).

[48] Capability demonstration, Oslo, March 2021.

[49] Anna Reading, 'Mobile Witnessing: Ethics and the Camera Phone in the "War on Terror"', *Globalizations* 6, 1 (2009), 61–76; Matthew Ford and Andrew Hoskins, *Radical War: Data, Attention and Control in the 21st Century* (London: Hurst, 2022).

situational awareness throughout the depth of operations. This is a mistake. Whilst the fusion of all of the available information would increasingly enable such a level of situational awareness, the ability to concentrate and interpret all of the relevant data in a single place within a relevant period is getting harder to achieve. Despite steady progress in increasing the bandwidth of data networks, the volume of data to be transmitted is increasing exponentially. Given that the problem is transferring data to be analysed by artificial intelligence, which is often proposed as a means of preventing headquarters from drowning in data,[50] this is unlikely to make total situational awareness feasible. Moreover—because of the threat of long-range precision fires—fixed infrastructure is far less survivable in the rear. In short, the aim to maintain total situational awareness for the land domain is likely illusory. This is exacerbated by the fact that to emit is to be detectable, so that many forward sensors will need to be highly selective in what and when they transmit their data. Rather than a transparent battlefield therefore, with the future command post a veritable panopticon, we should instead understand the trajectory of sensors as enabling a commander to find the answer to almost any question they choose to ask. Remaining concealed whilst moving, transmitting, or firing, is becoming harder and harder. But answering a commander's question will require the apportionment and assignment of sensors, and the synchronization of their data to be fused and analysed within a defined timeframe. There will therefore be a limit to the number of questions that a commander can ask. Furthermore, although the number of sensors is increasing, taskable penetrating ISTAR will not be inexhaustible. If ISTAR systems are overly attrited, the ability to interrogate the battlespace will diminish, and risks becoming uncompetitive with an adversary, which creates a rapid asymmetry in capability. The recce battle therefore has partially shifted from skirmishing by light screening forces, to a major line of effort coordinated by higher levels, and largely executed by troops controlled at divisional and corps echelons.[51]

The proliferation of sensors changes the requirements and methods for deception. Historically, armies have been confronted with a dense fog in battle and have had to discern what occurs behind the forward line of enemy troops (FLET) by assessing limited and fragmentary pieces of information. In this context, deception has depended upon minimizing the signature of the majority of the force and presenting a coherent picture through intentionally

---

[50] Keith Dear, 'Artificial Intelligence and Decision-Making', *RUSI Journal* 164, 5–6 (2019), 18–25.
[51] Jack Watling and Sean MacFarland, 'The Future of the NATO Corps', *RUSI Occasional Papers* (London: RUSI, 2021).

revealed elements that lead the adversary to the wrong conclusion.[52] Deception has enabled manoeuvre, by disorientating enemy forces. As penetrating ISTAR becomes more capable, however, the fog is increasingly penetrable, but its penetration are not even. Standoff ISTAR can penetrate the fog of war across a wide area, providing a large volume of returns representing potential enemy activity. These returns will result from movement or emissions and will be exceedingly difficult to conceal. However, the signatures captured by standoff ISTAR can also be mimicked. Adversaries can therefore bombard standoff ISTAR with false positives. Poor mimicry can be filtered out by analysis, but effective mimicry requires the assignment of penetrating stand-in ISTAR assets to confirm queries. These will take longer to put in place and be limited in number, with a deep but narrow field of regard. Because a force will be unable to shield itself from observation by standoff ISTAR, it is therefore necessary to deceive the adversary by either forcing them to waste effort interrogating false positives with their stand-in ISTAR, to ambush and attrit stand-in ISTAR assets, or else to present a narrative through the signatures captured by stand-off sensors, combined with the returns from stand-in sensors to cause the adversary to be satisfied with the answers to their questions but also to misconceive the meaning of the sensor data they have fused. In this sense, deception must become a more systematic component of operational manoeuvre to enable force protection. But the means for deception is also likely to shift from an activity where the foremost concern of units is signature reduction, to one where the aim is to make the mixture of true and false signatures present a misleading narrative.[53] This requires a much more conscious appreciation of what emissions look like and the story they tell across a force.[54] This coordination of emissions—rather than blanket suppression— likely requires a level of awareness across a force that will be greatly aided by artificial intelligence.

Artificial Intelligence

One of the problems with considering the trajectory of AI is that there is a strong perception of what a highly complex AI might look like—a cognitive machine—but the constituent processes and functions that contribute toward

---

[52] David Glantz, *Soviet Military Deception in the Second World War* (Abingdon: Frank Cass, 1989); J. C. Masterman, *The Double-Cross System, 1935–1945* (London: Pimlico, 1995).

[53] Alec Bane, Briefing in Warminster, January 2019.

[54] Requiring detailed mapping of the electromagnetic environment, see Dave Hewitt, 'Episode 29: Electronic Warfare and Cumulative Risk', *Western Way of War*, 17 December 2020, https://rusi.org/podcasts/western-way-of-war/episode-29-electronic-warfare-and-cumulative-risk.

such a machine are rarely considered AI in isolation.[55] Computer vision, object recognition, and other critical functions to having a machine that can contextually interact with its environment, have been around for some time, but are rarely thought of as AI. Yet increasingly sophisticated AI is likely to enter military service in stages, rather than bursting forth as a singularity. This is because some problems require more contextual understanding than others, and some tasks therefore lend themselves better to AI systems. AI systems are largely optimizers: they seek the most efficient means of achieving a defined end.[56] The clarity with which that end can be defined is critical to assuring the reliability of the system. AI systems are most effective within closed data sets, and whilst they can refine their understanding of a data set they struggle to extrapolate from it. Fundamentally AI are very effective at determining that $A = 1$ but struggle with problems that require to build upon $A \neq 1$ because to an artificial system all things that do not equal 1 are equally unalike, generating an infinite number of returns.

Within the constraints outlined above, it becomes possible to extrapolate as to which tasks AI systems are likely to take over. The need for assurance will likely see AI initially employed within a closed data system comprising data sets about friendly forces, or assured returns from a platform's sensors. Monitoring materiel consumption across a force, projecting supply needs, and optimizing route planning for rear echelon logistics is likely to be a task for which AI will soon be usable.[57] Support to military police in route management to maximize flow and force protection is a similar planning problem for which most of the data can be accessed within a force, whilst in the rear transfer of data can be much more readily assured. Another area where AI may more quickly become a critical tool is in support of lower echelon planning. Given the need to move between points, an AI system has the terrain data and speed of friendly forces that a staff would use to develop a scheme of manoeuvre. An AI system could interpret that data and plot and compare alternative routes, fields of fire, and optimal positions for radar and other assets far faster than a human team.[58] Its conclusions might be altered or disregarded based on the commander's wish to pursue an unorthodox rather

---

[55] Pamela McCorduck, *Machines Who Think* (London: Taylor & Francis, 2004), 204.

[56] Alan Brown, 'Session Twelve: Innovation and Adaptability', RUSI Land Warfare Conference 2019, https://www.youtube.com/watch?v=ZYzJIcS36Ls&list=PLFAgO2TZWpwCPUeJSx2M2WoWrbnY7d_Dj&index=12, accessed 15 May 2022.

[57] This is already increasingly the case in civilian logistics operations, see Matt Simon, 'Inside the Amazon Warehouse Where Humans and Machines Become One', Wired, 5 June 2019.

[58] Computers have long since been used to identify optimal firing positions, such as the air defence lay-down around the beachhead during the Falklands conflict, see Max Hasting and Simon Jenkins, *The Battle for the Falklands* (Oxford: Pan Macmillan, 2010), 253; however, today such tools can be used dynamically to plan operations, as observed by the authors in RNAS Yeovilton, January 2020.

than optimal approach, or because of new information regarding enemy activity. This would likely be left to a human planner. But by turning the laborious tabulation and calculations carried out by geo officers, engineering staff officers, and others into a coherent basic outline this could both reduce the length of the planning cycle, and the required size of lower echelon headquarters. This would help to improve survivability and tempo. Furthermore, by accelerating course of action generation it would give a commander more time to consider their options and therefore reduce the risks of cognitive overload. AI could enable tactical units to rapidly recalculate based on changes during the course of battle.[59] At the most tactical level, we may expect AI to also provide functions such as emissions analysis; helping a commander understand how they appear to the enemy, because an AI can analyse emissions data in real time in a way that a human would struggle to do without having an EW team focused on analysing friendly forces and reporting their findings.

A second highly likely area for AI to become increasingly prevalent is in the planning and coordination of fires. As precision strikes come to involve combinations of lethal and non-lethal munitions to bypass defences, and as route planning becomes critical to munitions reaching their target, the synchronization of salvos, the optimal ratio of munitions to saturate a given area, or bypass a defined density of defensive systems, is like to be increasingly plotted by AI. It is also reasonable to suppose that the management of counter-fires, and in particular point defence against UAVs and missiles, will become managed by AI.[60] This is because the latency between detection and impact leaves little room for human control, and so operators are liable to fall into a supervisory function. In this, defence of higher echelon infrastructure is likely to increasingly resemble defensive systems already afloat and aloft.

Whilst the application of counter-fires and point defence may become increasingly subordinated to AI, target selection and the control of offensive fires is far less likely to be entrusted to AI. This is less for ethical reasons than because of AI's limitations in the targeting process. A learning algorithm supporting target identification may become highly effective at confirming objects of interest within a data set.[61] However, it is also highly vulnerable to deception, and perhaps more importantly is less likely to pick up on anomalies that fall outside of its programmed concerns. For instance, a learning

---

[59] Bryan Clark, Dan Patt, and Harrison Schramm, 'Mosaic Warfare: Exploiting Artificial Intelligence and Autonomous Systems to Implement Decision-Centric Operations', CSBA, 2020.

[60] As already demonstrated by the US Navy Cooperative Engagement Capability and Naval Integrated Fire Control—Counter Air (NIFC-CA), see https://www.army.mil/article/175,940/navy_conducts_first_live_fire_nifc_ca_test_wtih_f_35_at_white_sands_missile_range, accessed 3 March 2021.

[61] Author briefing, US Army Futures Command, February 2021.

algorithm may be able to confirm the identification of mobile Surface-to-Air Missile (SAM) systems from reconnaissance photographs. However, it may not even think to flag the large number of buses that appear to be transiting the area. Moreover, SAM launchers covered to look like said buses could not only avoid its notice, but unlike a human operator may not even trigger a query requiring further analysis. Similarly, an AI may be able to map out the EW signatures of an enemy force to identify a formation but would likely struggle to notice the boundaries of the formation's operations. We may therefore expect targeting to remain a human led process. Once a human operator noted an unusual concentration of buses the tell-tale signs could be identified and an AI system used to rapidly separate decoys from targets, but it would be unlikely that an AI would make the initial discovery. We may therefore expect AI to enable smaller staffs to conduct more rapid targeting over a larger area, but the enterprise architecture is likely to require imagination and contextual judgement, for which humans are essential. For the foreseeable future, therefore, AI is likely to provide decision-support but not to be decision-making; it is likely to help to plan COAs, but not to conceive of them.

Combining Emerging Capabilities

The technologies outlined in this chapter—along with several that have been excluded given the constraints of space—will change how land warfare is conducted over the following three decades. The most notable omissions from this chapter include offensive electronic warfare, space-based assets, and cyber warfare. There is also little consideration of a range of potential novel weapons such as directed energy weapons. This is not because these are not important, but because their impact either falls largely outside the boundaries of 'land warfare' requiring a more joint analysis, or because they are likely to change the tools of fighting without fundamentally altering the concepts for doing so. Considering the technologies outlined in this chapter in combination points to some conclusions about future land operations.

The first clear conclusion is that densely networked sensors, feeding into targeting cells able to rapidly assess new information, connected to responsive precision fire at long range and able to coordinate high volumes of imprecise fire at medium range, will be able to rapidly destroy targets within the close battle area. Concentrating forces to attack across a narrow front is unlikely to effectively reduce the level of fire that can be brought to bear against them, because organic fires will have sufficient range to converge across unit boundaries. In this context the commitment of large ground manoeuvre formations to the close fight, whilst the capabilities outlined in this chapter remain available, appears to be a recipe for losing the force in short order.

Although the increasingly frenetic and lethal character of the close fight has been well established in military analysis, it must be noted that none of the capabilities outlined in this chapter diminish the complexity or level of capability required to conduct that close fight. It is simply that those forces will struggle to be brought to bear at a sufficient scale to deliver their intended effects. Even autonomous systems, which enable the rapid isolation of a sector and limit the force concentrations required to assault terrain, will need to be emplaced. Their emplacement will take time and will not likely be feasible if the full combination of stand-off and stand-in sensors along with accompanying fires are in place. We may also note that the extent to which a tactical action can be reinforced will also be limited because increasing sensor fidelity and the reach of their fires will limit the endurance of units in the close fight by the threat posed to their logistics. Nevertheless, once ground is taken, the same issue of bringing to bear enough materiel to dislodge troops whilst under persistent threat in depth will be imposed upon the adversary.

We may therefore conceive of future conflict as becoming increasingly disjointed whereby tactical tempo is accelerating, but operational tempo is slowing. Whilst under favourable conditions, a force may take considerable ground against comparably sized forces and seize positions of operational significance, this will require extensive shaping. The most significant shaping activities will likely be the dislocation of enemy stand-off and stand-in sensors to enable deception, the use of deception to deplete adversary stocks of precision munitions and attrit ISR functions, the setting of patterns to teach adversary AI the wrong lessons, to be broken once the adversary has lost the ability to observe changes, and thence the commitment of manoeuvre elements.

We may also expect to see a considerable realignment of the current distribution of tasks across echelons. Brigades are likely to need to be more manoeuvrable, capable of dispersing for protection, going static for concealment, and then concentrating at speed from multiple axes when conditions enable them to access the close fight. Brigades, supported by AI to slim down their staffs, will likely need to practise high levels of mission command and hold a range of capabilities organically. The Division is liable to become less important as a command function and instead be pivotal in supporting its brigades through the assurance of sustainment, force protection, ISR, and fires. The Division is likely to be put under the greatest threat from enemy precision fires having a great deal of infrastructure that is hard to conceal and protect, whilst simultaneously being in range of both enemy sensors and shooters. The Corps by contrast is likely to take over much of the responsibility currently held at Division for the deep fight, being far enough from the front to receive, fuse, and process a sufficient volume of sensor data to understand the battlefield and being far enough from contact to plan and execute deep effects that are sufficiently coordinated to penetrate layered defences. In this sense it is the Corps that is likely to become the echelon that must win the deep fight to enable success in the close.

There will, undoubtedly, be deviations from the course charted above. New discoveries may drastically alter the offence–defence balance at difference ranges. What is clear, however, is that whilst there are tipping points in capability development that cause a substantial shift in how war is prosecuted, many of the prophets of a technological nirvana will be disappointed by the iterative, messy, and piecemeal development of military capability over the next three decades. Furthermore, critics and campaigners, seeking to prevent the erection of certain defined end states will be frustrated by the differences between what emerges and what they had envisaged in their regulations. Finally, whilst autonomous systems, layered precision fires, pervasive sensors, and AI will alter where humans are most important on the battlefield, and what they do, it will not see the need for mass or personnel diminish in the foreseeable future.

9

Interoperability Challenges in an Era of Systemic Competition

Andrew Curtis

Introduction

Admiral Ernest King, United States (US) Chief of Naval Operations during the Second World War, is purported to have said: 'I don't know what the hell this "logistics" is that [General George] Marshal is always talking about, but I want some of it'. This depressing lack of knowledge and associated desire could equally describe the approach of some Western policymakers to interoperability. Notwithstanding this, interoperability has been ever-present in the lexicon of defence for over 50 years. Moreover, despite the lack of understanding in some quarters, it could be argued that interoperability is actually a sine qua non of developing and maintaining military capability. For example, since the decision in 1968 to withdraw British forces from East of Suez,[1] and the recognition that NATO should remain the first and overriding charge on the resources available for defence,[2] UK defence policy has always emphasized the importance of defence cooperation with its allies and, in particular, within the North Atlantic Alliance.[3] Nonetheless, it is not always clear why interoperability is so popular with policymakers, or what they expect to gain from pursuing it.

Arguably the most influential factors in the decision-making process for Western policymakers are the extant and anticipated future global strategic environment. Over the last decade, the global strategic environment has

---

[1] Phillip Darby, *British Defence Policy East of Suez 1947–1968* (London: Oxford University Press, 1973), 325.
[2] *Statement on the Defence Estimates 1975, Cm 5976* (London: Ministry of Defence (MoD), Her Majesty's Stationery Office, 1975), 7.
[3] See, for example, MoD, 'Statement on the Defence Estimates 1975, 29; *Statement in the Defence Estimates, Britain's Defence for the 90s, Cm 1559-I* (London: MoD, Her Majesty's Stationery Office, 1991), 29; and *Securing Britain in an Age of Uncertainty: The Strategic Defence and Security Review, Cm 7948* (London: HM Government, The Stationery Office, 2010), 59–63.

Andrew Curtis, *Interoperability Challenges in an Era of Systemic Competition*. In: *Advanced Land Warfare*.
Edited by Mikael Weissmann and Niklas Nilsson, Oxford University Press. © Andrew Curtis (2023).
DOI: 10.1093/oso/9780192857422.003.0009

changed considerably. In 2003, the UK government set out its analysis of the future security requirement in a Defence White Paper.[4] Specifically, it confirmed 'There are currently no major military threats to the UK or NATO'.[5] By 2017, however, the UK's National Security Advisor believed it necessary to undertake a National Security Capability Review (NSCR)[6] outwith the recently established quinquennial defence and security review cycle[7] 'to deal with the evolving threat picture'.[8] This was corroborated by the Parliamentary Joint Committee on the National Security Strategy, which, in a report published shortly after the NSCR was completed, painted a picture of an increasingly unstable and unpredictable global context.[9] The UK's latest defence and security review—the 2021 Integrated Review—built on these themes, suggesting that 'the nature and distribution of global power is changing as we move toward a more competitive and multipolar world'.[10] One of the four overarching trends it judged would be of particular importance to the UK in the next decade was systemic competition, which it defined as:

> The intensification of competition between states and with non-state actors, manifested in: a growing contest over international rules and norms; the formation of competing geopolitical and economic blocs of influence and values that cut across our security, economy and the institutions that underpin our way of life; the deliberate targeting of the vulnerabilities within democratic systems by authoritarian states and malign actors; and the testing of the boundary between war and peace, as states use a growing range of instruments to undermine and coerce others.[11]

This view is not limited to the UK. The 2018 US National Defense Strategy confirmed that strategic competition, not terrorism, is now the primary concern in US national security.[12] In addition, NATO Heads of State and

---

[4] *Delivering Security in a Changing World Defence White Paper, Cm 6041-I* (London: MoD, The Stationery Office, 2003).

[5] *Delivering Security in a Changing World*, 7.

[6] *National Security Capability Review* (London: HM Government, The Stationery Office, 2018).

[7] *A Strong Britain in an Age of Uncertainty: The National Security Strategy* (London: HM Government, The Stationery Office, 2010), 35. In the 2010 National Security Strategy report, the UK government committed to undertaking a Strategic Defence and Security Review every five years.

[8] *Oral Evidence: Work of the NSA*, Joint Committee on the National Security Strategy, Evidence Session No. 1, HC 625, 18 December 2017, 3.

[9] *National Security Capability Review: A Changing Security Environment*, Joint Committee on the National Security Strategy, First Report of Session 2017–2019, HL Paper 104/HC 756, March 2018, 8.

[10] *Global Britain in a Competitive Age: The Integrated Review of Security, Defence, Development and Foreign Policy, CP 403*, (London: HM Government, The Stationery Office, 2021), 24.

[11] *Ibid.*

[12] Secretary of Defense, *Summary of the 2018 National Defense Strategy of the United States of America: Sharpening the American Military's Competitive Edge* (Washington, DC: US Department of Defense, 2018), 1.

Government included the following text in the communique that followed their June 2021 summit in Brussels: 'We face multifaceted threats, *systemic competition* [italics added] from assertive and authoritarian powers, as well as growing security challenges to our countries and our citizens from all strategic directions.'[13] This worsening global strategic picture has, unsurprisingly, forced policymakers to review military force structures as well as the concepts for their use. NATO members have also reacted collectively. Following the 2014 Wales Summit, NATO launched its Readiness Action Plan, a package of assurance measures for NATO allies in central and eastern Europe that it sees as 'an essential driver ... to the changed and evolving security environment.'[14] Two years later, Alliance members further agreed to strengthen their deterrence and defence posture through the establishment of an enhanced forward presence in Poland and the Baltic States.[15] This military response to the threat of systemic competition is having, and will continue to have, significant implications for all aspects of interoperability.

Against this backdrop, this chapter explores the future challenges for interoperability in an era of systemic competition. It begins with an assessment of what interoperability is, its characteristics, and its benefits. This analysis is centred on NATO's approach to interoperability and how that has influenced the actions and activities of its member states. Thereafter, the chapter examines the issues surrounding the pursuit of interoperability in an emerging era of systemic competition. Recognizing the impact that the latest evolution of the American way of war—multi-domain operations (MDO)—will have on the development of Western military capability in the coming decade, it will consider what the future may hold for the various characteristics of interoperability. Finally, the chapter considers the UK's approach to interoperability, driven as it has been by the demands of the Cold War, expeditionary operations, and now the outcomes of its recent Integrated Review. Whilst this final section touches on all five operational domains,[16] the emphasis is firmly on the British Army and interoperability in the land environment.

---

[13] 'Brussels Summit Communiqué', NATO, 14 June 2021, https://www.nato.int/cps/en/natohq/news_185000.htm?selectedLocale=en.

[14] 'Readiness Action Plan', NATO, accessed 18 August 2021, https://www.nato.int/cps/en/natohq/topics_119353.htm.

[15] 'Boosting NATO's presence in the east and southeast', NATO, accessed 18 August 2021, https://www.nato.int/cps/en/natohq/topics_136388.htm.

[16] Maritime, land, air, space, and cyber.

Understanding Interoperability

The 2020 RAND study, *Chasing Multinational Interoperability*, concluded that achieving interoperability is an ongoing challenge.[17] It argued that it is a buzzword, often touted as a solution to an unexplained problem, and that policymakers do not have a precise enough understanding of why more and better interoperability is needed. It also suggested that the drive to be interoperable is predicated on military forces having a poor track record in interoperability.[18] Probably the clearest reason for this situation is that there are so many different definitions of interoperability. For example, the current NATO Glossary of Terms and Definitions[19] includes three different definitions: one for force interoperability;[20] one for military interoperability;[21] and one simply for interoperability.[22] A 2007 survey[23] identified thirty-four definitions for interoperability and concluded that many of them could be traced back to the following US Department of Defense definition, which was believed to have been first used in 1967:

> The ability of systems, units, or forces to provide services to and accept services from other systems, units, and forces and to use the services so exchanged to enable them to operate effectively together.[24]

This definition probably most closely describes what interoperability is; however, there remains some ambiguity around the use of the word *services*. This was recognized by Christopher G. Pernin et al. Their solution was to align the services in the definition with the US Army's seven warfighting functions,[25] which, in turn, are synonymous with tasks that the US military

---

[17] Christopher G. Pernin, Angela O'Mahony, Gene Germanovich, and Matthew Lane, *Chasing Multinational Interoperability: Benefits, Objectives, and Strategies* (Santa Monica, CA: RAND Corporation, 2020), ix.

[18] *Ibid.*

[19] *AAP-06 Edition 2020: NATO Glossary of Terms and Definitions* (Brussels: NATO Standardization Office, 2020).

[20] *AAP-06 Edition 2020*, 55. The ability of forces of two or more nations to train, exercise, and operate effectively together in the execution of assigned missions and tasks.

[21] *AAP-06 Edition 2020*, 82. The ability of military forces to train, exercise, and operate effectively together in the execution of assigned missions and tasks.

[22] *AAP-06 Edition* 2020, 70. The ability to act together coherently, effectively, and efficiently to achieve Allied tactical, operational, and strategic objectives.

[23] Thomas C. Ford, John M. Colombi, Scott R. Graham, and David R. Jacques, *A Survey on Interoperability Measurement* (Wright-Patterson AFB, OH: Air Force Institute of Technology, 2007).

[24] *A Survey on Interoperability Measurement*, 4.

[25] Mission Command, Intelligence, Movement and Manoeuvre, Fires, Protection, Sustainment, and Engagement.

might provide to, or accept from, another force.[26] They went on to conclude that:

> Interoperability is done to enable the provision of services from one or many other nations, and if those services are so desired to meet overall national or military objectives, should directly connect to the multinational force's ability to effectively deter and defeat an adversary.[27]

The definition and amplification above provide an excellent foundation to develop an understanding of interoperability and the benefits that can be leveraged from its successful pursual.

Armed with an appreciation of interoperability, the next step is to identify its benefits. From the outset, it is important to acknowledge that interoperability is not an end in itself. Instead, it is a means to an end. It enables multi-national forces to achieve their objectives more efficiently and/or effectively than they would otherwise have done. To that end, if interoperability in a given circumstance does not add value, it should not be pursued. Interoperability is not a free good; therefore, some form of cost–benefit analysis should always be undertaken before committing to it at any level.

An examination of the literature reveals a surprising lack of analysis regarding the benefits of interoperability. Journal articles and monographs, especially those written by practitioners, tend to focus on ways to achieve or improve interoperability, rather than the benefit it will bring.[28] On its website, NATO identifies the components, mechanisms, and evolution of interoperability, but makes no mention of benefits.[29] One notable exception to the paucity of analysis around benefits is the *Chasing Multinational Interoperability* RAND study. Its authors conducted wide-ranging interviews across the US Army to identify discrete benefits that might accrue through interoperability.[30] Although obviously US Army centric, the resultant list, at Table 9.1, has relevance for most multi-national forces:

---

[26] Christopher G. Pernin, Jacob P. Hlavka, Matthew E. Boyer, Hohn Gordon IV, Michael Lerario, Jan Osburg, Michael Shurkin, and Daniel C. Gibson, *Targeted Interoperability: A New Imperative for Multinational Operations* (Santa Monica, CA: RAND Corporation, 2019), 16.

[27] *Ibid.*

[28] See, for example, Douglas M. Chalmers, *British Units under US Army Control: Interoperability Issues* (Fort Leavenworth, KS: US Army Command and General Staff College, 2001); Marc Bouchard, *Interoperability: A Must for the Canadian Forces* (Toronto: Canadian Forces College, 2010); and Paul W Fellinger, *Enhancing NATO Operability* (Carlisle, PA: US Army War College, 2013).

[29] 'Interoperability: Connecting NATO Forces, NATO, accessed 11 August 2021, https://www.nato.int/cps/en/natolive/topics_84112.htm.

[30] Pernin et. al., *Chasing Multinational Interoperability*, 9.

Table 9.1 Benefits of interoperability.

Benefits	Explanation
Enabling access to locations and populations	There is uncertainty in where US forces might operate for future operations. Interoperability can make it easier to work out operational details of access.
Leveraging partner capabilities	Some partners have valuable niche capabilities that can bolster US Army performance.
Filling capability gaps in force structure	The US Army has force structure and capability gaps in key scenarios that partners could help bridge.
Increasing legitimacy of operations	The US Army often seeks involvement from partners to show commitment and enhance the legitimacy of its operations.
Increasing operational safety	The US will inevitably work together with partners and thus needs to reduce downside effects of operating with disparate forces, such as fratricide and collateral damage.
Deterring adversaries	By increasing capabilities and demonstrating commitment, interoperability can deter adversaries.
Meeting treaty obligations	Interoperability increases multi-national capabilities to meet treaty obligations.
Reassuring partners	Working closely with partnerships partners to understand US Army capabilities and demonstrates US Army commitments.
Reducing costs of operations	Global commitments over long periods entail finding ways of reducing overall costs of operations. Interoperability can help efforts to maintain readiness whilst meeting current demands.
Shaping partner purchases	Interoperability increases purchases of shared materiel and training.
Sharing burdens for operations	Interoperability provides a mechanism for burden sharing.
Supporting partner-led missions	The USA is committed to supporting partners in maintaining stability and sovereignty.

Source: Pernin et al., *Chasing Multinational Interoperability*, p. 10.

An understanding of interoperability is incomplete without an exploration of its characteristics. Michael Codner argued that there are many different sorts of interoperability and, therefore, it should be considered a multidimensional concept.[31] He offered three ways to describe interoperability. In the first instance, it can be described by reference to the organizational level[32] at which it is being attempted. It can also be described with reference to the actors among whom interoperability is being undertaken. Finally, it can be described with reference to the services that are provided for which interoperability is required.[33] In a different vein, NATO describes interoperability by dimensions. First is the procedural dimension that covers concepts and doctrine, plus their associated tactics, techniques, and procedures (TTPs). Second, and probably best known, is the technical dimension, which includes weapons and communications systems and armaments. Finally, NATO recognizes the human dimension, comprising of behavioural, terminology, and training. NATO interoperability dimensions provide the ideal framework within which to examine interoperability in an emerging era of systemic competition.

Interoperability in an Era of Systemic Competition

Procedural Dimensions

The procedural dimensions of interoperability include concepts and doctrine, and their associated TTPs. By far the most significant reset of Western concepts and doctrine in the last decade is the USA's adoption of MDO. Whilst each of the service branches has its own ideas on MDO,[34] it is the US Army's approach that is most likely to influence interoperability within the NATO area of operations. Its concept focuses on China and Russia, although the ideas therein also apply to other threats.[35] In the emerging operational environment, the US Army recognizes that four interrelated trends are shaping competition and conflict: adversaries are contesting all domains, and US dominance is not assured; smaller armies fight on an expanded battlefield that is becoming increasingly lethal; nation-states have more difficulty in

[31] Michael Codner, *Hanging Together: Interoperability within the Alliance and with Coalition Partners in an era of Technological Innovation* (London: Royal United Services Institute, 1999), 13.
[32] The organizational levels of war accepted throughout NATO are: grand strategic; military strategic; operational; and tactical.
[33] Codner, *Hanging Together*, 13.
[34] Grant J. Smith, 'Multi-Domain Operations: Everyone's Doing It, Just Not Together', Over the Horizon: Multi-domain operations and strategy, 24 June 2019, https://othjournal.com/2019/06/24/multi-domain-operations-everyones-doing-it-just-not-together/.
[35] *The US Army in Multi-Domain Operations 2028* (Fort Leavenworth, KS: US Army Training and Doctrine Command (TRADOC), 2018), 5.

imposing their will within a politically, culturally, technologically, and strategically complex environment; and near-peer states more readily compete below armed conflict making deterrence more challenging.[36] In a state of systemic competition, it confirms that China and Russia are exploiting the conditions of this emerging operational environment to achieve their objectives without resorting to armed conflict by fracturing the USA's alliances, partnerships, and resolve.[37] It also predicts that, in armed conflict, China and Russia will seek to achieve physical standoff[38] by employing layers of anti-access and area denial systems designed rapidly to inflict unacceptable losses on US and partner military forces, in order to achieve campaign objectives faster than the USA can effectively respond.[39] MDO is the US Army's solution to these problems. It is underpinned by the following three interrelated tenets:[40]

- *Calibrated Force Posture* is the combination of position and ability to manoeuvre across strategic distances.
- *Multi-Domain Formations* possess the capacity, capability, and endurance necessary to operate across multiple domains in contested spaces against a near-peer adversary.
- *Convergence* is rapid and continuous integration of capabilities in all domains, and the information environment, that optimizes efforts to overmatch an enemy through cross-domain synergy and multiple forms of attack.

The US Army is committed to delivering a multi-domain force by 2035 (specifically, its Aimpoint 2035 is a multi-domain army that will be modernized and prepared to dominate adversaries in sustained large scale combat operations).[41] Here is the first, and most significant challenge to future interoperability—the rest of NATO members are not. The UK, for example, is underpinning its military modernization plans through the exploratory concept of *multi-domain integration* (MDI), which it defines as 'an ambitious vision for maintaining advantage in an era of persistent competition'.[42] As

---

[36] *The US Army in Multi-Domain Operations 2028*, 6–8.
[37] Ibid., 9–11.
[38] Standoff is the political, temporal, special, and functional separation that enables freedom of action in any, some, or all domains, and the information environment, to achieve strategic and/or operational objectives before an adversary can adequately respond.
[39] *The US Army in Multi-Domain Operations 2028*, 11–13.
[40] Ibid., 17–24.
[41] *Army Multi-Domain Transformation: Ready to Win in Competition and Conflict*, (Washington DC: US Army, US Department of the Army, 2021), 29.
[42] *Joint Concept Note 1/20: Multi-Domain Integration* (Shrivenham: MoD Development, Concepts and Doctrine Centre, 2020), v.

Chris Tuck points out, this is not the same as MDO; instead, MDI is a far broader concept that 'explicitly attempts to move the British military's thinking beyond jointery toward a new and more holistic approach to meeting threats and maximising British influence'.[43] By contrast, NATO is forging ahead with its own Warfighting Capstone Concept (the NWCC), which seeks to detail how member nations must 'develop their militaries to maintain advantage for the next twenty years'.[44] Aligned with this is an initiative to evolve the concept of MDO—NATO Joint All Domain Operations (JADO). The aim of NATO JADO is to identify and propose solutions to the problems associated with fully utilizing the collective capabilities of all assets assigned to a NATO-led effort.[45]

This raft of new thinking about how best to confront the threat of systemic competition is welcome; however, the dangers are obvious. Any divergence among NATO member states in concepts, doctrine, and TTPs will only make it harder to pursue interoperability and, thus, reduce the effectiveness of the Alliance as a fighting force. During the Cold War, NATO mandated the high-level warfighting concepts employed to counter the Soviet threat, even if many were originally conceived by the US military, for example the adoption of AirLand Battle.[46] After 1991, the imperative for a single, NATO-led approach disappeared and member states, in particular the USA, modernized their militaries at a pace and complexity to suit their own policies and budgets. That now needs to change. To rebuild a credible collective defence posture, NATO must once again take the conceptual lead. Whether that is through its own NWCC or the acceptance of the US Army's MDO concept is a big decision, but one that needs to be made quickly. Jack Watling and Daniel Roper suggest the biggest barrier to interoperability is the lack of a common language across the Alliance to describe the multi-domain environment.[47] That barrier will only be removed by the adoption of a single, NATO-wide approach to which all member states can then align their modernization plans. Furthermore, until that barrier is removed, it is hard to see how the technical and human dimensions exposed below can be addressed.

[43] Chris Tuck, 'What is Multi-Domain Integration?', Defence-In-Depth, 16 May 2021, https://defenceindepth.co/2021/05/14/what-is-multi-domain-integration/.

[44] 'NWCC: NATO Warfighting Capstone Concept', NATO Allied Command Transformation, accessed 24 August 2021, https://www.act.nato.int/nwcc.

[45] 'NATO JADO: A Comprehensive Approach to Joint All Domain Operations in a Combined Environment', Joint Air Power Competence Centre, accessed 24 August 2021, https://www.japcc.org/portfolio/nato-joint-all-domain-operations/.

[46] *FM 100–5 Operations* (Washington, DC: US Army, US Department of the Army, 1993).

[47] Jack Watling and Daniel Roper, *European Allies in US Multi-Domain Operations* (London: Royal United Services Institute, 2019), 31.

Technical Dimensions

The technical dimensions of interoperability are all about hardware. Whilst thirty member states will never all field the same weapons systems, NATO has always promoted a number of simple ways to improve collaborative working, the most obvious of which is its standardization programme. However, the post-Cold War expansion of the Alliance introduced considerable amounts of Soviet-era military equipment into the NATO ORBAT at a time when the focus on interoperability was waning. Retrofitting standardization was never a priority and, as a result, the challenges of NATO forces operating together are now probably greater than ever. For example, in a 2017 joint exercise in Poland, US Army soldiers discovered that their fuel nozzles did not fit the fuel tanks of Polish armoured vehicles.[48] Whilst this problem was swiftly overcome through the procurement of adaptors, it would never have arisen during the Cold War because of the rigid application of NATO STANAGS. Since 2014, European member states and Canada have increased annual defence expenditure by an average of 3.7 per cent.[49] As this military re-capitalization continues, it is vital that interoperability becomes a default consideration in all nations' procurement decision-making.

Whilst low-level equipment interface problems can be easily overcome, technological disparities at the weapons system level between NATO forces remain challenging. This has the greatest impact in the areas of communication and situational awareness. To prevail in future competition against a peer, or near-peer, adversary, Watling and Roper posit the criticality of the following two elements: timely and verified information describing the operational environment across all domains; and commanders who can understand the multi-domain battlespace and shape their operations to maximize their contribution to the fight across them.[50] In 2015 there were at least thirteen different systems for battle tracking within NATO, many with different technical standards.[51] Not knowing what the commander on your flank knows, and not being able to tell them what you know, is not a good starting

---

[48] Hans Binnendijk and Elisabeth Braw, 'For NATO, True Interoperability Is No Longer Optional', Defense One, December 18 2017, https://www.defenseone.com/ideas/2017/12/nato-true-interoperability-no-longer-optional/144650/.
[49] 'Defence Expenditure of NATO Countries (2014–2021)' (Brussels: NATO Public Diplomacy Division, 2021), 2.
[50] Watling and Roper, *European Allies in US Multi-Domain Operations*, 15.
[51] James Derleth, 'Enhancing Interoperability: The Foundation for Effective NATO Operations', NATO Review, 16 June 2015, https://www.nato.int/docu/review/articles/2015/06/16/enhancing-interoperability-the-foundation-for-effective-nato-operations/index.html.

point for maximizing each other's capabilities, exploiting opportunities, and mitigating vulnerabilities.[52]

Human Dimensions

The final dimension considers behavioural, terminological, and training aspects of interoperability. Here too, there are significant obstacles to overcome. For instance, whilst the multi-national battlegroups of NATO's enhanced forward presence generate regular training opportunities for the armies of over two-thirds of member states, the numbers involved are extremely small (the approximate total troop number for all four battle-groups is under 5,000, with some nations' contributions no more than single figures).[53] What is needed are regular and demanding division- or corps-level training events akin to the annual Cold War Reforger (Return of Forces to Germany) exercises that tested NATO's ability swiftly to deploy ground forces, mainly from the USA, into West Germany.[54] Even though the COVID-19 pandemic forced a reduction in scale and scope, Exercise Defender-Europe 20 was a good first step. It was the third-largest military exercise in Europe since the Cold War, and exercised the large-scale movement of forces across the Atlantic and into training areas in Germany and Poland.[55] Frequent large-scale exercises are vital to ensure that procedural and technical interoperability problems are both exposed and solved. For example, integrating allies into a future joint force is a key tenet of MDO. To that end, the US Army will undoubtedly be expected to solve the myriad sustainability issues arising from the coming together of allied ground forces in a multi-national corps.[56] Knowing in advance the associated sustainment challenges, and documenting options to resolve them through the post-exercise lessons process, are essential if commanders are to mitigate the worst logistic frictions of a mobilization phase that, in an era of systemic competition, is sure to be contested.

---

[52] Watling and Roper, *European Allies in US Multi-Domain Operations*, 15.

[53] NATO's Enhanced Forward Presence, NATO, March 2021, https://www.nato.int/nato_static_fl2014/assets/pdf/2021/3/pdf/2103-factsheet_efp_en.pdf.

[54] 'Reforger', GlobalSecurity.org, accessed 31 August 2021 https://www.globalsecurity.org/military/ops/reforger.htm.

[55] Gareth Thomas, Peter Williams, and Yanitsa Dyakova, 'Exercise Defender-Europe 20: enablement and resilience in action', NATO Review, 16 June 2020, https://www.nato.int/docu/review/articles/2020/06/16/exercise-defender-europe-20-enablement-and-resilience-in-action/index.html.

[56] Rodney Fogg, Simon Heritage, Thierry Balga, and Mark Stuart, 'Interoperability: Embrace it or Fail!', US Army, February 10 2020, https://www.army.mil/article/231653/interoperability_embrace_it_or_fail.

As member states develop new concepts to meet the changing character of warfare, behavioural and terminological dimensions will demand more attention. In particular, the US Army's journey toward Aimpoint 2035 will include the consolidation of some existing MDO ideas as well as the disinvestment of others. It may also accommodate the formulation of ideas not even being thought about today. It will move rapidly and experiment aggressively to ensure that its final ways of working can meet the threat of systemic competition. Most NATO nations will struggle to keep up and even those that do may take issue with some of the changes the US Army may want to make. For example, most member states recognize the mission command approach to command and control, yet the decentralization principles of MDO could drastically increase the need further to empower subordinate commanders to a level beyond which they are comfortable.[57] In a similar vein, the development of future land systems will create ethical issues around the employment of capabilities such as artificial intelligence and robotics. Not all NATO members are as comfortable as the USA regarding the development of autonomous weapons systems.[58] Finally, notwithstanding the issues regarding the technical ability to share situational data outlined above, MDO's demands for full integration will also create behavioural difficulties. To ensure success in warfighting operations, most allies would likely consider complete data transparency to be a price worth paying. In an era of systemic competition, however, the US Army will want that 24/7. As Watling and Roper recognize, many NATO members might be reluctant to sign up to the ubiquitous exportation of large amounts of their data to a third party.[59]

The UK's Approach to Interoperability

Interoperability during the Cold War

The UK's approach to interoperability during the Cold War was relatively straightforward. As Michael Codner pointed out, 'the threat to NATO [from the Soviet Union and the Warsaw Pact] was so immediate and the perceived balance of advantage was so unfavourable that there was a premium on any mechanisms for achieving greater military efficiency'.[60] To that end, the

---

[57] Mark Balboni, John Bonin, Robert Mundell, Doug Orsi, Craig Bondra, Antwan Dunmyer, Lafran Marks, and Daniel Miller, *Mission Command of Multi Domain Operations* (Carlisle, PA: US Army War College, 2019).

[58] Watling and Roper, *European Allies in US Multi-Domain Operations*, 16.

[59] *Ibid.*

[60] Codner, *Hanging Together*, 7.

UK was a strong proponent of NATO standardization efforts;[61] in particular, leading the development and implementation of multi-national concepts and doctrine, and adhering to standardization agreements (STANAGS).[62] During this time, the single services were predominantly focused on operations within their own environment. Thus, by the 1980s, the Royal Navy was concentrated on maritime operations in the Eastern Atlantic and English Channel; the Army was fixed on the forward defence of the Federal Republic of Germany through the British Army on the Rhine; and the RAF was invested in the air defence of the UK homeland and the provision of a Tactical Air Force for operations in the Central Region.[63] Within these environments, considerable amounts of interoperability were achieved. Good examples are the Royal Navy's contribution to STANAVFORLANT,[64] the Army's commitment to the Allied Command Europe (ACE) Mobile Force,[65] and the RAF's support to the NATO Integrated Air Defence System, through the UK Air Defence Ground Environment (UKADGE).[66] At this time, whilst policymakers also recognized the need to maintain the ability to operate beyond the NATO area,[67] this was not expected to attract a heavy interoperability burden. And so it proved, as the only out of area operation conducted by UK forces after the East of Suez withdrawal was Operation *Corporate*,[68] which was conducted without the overt assistance from any NATO partners. Following the fall of the Berlin Wall in 1989, NATO members, including the UK, were eager to capitalize on the perceived 'peace dividend' and quick to disinvest in defence.[69] As early as 1991, UK policymakers confirmed that 'the capability to mount a timely defence against such a massive threat

---

[61] 'Standardization', NATO, accessed 13 August 2021, https://www.nato.int/cps/en/natolive/topics_69269.htm.

[62] A STANAG is a NATO standardization document that specifies the agreement of member nations to implement a standard, in whole or in part, with or without reservation, in order to meet interoperability requirements.

[63] Clearly there was some overlap of the three services' missions and tasks, for example the RAF's provision of close air support and support helicopters to the First British Corps, and its maintenance of a maritime strike capability.

[64] NATO's Standing Naval Force Atlantic (STANAVFORLANT) was a multinational squadron of frigates and destroyers, established in 1968. Ships were permanently committed to the squadron by Canada, the Federal Republic of Germany (then Germany), Netherlands, the UK, the USA and, from 2000, Spain.

[65] The ACE Mobile Force was a brigade-sized quick reaction force, composed of force elements from up to fourteen NATO members.

[66] The UKADGE was the RAF's ground-controlled interception system for the British Isles that linked ground-based radar sites, airborne early warning aircraft, and RN warships.

[67] See, for example, *The United Kingdom Defence Programme: The Way Forward*, Cm 8288 (London: MoD, Her Majesty's Stationery Office, 1981), 11.

[68] The re-capture of the Falkland Islands following an Argentinian invasion in 1982.

[69] See, for example, David Greenwood, 'Expenditure and Management', in *British Defence Policy: Thatcher and beyond*, edited by Peter Byrd (Hemel Hempstead: Philip Allen, 1991), 63.

[from the Warsaw Pact] is no longer the main focus of our concern'[70] The imperative for interoperability among NATO members slowly withered on the vine.

The Impact of Expeditionary Operations

In the final decade of the twentieth century, UK defence policy shifted markedly toward a posture of expeditionary operations. This culminated in the 1998 Strategic Defence Review (SDR), in which the then Secretary of State for Defence George Robertson confirmed 'in the post-Cold War world, we must be prepared to go to the crisis, rather than have the crisis come to us'.[71] The SDR was underpinned by a series of initiatives across defence to coordinate the activities of the three Services more closely.[72] This new 'joint' approach forced the Royal Navy, British Army, and RAF to work together to a degree not seen during the Cold War, and generated a considerable increase in intra-service interoperability. For example, joint doctrine was published;[73] joint headquarters were established;[74] and joint logistics processes were introduced.[75] After 9/11 the UK doubled down on its expeditionary defence policy. Whilst at times this did include operating under a NATO command structure,[76] during the first decade of the twenty-first century, UK forces predominantly operated bilaterally with the US military.[77] Indeed, the 2010 National Security Strategy listed the USA ahead of both NATO and the European Union in 'its unique network of alliances and relationships'.[78] Understandably, the UK armed forces stepped up their interoperability efforts with the US military. For instance, the USA's adoption of Network Centric Warfare was closely followed by a similar UK initiative—Network Centric Capability.[79] However, in its efforts to keep up

---

[70] Britain's Defence for the 90s, 31.

[71] The Strategic Defence Review: Modern Forces for the Modern World (London: MoD, The Stationery Office, 1998), 2.

[72] Ibid.

[73] The first joint doctrine publication—British Defence Doctrine: Joint Warfare Publication 0-01—was published in 1996.

[74] For example, operational command of UK deployments overseas was vested centralized in a single organization—the Permanent Joint Headquarters—at Northwood in 1996.

[75] The single services logistics departments and MoD central logistics agencies were amalgamated to form the Defence Logistics Organisation in 2000.

[76] From August 2003 until December 2014, UK forces were part of the NATO led, UN-mandated International Security Assistance Force in Afghanistan.

[77] For example, in 2001 on Operation Enduring Freedom—the USA's global war on terrorism—in Afghanistan, and in 2003 as part of Operation Iraqi Freedom.

[78] A Strong Britain in an Age of Uncertainty, 11.

[79] The Strategic Defence Review: A New Chapter, Cm 5566 Vol 1 (London: MoD, The Stationery Office, 2002), 15.

with USA's latest technology-driven operating concepts, the UK paid less and less attention to its interoperability links with other NATO partners.

Rediscovering the Benefits of Interoperability

In its 2021 Integrated Review, the UK government recognized that the international order had become more fragmented and was now characterized by intensifying competition between states over interests, norms, and values.[80] It also suggested that an era of systemic competition was emerging, in which the distinctions between peace and war; home and away; state and non-state; and virtual and real were becoming increasingly blurred.[81] In response, the review reaffirmed the UK's desire to be able to influence the international agenda and recognized that this aspiration must continue to be underpinned by a global power projection capability. It also stated the UK's unequivocal commitment to European security,[82] whilst identifying Russia as the most acute threat to it.[83] Given this, the Defence Command Paper that accompanied the Integrated Review balanced the need to maintain modern expeditionary forces capable of operating world-wide, with a re-emerging requirement to sustain a credible deterrence posture within NATO's traditional area of operations. It also recognized that the character of warfare is rapidly evolving and explained the changes the UK's armed forces must make to keep up. These changes were first articulated in a new Integrated Operating Concept (IOpC),[84] which was published in 2020. It provides the conceptual north star for the UK's future force structure and was heavily influenced by the US Army's thinking around MDO.

By contrast, the UK's current capstone doctrine publication—Joint Doctrine Publication 0-01: UK Defence Doctrine[85]—has little to say about interoperability. Other than recognizing it is necessary to employ military capability across a coalition force, it simply confirms that interoperability may be expensive to achieve and sustain and may also require adherence to a common standard.[86] However, following the Integrated Review, UK Defence Doctrine is being revised to reflect the MoD's new approach to the

---

[80] *Global Britain in a Competitive Age*, 11–22.
[81] *Defence in a Competitive Age, CP 411* (London: MoD, The Stationery Office, 2021), 5.
[82] *Global Britain in a Competitive Age*, 11.
[83] Ibid., 19.
[84] *Introducing the Integrated Operating Concept* (London: MoD, The Stationery Office, 2020).
[85] *Joint Defence Publication 0–01: UK Defence Doctrine* (Shrivenham: MoD, Development, Concepts and Doctrine Centre, 2014).
[86] *Joint Defence Publication 0-01*, 27.

utility of armed force in an evolving era of systemic competition. This revision is expected to include a description of Defence Lines of Development (DLODs), a list of nine essential factors that shape the development and maintenance of military capability.[87] Significantly, the DLODs are supported by two cross-cutting themes: resilience and interoperability. Within the DLOD construct, interoperability is seen as providing the capability for the Integrated Force[88] to train, exercise, and operate effectively together when executing assigned missions and tasks. It also acknowledges the following three levels of interoperability:

- *Integrated.* Operates seamlessly and interchangeably as a single force.
- *Compatible.* Complement and work alongside each other as separate forces.
- *Deconflicted.* Coexist, but separated.

The value of interoperability is clearly a feature of the new IOpC. It confirms that the UK must respond to systemic competition by recognizing and continuing to resource its strengths.[89] Second on that list of strengths is allies and partners. However, whilst acknowledging the centrality of NATO, the IOpC also urges military planners to look to other alliances, and to give real meaning to multi-national cooperation. Similarly, it stresses that future military activity should be constructed with allies in mind.[90] The central idea of the IOpC is to drive the conditions and tempo of strategic activity, rather than respond to the actions of others.[91] This, it is claimed, can only be realized through being more integrated and, specifically, integrated within the military instrument, across government and with allies. A quarter of a century of joint operations means the UK armed forces are both experienced and competent at operating together across the traditional environments of maritime, land, and air. The recent additions of cyber and space as operational domains, however, bring fresh challenges for intra-service interoperability. Add to that the difficulties associated with closer pan-government cooperation, and the need for the UK to re-invigorate working relationships with all its allies, except possibly the USA and France, there is much to do before integrated by advantage moves from being a strapline to a reality. Nevertheless, the IOpC's commitment to allies and partners, as well as committing to

---

[87] The DLODs are: training; equipment; personnel; information; doctrine and concepts; organization; infrastructure; logistics; and security. They are known by the acronym TEPIDOILS.

[88] The UK's future force structure, detailed in the IR, is referred to as the Integrated Force.

[89] *Integrated Operating Concept*, 7.

[90] *Ibid.*

[91] *Integrated Operating Concept*, 8.

integration at every level,[92] leaves the reader in no doubt that interoperability is seen as a cornerstone of the UK's future operating concept.

One level down, the British Army uses the concept of fighting power to describe its operational effectiveness;[93] moreover, it recognizes fighting power varies depending on the level of interoperability that a force can achieve with other military formations and with other actors.[94] UK Land Power doctrine also acknowledges that 'achieving high levels of interoperability takes time and resources to develop and maintain, and must be honed through training and by lessons identified during operations.'[95] To that end, the British Army coordinates its approach to interoperability through a dedicated Army Command Standing Order and maintains an interoperability programme with a number of the UK's strategic partners.[96] Its highest priority remains cooperation with the US Army, with the goal of reaching an integrated level of interoperability for a UK division in a US corps by 2025.[97] Using key leader engagement, personnel exchange, and doctrine, training, and education as main lines of effort, its plans are already well advanced. As an example, after Exercise Warfighter 19-4, the US Army's premier Command Post Exercise at Fort Hood, Texas in 2019, Lieutenant General Paul Funk, Commanding General III (US) Corps stated that 'working with [Headquarters] 3 (UK) Division was just like having a US division under command'.[98] But it is not all about transatlantic partnerships. Under the auspices of the Joint Expeditionary Force, a multi-national force of 10 northern European nations,[99] the British Army is also committed to integrating a Danish battlegroup into a UK brigade by 2025.[100] It has also already achieved compatible interoperability with a French brigade as part of the Combined Joint Expeditionary Force.[101] In his speech to the 2021 Royal United Services Institute Land Warfare Conference, General Sir Mark Carleton Smith, the then Chief of the

---

[92] *Ibid.*
[93] Fighting power recognizes the fact that forces do not simply consist of such tangibles as people and equipment, they also have intangible conceptual and moral properties that can play a decisive role in shaping their effective employment.
[94] *Joint Doctrine Publication 0-20, UK Land Power* (Shrivenham: MoD, Development, Concepts and Doctrine Centre, 2017), 37.
[95] *Joint Doctrine Publication 0-20*, 54.
[96] United States, France, Germany, and Denmark.
[97] 'In Front', *British Army Newsletter* 3, Summer 2019, 22.
[98] 'In Front', 19.
[99] 'Joint Expeditionary Force Policy Direction—July 2021', Ministry of Defence, Gov.uk, 12 July 2021, https://www.gov.uk/government/publications/joint-expeditionary-force-policy-direction-july-2021.
[100] 'In Front', 22.
[101] 'UK and France able to deploy a 10,000 strong joint military force in response to shared threats', Ministry of Defence and Rt Hon Ben Wallace MP, Gov.uk, 2 November 2020, https://www.gov.uk/government/news/uk-and-france-able-to-deploy-a-10000-strong-joint-military-force-in-response-to-shared-threats.

General Staff, highlighted the need to place more emphasis on a wider coalition of partners and allies to combat the growing cycle of competition.[102] There is increasing evidence that the British Army is already taking that direction to heart.

Conclusion

Interoperability has been a staple in the defence policies of Western nations for the last 50 years, even though the reasons why are not always obvious. During the Cold War, NATO did much to facilitate interoperability across the force structures of its member states through the development of Alliance-wide concepts and doctrine, the maintenance of STANAGS and the implementation of a rigorous exercise programme. Whilst NATO-led operations and missions continued to drive some tactical-level cooperation among member states and partners, for example in Afghanistan and Kosovo,[103] the urgency to maintain intra-NATO interoperability in the immediate post-Cold War period no longer existed. In its place, the UK concentrated on joint warfare and invested heavily in capability that supported bilateral operations with the USA.

In the last decade, however, the global strategic environment has undergone a considerable change and the multipolar world in 2021 is far more competitive. Significantly, the USA now recognizes that its military dominance is no longer assured. In response, the US Army is committed to a transformation programme underpinned by its new concept of multi-domain operations. This approach has close relations in the UK's concept of multi-domain integration and NATO's Joint All Domain Operations concept. Procedurally, the lack of a single, NATO-wide unifying concept will mean there is no common approach around which future interoperability efforts can coalesce. Technically, although the introduction of a more regular and demanding NATO-wide exercise programme will drive out minor equipment interface problems, re-capitalizing member states will have to work much harder to overcome potential technological disparities in future weapons systems. Finally, behavioural differences among NATO allies will become more and more prevalent, especially as some nations aggressively press ahead with

---

[102] Chief of the General Staff RUSI Land Warfare Conference 2021, Ministry of Defence and General Sir Mark Carleton-Smith, Gov.uk, 2 June 2021, https://www.gov.uk/government/speeches/chief-of-the-general-staff-rusi-land-warfare-conference-2021.

[103] Operations and missions: past and present, NATO, accessed 1 September 2021, https://www.nato.int/cps/en/natohq/topics_52060.htm.

the exploitation of new technologies whilst others follow more cautiously behind.

At the height of the Cold War, the threat to NATO was immediate and the balance of advantage was perceived to be critically unfavourable. Accordingly, member states considered investment in interoperability to be incontrovertible. Today, the members of an increased North Atlantic Alliance are rushing to develop their militaries to maintain advantage in an era of systemic competition. Interoperability will be as important to meeting that challenge as it has ever been.

10

The Moral Component of Fighting

Bringing Society Back In

Tua Sandman

Introduction

Theories of combat tactics and battle victory often include a moral dimension.[1] The question of how to win in battle or war cannot merely centre on the physical means to fight or conceptual problems of how to fight. To understand and shape the outcome of ground warfare, one must also take into consideration the moral component of fighting, essentially the will to fight.[2] As Friedman notes, moral power is 'too intangible to be reduced to strict codification';[3] however, it is typically considered to concern motivation, morale, and moral cohesion.[4] These aspects, although rarely specified in the literature, are considered integral and critical aspects of combat effectiveness and how to achieve advantage in battle or war.

But what generates moral power and fighting will? The scholarly debates that speak to the moral component of fighting reflect diverging and, at times, opposing views on what truly enhances, maintains, or disrupts soldiers' willingness to engage in combat and war. As shown in recent publications on how to expand and advance the debates on combat motivation,[5]

[1] E.g. Randall Collins, 'A Dynamic Theory of Battle Victory and Defeat', *Cliodynamics* 1 (2010): 3–25; B. A. Friedman, *On Tactics: A Theory of Victory in Battle* (Annapolis, MD: Naval Institute Press, 2017); J. F. C. Fuller, *The Foundations of the Science of War* (London: Hutchinson & Co, 1926); Jim Storr, *The Human Face of War* (Cornwall: Continuum, 2009).

[2] E.g. Jonathan Fennell, 'Morale and Combat Performance: An Introduction', *The Journal of Strategic Studies* 37, 6–7, (2014): 796–798; Storr, *The Human Face of War*; Christopher Tuck, *Understanding Land Warfare* (Abingdon, Oxon: Routledge, 2014), 24.

[3] Friedman, *On Tactics*, 23.

[4] E.g. Friedman, *On Tactics*, 89; Storr, *The Human Face of War*, 8; Tuck, *Understanding Land Warfare*, 24.

[5] Tarak Barkawi, 'Subaltern Soldiers: Eurocentrism and the Nation-State in the Combat Motivation Debates', in *Frontline: Combat and Cohesion in the Twenty-First Century*, edited by Anthony King (Oxford: Oxford University Press, 2015); Ilya Berkovich, *Motivation in War: The Experience of Common Soldiers in Old-Regime Europe* (Cambridge: Cambridge University Press, 2017); Michal Pawinski and Georgina Chami, 'Why They Fight? Reconsidering the Role of Motivation in Combat Environments', *Defence Studies* 19, 3 (2019): 297–317.

Tua Sandman, *The Moral Component of Fighting*. In: *Advanced Land Warfare*. Edited by Mikael Weissmann and Niklas Nilsson, Oxford University Press. © Tua Sandman (2023). DOI: 10.1093/oso/9780192857422.003.0010

morale,[6] and cohesion,[7] it is important to push these debates forward and problematize taken-for-granted perspectives and assumptions. In light of various contexts and changes in the character of war and battle, a critical interrogation of the theoretical explanations that have come to dominate the field is reasonably warranted, especially as some of these explanations are now virtually regarded as universal.

Essentially, this chapter contends that the literature on combat motivation, morale, and cohesion at large reproduces certain frames of intelligibility that potentially delimit our field of perception and comprehension in regard to the question of why soldiers fight. The section called *How society slips from view* seeks to problematize and uncover how the literature tends to overlook or diminish the role of society and broader socio-political discourses that soldiers inevitably are embedded in. It argues that attention has been skewed toward the 'micro' dimension of the postulated micro–macro dichotomy, the analytical focus narrowed down to the *here and now* of combat, and an idea of the military unit as detached from society has been reproduced; as a result, the home front's or society's role with regard to the will to fight has largely been underexplored and undertheorized. Subsequently, the section called *Considering the role of society* contends that when so-called macro-level forces are considered, when societal factors are acknowledged and explored, we need to look more closely at and appreciate the *contingent* character of shared beliefs and ideas, and how the will to fight on the ground *and* at the home front are interlinked and continuously in the making, susceptible to change. To account for the element of contingency and explore how the will to fight among soldiers and in society may interweave, the section suggests that perceived legitimacy and righteousness, essentially the notion of a 'just cause', are concepts worth foregrounding. Whilst the notion of legitimacy among the broader public is affected by the representation of experiences on the ground, the home front's (possibly shifting) sense of legitimacy and righteousness in turn reasonably affects those performing violence on the ground.

As the chapter critically engages with the literature on combat motivation, morale, and moral cohesion, it will initially provide a brief overview of the most central points of debate in the scholarly endeavour set out to grasp *Why*

---

[6] Jonathan Fennell, 'In Search of the "X" Factor: Morale and the Study of Strategy', *The Journal of Strategic Studies* special issue; 'Morale and Combat Performance' 37, 6-7 (2014): 799–828; Hew Strachan, 'Training, Morale and Modern War', *Journal of Contemporary History* 41, 2 (2006): 211–227.

[7] Ilmari Käihkö, 'Broadening the Perspective on Military Cohesion', *Armed Forces & Society* 44, 4 (2018): 571–586; Anthony King, *The Combat Soldier: Infantry Tactics and Cohesion in the Twentieth and Twenty-First Centuries* (Oxford: Oxford University Press, 2013); Anthony King, 'On Cohesion', in *Frontline: Combat and Cohesion in the Twenty-First Century*, edited by Anthony King (Oxford: Oxford University Press, 2015); Siniša Malešević, *The Rise of Organised Brutality: A Historical Sociology of Violence* (Cambridge: Cambridge University Press, 2017).

soldiers fight. This section will not, nor will the chapter at large, account for these research fields as such or in their entirety, but focus specifically on how the question of why soldiers fight is projected and discussed in these overlapping debates. The purpose of the chapter is not to evaluate or dismiss prevalent explanations for why soldiers fight or how they come to fight well, or to set the dispute between diverging explanatory frameworks; ultimately, the chapter rather wishes to highlight aspects that are worth exploring and examining further as the field moves forward.

Why Soldiers Fight

What makes soldiers engage the enemy and in combat? Why do people enlist and sign up for war in the first place? What brings veterans to redeploy? Questions of this kind have been widely debated throughout the last century. According to Berkovich, the 'genuine interest in common soldiers and the forces influencing them' truly began during the First World War, but it was not until the Second World War and its aftermath that the issue of combat motivation became a subject of systematic review.[8] As of then, several strands of literature have emerged that in various ways confront the enigma of willingness and motivation. However, with few exceptions, these questions have seldom been studied empirically, and the scholarly discussion to a great extent still relies on observations from the past.[9]

Based on experiences and observations during the Second World War, and originally advanced by Marshall, Shils and Janowitz, and Stouffer et al.,[10] the thesis of primary group solidarity, or the primary group model of social cohesion, has heavily influenced the scholarly debate on combat motivation, morale, and cohesion. The thesis states that soldiers essentially fight for their buddies. 'The element of self-concern in battle' is minimized, notes Shils and Janowitz, when the surrounding group is considered to satisfy one's

[8] Berkovich, *Motivation in War*, 22. For further elaboration on the origins and evolution of the study of why soldiers fight, see e.g. Berkovich, *Motivation in War*; Käihkö 'Broadening the Perspective on Military Cohesion'; Malešević, *The Rise of Organised Brutality*; Simon Wessely, 'Twentieth-century Theories on Combat Motivation and Breakdown', *Journal of Contemporary History* 41, 2 (2006): 269–286.

[9] Peter van den Aker, Jacco Duel and Joseph Soeters, 'Combat Motivation and Combat Action: Dutch Soldiers in Operations since the Second World War; A Research Note', *Armed Forces & Society* 42, 1 (2016): 211–225; see also Fennell, 'Morale and Combat Performance'.

[10] S. L. A. Marschall, *Men Against Fire* (New York: William Morrow, 1947); Edward A. Shils and Morris Janowitz, 'Cohesion and Disintegration in the Wehrmacht in World War II', *Public Opinion Quarterly* 12, 2 (1948): 280–315; Samuel A. Stouffer, Edward A. Suchman, Leland C. Devinney, Shirley A. Star, and Robin M. Williams Jr., *The American Soldier: Adjustment during Army Life.* (Studies in social psychology in World War II) (Princeton, NJ: Princeton University Press, 1949).

basic needs and provide the moral support to push forward.[11] Thus, soldiers do not fight for 'abstractions' and are not primarily driven by shared beliefs and national convictions; as soon as combat begins, it is rather the loyalty to one's immediate primary group that keeps one going.[12] The primary group theory has become 'almost dogma in the current practices of western armies'[13] and has remained 'standard doctrine' to this day.[14] Among the more contemporary proponents of peer bonding and social cohesion as the most critical motivating factor in combat, one may look to Collins, Siebold, or Wong et al. among others.[15] This notion of what essentially matters for soldiers in the field has also been popularized through war movies and other cultural expressions, with depictions of camaraderie and 'leave no man behind' as now familiar tropes. Still, as Käihkö points out, tactical developments have deeply influenced our understanding of combat motivations; whereas the First World War has been associated with mass armies, mass frontal assaults, and thus mass ideology and patriotism, the Second World War—and the turn to smaller group formations—drew attention to the significance of interpersonal solidarity.[16]

Although recognizing the merits of strong unit cohesion, many scholars have emphasized the possible counterproductive effects that unit cohesion and primary group solidarity may have. Under certain conditions, tight-knit groups and strong in-group loyalty could serve to undermine the army and combat effectiveness.[17] It could lead to mutiny or soldiers refusing to harm

---

[11] Shils and Janowitz, 'Cohesion and Disintegration in the Wehrmacht in World War II', 281. Note that their observations are based on interviews with German prisoners of war. Similarly, the results of Leonard Wong, Thomas A. Kolditz, Raymond A. Millen, and Terrence M. Potter, *Why They Fight: Combat Motivation in the Iraq War* (Carlisle Barracks, PA: Strategic Studies Institute US Army War College, 2003) are based on interviews with Iraqi prisoners of war.

[12] E.g. Neal A. Puckett and Marcelyn Atwood, 'Crime on the Battlefield', in *The Oxford Handbook of Military Psychology*, edited by Janice H. Laurence and Michael D. Matthews (Oxford: Oxford University Press, 2012), 81.

[13] Strachan, 'Training, Morale and Modern War', 211.

[14] Wessely, 'Twentieth-century Theories on Combat Motivation and Breakdown'.

[15] Randall Collins, 'Does Nationalist Sentiment Increase Fighting Efficacy? A Sceptical View from the Sociology of Violence', in *Nationalism and War*, edited by John A. Hall and Siniša Malešević (Cambridge: Cambridge University Press, 2013): 31–43; Guy L. Siebold, 'The Essence of Military Group Cohesion', *Armed Forces & Society* 33, 2 (2007): 286–295; or Leonard Wong et al., 'Why They Fight: Combat Motivation in the Iraq War'.

[16] Käihkö, 'Broadening the Perspective on Military Cohesion', 574.

[17] Berkovich, *Motivation in War*, 25; Elliot Chodoff, 'Ideology and Primary Groups', *Armed Forces and Society* 9, 4 (1983), 570; King, *The Combat Soldier*, 73; Fennell 'In Search of the "X" Factor: Morale and the Study of Strategy', 804; Jonathan Fennell, 'Re-evaluating Combat Cohesion: The British Second Army in the Northwest Europe Campaign of the Second World War', in *Frontline: Combat and Cohesion in the Twenty-First Century*, edited by Anthony King (Oxford: Oxford University Press, 2015), 138; Strachan, 'Training, Morale and Modern War', 213; Malešević, *The Rise of Organised Brutality*, 284; Wong et al., 'Why They Fight: Combat Motivation in the Iraq War', 3–4.

the enemy.[18] Or it could facilitate unethical behaviour and cover-ups[19] or even fragging.[20] Thus, as suggested, the enhancement of small group cohesion does not necessarily, or automatically, benefit the military mission or operation as such.

Over the years, the primary group thesis and the emphasis on camaraderie and solidarity have also been outright challenged. Bartov, for one, has questioned the conclusions drawn regarding the cohesion of *Wehrmacht* soldiers; in his study of the Eastern front, he argues that primary group solidarity was unlikely to be strong or possible to sustain, given the casualty figures and the frequent disintegration of units.[21] Significant obstacles to building unit cohesion have also been noted by Rush in his study on the German LXXIV Infantry Corps on the Western front in 1944. The combat motivation of the *Wehrmacht* could thus not exclusively be attributed to strong small group loyalty.[22] Similarly, in his study on combat motivation and the combat experience of US soldiers in Vietnam, Moskos calls for a modification of the primary group theory; essentially, he argues that the US rotation system in Vietnam reinforced a privatized view of the war and that primary group ties are 'best viewed as mandatory necessities arising from immediate life and death exigencies'. Although soldiers essentially rely on the group's physical, technical, and moral support to survive, one must also take into consideration the 'salient ideological factors', he argues—or 'latent ideology'—which 'serve as preconditions supporting the soldier in dangerous situations'.[23] In a similar manner, Bartov points to ideological internalization in the case of the *Wehrmacht*, not simply social ties; instead of ascribing meaning to primary groups as in the original understanding of the term, he rather foregrounds the attachment to 'an *ideal* primary group', meaning the projection of Self as

[18] Randall Collins, *Violence: A Micro-sociological Theory* (Princeton, NJ: Princeton University Press, 2008): 57; J. G. Fuller, *Troop Morale and Popular Culture in the British and Dominion Armies 1914–1918* (Oxford: Clarendon, 1991): 22–24; Paul G. Kooistra and John S. Mahoney, 'The Road to Hell: Neutralization of Killing in War', *Deviant Behavior* 37, 7 (2016): 768; Chiara Ruffa, 'Cohesion, Political Motivation, and Military Performance in the Italian Alpini', in *Frontline: Combat and Cohesion in the Twenty-First Century*, edited by Anthony King (Oxford: Oxford University Press, 2015): 250.

[19] E.g. James Connor, Dia Jade Andrews, Kyja Noack-Lundberg, and Ben Wadham, 'Military Loyalty as a Moral Emotion', *Armed Forces & Society* (2019): 1–21; Kooistra and Mahoney, 'The Road to Hell: Neutralization of Killing in War', 769.

[20] Rune Henriksen, 'The Character of War and the Nature of Combat', in *The Character of War in the 21st Century*, edited by Caroline Holmqvist-Jonsäter and Christopher Coker (London: Routledge, 2010): 21; Charles C. Moskos, 'The American Combat Soldier in Vietnam', *Journal of Social Issues* 31, 4 (1975): 35.

[21] Omer Bartov, 'Soldiers, Nazis, and War in the Third Reich', *The Journal of Modern History* 63, 1 (1991): 44–60.

[22] Robert S. Rush, 'A Different Perspective: Cohesion, Morale and Operational Effectiveness in the German Army, Fall 1944', *Armed Forces & Society* 25, 3 (1999): 477–508.

[23] Charles C. Moskos, 'The American Combat Soldier in Vietnam', 27–29.

opposed to an Other.[24] Needless to say, this notion of Self was permeated with racism and built on conceptions of the enemy as *untermenschen*.[25] In a similar fashion, Shils and Janowitz also acknowledge that ideological factors *strengthened* primary group solidarity; for instance, they point to the presence of Nazi 'hard cores' who embodied a particular notion of masculinity.[26] Masculinity is often foregrounded in discussions on combat motivation, although perhaps rarely acknowledged as an ideological influence in itself, or as an integral part of nationalism and militarism. Most notably, King has accounted for the appeal to masculinity and the male ideal throughout the Second World War and the Vietnam War.[27]

The debates on combat motivation, morale, and cohesion largely sprang out of the observed passivity and lack of offensive spirit among many US soldiers during the Second World War. And for the military at the time, this was a cause for great concern.[28] Whilst some might suggest that certain soldiers, or warriors,[29] are 'naturally born killers', most would agree that the reluctance to kill or to actively fight is something that so-called warriors are trained and/or forced to overcome.[30] As Henriksen emphasizes: 'the encounter with combat is a frightening and shocking experience for most'.[31] Historically, the use of force has been arranged in ways that help soldiers overcome their resistance to killing; consider for instance how crew-served weapons facilitate group anonymity,[32] how artillery—that is: firing at a distance—has served to battle confrontational tension/fear,[33] or how modern day drone technologies offer 'powerful means of distancing', despite the ocular proximity.[34] Fundamentally, violence represents an act of extreme deviance, and a vast majority of all people find killing extremely difficult.[35] Thus, to make soldiers prepared, willing, and motivated to fight is at the very heart of military

[24] Omer Bartov, 'Soldiers, Nazis, and War in the Third Reich'.
[25] See also Dave Grossman, *On Killing: the Psychological Cost of Learning to Kill in War and Society* (London: Little, Brown and Company, 1996): 162.
[26] Shils and Janowitz, 'Cohesion and Disintegration in the Wehrmacht in World War II', 286.
[27] King, *The Combat Soldier*, 63–73.
[28] E.g. Joanna Bourke, *An Intimate History of Killing: Face-to-face Killing in the Twentieth-Century Warfare* (New York: Basic Books, 1999): 61–65; Grossman, *On Killing*; Marschall, *Men Against Fire*.
[29] For a conceptual discussion on the difference between soldiers and warriors, see Rune Henriksen, 'Warriors in Combat—What Makes People Actively Fight in Combat?', *The Journal of Strategic Studies* 30, 2 (2007): 187–223; Christopher Coker, *The Warrior Ethos: Military Culture and the War on Terror* (Abingdon: Oxon: Routledge, 2007); or for a more advanced soldier typology, see Iselin Silja Kaspersen, 'New Societies, New Soldiers? A Soldier Typology', *Small Wars & Insurgencies* 32, 1 (2020): 1–25.
[30] E.g. Bourke, *An Intimate History of Killing*; Grossman, *On Killing*; Henriksen, 'Warriors in Combat'; Malešević, *The Rise of Organised Brutality*; Bruce Newsome, 'The Myth of Intrinsic Combat Motivation', *Journal of Strategic Studies* 26, 4 (2003): 24–46.
[31] Henriksen, 'Warriors in Combat', 188.
[32] Grossman, *On Killing*, 149–155.
[33] Collins, 'Does Nationalist Sentiment Increase Fighting Efficacy?', 37.
[34] Grégoire Chamayou, *Drone Theory* (London: Penguin Books, 2015), 119.
[35] E.g. Kooistra and Mahoney, 'The Road to Hell: Neutralization of Killing in War'; see also Collins, 'Does Nationalist Sentiment Increase Fighting Efficacy?'.

training.[36] Accordingly, training has in itself been brought forward as a key explanation to how morale is sustained,[37] and to high combat performance.[38] As King suggests, in the context of the professional army, 'primary-group theory and ideological explanations have been displaced by training, drills and preparation'.[39]

The debates on combat motivation, morale, and cohesion are largely based on historical accounts of battles fought with conscription armies. Today, however, the conscription system has formally or de facto been abolished in most Western democracies, and armed forces are largely made up of professional soldiers. This transition has implied a shift in the self-image of the so-called democratic warrior,[40] and it reasonably raises new questions concerning the will to fight. As related by King: in the last two decades, there has been a 'practical or performative turn' in military scholarship, a turn from comradeship to competence.[41] The debate on cohesion, he argues, has reoriented itself from interpersonal bonds to practical military teamwork—that is: from social cohesion to *task* cohesion, as in a shared commitment.[42] However, King largely reconceptualizes cohesion, disassociates it from motivation, and defines it as 'the successful coordination of actions on the battlefield'.[43] With such a definition, cohesion no longer serves as an *explanation* to combat performance but is made synonymous with combat performance itself. The questions of what soldiers *fight for* and what makes them *fight effectively*—as in cohesively and ultimately successfully—are thus becoming increasingly blurred. As Malešević points out, the focus on military utility in these debates often obscures key sociological questions of, say, how group ties and commitments develop, transform, or collapse. It is thus necessary, he argues, 'to move away from the obsession with the military performance'.[44]

On principle, we need to be cautious about terms such as 'will' and 'motivation'. To ask, say, what forms soldiers' willingness to engage in combat is to

[36] Maya Eichler, *Militarizing Men: Gender, Conscription, and War in Post-Soviet Russia* (Stanford, CA: Stanford University Press, 2012): 112.

[37] Strachan, 'Training, Morale and Modern War'.

[38] Anthony King, *The Transformation of Europe's Armed Forces: From the Rhine to Afghanistan* (New York: Cambridge University Press, 2011); Anthony King, 'Discipline and Punish', in *Frontline: Combat and Cohesion in the Twenty-First* Century, edited by Anthony King (Oxford: Oxford University Press, 2015); King, *The Combat Soldier*.

[39] King, 'Discipline and Punish', 94.

[40] Andreas Herberg-Rothe, 'The Democratic Warrior and World-Order Conflicts', in *Heroism and the Changing Character of War: Toward Post-Heroic Warfare?*, edited by Sibylle Scheipers (Basingstoke: Palgrave Macmillan, 2014), 290.

[41] King, *The Transformation of Europe's Armed Forces*; King, 'On Cohesion', 10, 12.

[42] King, *The Combat Soldier*, 34. See also Robert J. MacCoun, Elizabeth Kier and Aaron Belkin, 'Does Social Cohesion Determine Motivation in Combat? An Old Question with an Old Answer', *Armed Forces & Society* 32, 4 (2006): 646–654.

[43] King, *The Combat Soldier*, 36.

[44] Malešević, *The Rise of Organised Brutality*, 285.

suggest that there essentially *is* a will behind it. For the soldier, fighting has, throughout the history of warfare, constituted a necessity. As been pointed out by many: usually, soldiers simply fight for survival[45] and soldiers thus engage in their own 'very private war'.[46] This has occasionally been termed situational motivation.[47] The instinct for personal survival is a fundamental part of 'the soldier's dilemma', as when one's sense of duty and sacrifice essentially is to be balanced against one's will to *live*.[48] So, in regard to the moral component of fighting, the sense of having no alternative[49] seems to be an aspect to be taken seriously; and so does the element of intimidation and oppression. For time and again, soldiers have been coerced or disciplined into action.[50] Often, they have been outright *forced* to fight, or they are made to fight, just as young men (and occasionally, women) have been legally obligated to serve through conscription. 'Punishment and deterrence', Strachan argues, are often forgotten in debates on motivation, despite their central place in warfare.[51] In fact, Rush ascribes intimidation and threat the single most important explanation for why *Wehrmacht* soldiers kept on fighting toward the end of 1944, although the war was clearly lost.[52] During the Second World War, thousands upon thousands of German and Soviet servicemen were executed.[53] One may also note how it was common for soldiers during the First World War to refuse to 'go over the top' of the trenches unless forced at gunpoint.[54] When recounting the 1916 battle of the Somme, Keegan similarly emphasizes the impossibility of running away or refusing to engage the enemy.[55] As testimonies from the battlefield of the Western front attest to, fighting was hardly optional:

---

[45] Henriksen, 'The Character of War and the Nature of Combat', 11.

[46] Moskos, 'The American Combat Soldier in Vietnam', 37.

[47] E.g. Berkovich, *Motivation in War*, 29.

[48] Christopher Dandeker and Simon Wessely, 'Beyond the Battlefield', in *Frontline: Combat and Cohesion in the Twenty-First Century*, edited by Anthony King (Oxford: Oxford University Press, 2015): 293.

[49] See Herberg-Rothe, 'The Democratic Warrior and World-Order Conflicts' 289–290;. Rush, 'A Different Perspective', 497.

[50] Fennell, 'Re-evaluating Combat Cohesion', 138; Fennell, 'In Search of the "X" Factor', 805; John Keegan, 'Towards a Theory of Combat Motivation', in *Time to Kill: The Soldier's Experience of War in the West 1939–1945*, edited by Paul Addison and Angus Calder (London: Pimlico, 1997): 3–11; Rush, 'A Different Perspective', 488.

[51] Hew Strachan, 'The Soldier's Experience in Two World Wars: Some Historiographical Comparisons', in *Time to Kill: The Soldier's Experience of War in the West 1939–1945*, edited by Paul Addison and Angus Calder (London: Pimlico, 1997), 369–378, 374. See also King, 'Discipline and Punish'.

[52] Rush, 'A Different Perspective'.

[53] Strachan, 'Training, Morale and Modern War', 215.

[54] Bourke, *An Intimate History of Killing*, 61.

[55] John Keegan, *The Face of Battle: A Study of Agincourt, Waterloo and the Somme* (Reading, Berkshire: Pimlico, 2004), 278: 'To surrender was dishonourable and might be dangerous. To run away was impossible (for the Germans, of course, had their own battle police farther down the trenches). To kill the British was, therefore, a necessity—though the majority would have called it a duty and, to the British on the wrong side of the wire, it may have seemed that they found it a pleasure'.

Sergeant Moore, he was standing behind the trench. He got a revolver in his hand and said 'Anybody goes back, I shoot them'. So that if we didn't go one way, we wouldn't go the other.[56]

These coercive dynamics are not restricted to our past, but very much part of the reality and logic of war still today. For instance, in their interview study with Iraqi Regular Army prisoners of war, Wong et al. report that fear of retribution and punishment was a prime motivating force; in fact, it was a near universal response.[57] And in the early 2000s, the Israeli Defence Forces toughened its policies on conscious objectors and, for the first time since the 1970s, brought many so-called 'refuseniks' before court martial.[58] However, whether coercion is a successful formula to uphold motivation and combat performance has largely been disputed.[59]

How Society Slips from View

The question of why one fights in war and engages in combat is complex and, reasonably, impossible to pin down. As Wessely has pointed out: 'there is no universal explanation why men fight, or why they break down in battle';[60] 'science may tell us one day', Keegan notes—'though I doubt it'.[61] Thus, rather than viewing the interpretations of why people fight in battle or war as universal truths, the theories which have dominated the scholarly debate over the twentieth century should be considered 'historical material in their own right'.[62] These theories simplify an experience that reasonably is profoundly complex and they thus potentially delimit our field of perception and comprehension;[63] how historical explanations are presented and reiterated, how they structure and colour contemporary debate and practice, and what notions of war's dynamics they reproduce and reinforce, are thus vital questions to critically interrogate. As demonstrated, previous research has predominantly centred on the following to account for why soldiers fight:

---

[56] *They shall not grow old* [Documentary film]. Director: Peter Jackson. United Kingdom/New Zealand: WingNut Films, House Productions, Warner Bros, 2018, 1:00:53.

[57] Wong et al., 'Why They Fight: Combat Motivation in the Iraq War', 6. For a methodological critique of their study, see MacCoun et al., 'Does Social Cohesion Determine Motivation in Combat?'.

[58] Tamar S. Hermann, *The Israeli Peace Movement: A Shattered Dream* (New York: Cambridge University Press, 2009): 260; Baruch Kimmerling, *Politicide: Ariel Sharon's War Against the Palestinians* (New York: Verso Books, 2003).

[59] E.g. Benjamin Barber IV and Charles Miller, 'Propaganda and Combat Motivation: Radio Broadcasts and German Soldiers' Performance in World War II', *World Politics* 71, 3 (2019): 462.

[60] Wessely, 'Twentieth-century Theories on Combat Motivation and Breakdown', 286.

[61] Keegan, 'Towards a Theory of Combat Motivation', 11.

[62] Wessely, 'Twentieth-century Theories on Combat Motivation and Breakdown', 286.

[63] See also Fennell, 'In Search of the "X" Factor', 810.

social cohesion or in-group solidarity; 'latent' ideology, masculinity, or ideological indoctrination; training or task cohesion/commitment; and survival or sheer necessity or coercion. In an attempt to shed further light on the potentially dynamic role of society in regard to the will to fight, this section will seek to uncover the ways in which the role of society has largely slipped from view and ultimately been rendered irrelevant for the question of why soldiers fight.

First, the literature on combat motivation, morale, and cohesion tends to reproduce a dichotomic understanding of why soldiers fight; in short, it continuously reflects the idea that there are two distinct and separate levels of influence. In large part, the scholarly debate on the will to fight has been structured around a micro–macro dichotomy. The micro-level is typically associated with primary groups and interpersonal solidarity, whereas the macro-level encompasses factors related to state and society.[64] These levels are not only depicted as distinct and separate, but are continually set in opposition. In many empirical or conceptual accounts, isolated explanations to combat motivation or cohesion are typically 'tested' and various influencing factors are often contrasted and weighted against another—as in micro *or* macro. Consider for instance Wong et al. who clearly pursue an 'either/or' line of reasoning, largely contrasting 'fighting for my buddies' against 'the cause'.[65] Another case in point is Collins who dismisses ideational factors altogether, in support of small-group loyalty.[66] Fundamentally, the framing of 'either/or' reflects an idea of these explanations as mutually exclusive. Although many acknowledge that fighting will is influenced by a number of factors[67]—that motivational factors such as ideology or primary group cohesion may overlap,[68] or be more or less significant in different situations[69]—the idea of macro and micro, ideology and primary group solidarity, as distinct variables is continually reproduced. Just consider how this dichotomy also structures and governs contemporary discussions and literature reviews such as the one presented in this chapter.[70] Micro-level explanations have come to

---

[64] E.g. Käihkö, 'Broadening the Perspective on Military Cohesion'.

[65] Wong et al., 'Why They Fight: Combat Motivation in the Iraq War'.

[66] Collins, 'Does Nationalist Sentiment Increase Fighting Efficacy?'.

[67] E.g. Fuller, *Troop Morale and Popular Culture in the British and Dominion Armies 1914–1918*; King, *The Transformation of Europe's Armed Forces*, 206; Moskos, 'The American Combat Soldier in Vietnam'.

[68] Chodoff, 'Ideology and Primary Groups'.

[69] Anthony Kellett, *Combat Motivation: The Behavior of Soldiers in Battle* (Boston, MA: Kluwer, 1982), 319.

[70] See e.g. Fennell, 'In Search of the "X" Factor'; Wessely, 'Twentieth-century Theories on Combat Motivation and Breakdown'; Berkovich, *Motivation in War*.

dominate the debate on combat motivation, morale, and military cohesion;[71] as it appears, it has generally been deemed a more plausible—even intuitive—answer to why soldiers fight. The significance of comradeship for combat motivation is not uncommonly projected as a self-evident fact, perhaps on the grounds of Marschall's claim that 'one of the simplest truths of war' is that infantry soldiers keep going for their comrades.[72] In line with the discourse of micro *versus* macro, influencing forces associated with societal beliefs and identifications are often disregarded as 'abstractions', which allegedly has little explicit significance to soldiers in combat. Naturally, the micro–macro dichotomy, and other such categorizations, serve to order and make the overload of potential factors influencing morale and combat motivation manageable and comprehensible. However, the will to fight is reasonably more complex and layered than for example the micro–macro dichotomy suggests or than the positivist quest to isolate variables allows us to appreciate.

Second, discussions on combat motivation, morale, and cohesion tend to focus primarily on specific combat situations. Yet, with an exclusive focus on the *here and now* of combat, our understanding of the forces influencing soldiers in the field reasonably becomes unnecessarily narrow. It might be true, as Henriksen argues, that combat is not *about* something, but that it simply *is*; and accordingly, it does not 'inspire questions' of its meaning.[73] The intensity of fighting, Malesevic relates, does not allow for reflection; one rather tends to focus on the practical and the technical.[74] As in Gray's philosophical memoir of his four years of service during the Second World War, when he seeks to capture the transition from soldier to fighter. This *becoming*, as it were, has seemingly little to do with will as such:

> The soldier who has yielded himself to the fortunes of war, has sought to kill and to escape being killed, or who has even lived long enough in the disordered landscape of battle, is no longer what he was. He becomes in some sense a fighter, whether he wills it or not—at least most men do. His moods and disposition are affected by the presence of others and the encompassing environment of threat and fear. He must surrender in a measure to the will of others and to superior force. In a real sense he becomes a fighting man, a *Homo furens*.[75]

---

[71] See e.g. Fennell, 'In Search of the "X" Factor', 807; Käihkö, 'Broadening the Perspective on Military Cohesion'.

[72] Marschall, *Men Against Fire*, 42.

[73] Henriksen, 'The Character of War and the Nature of Combat', 17.

[74] Siniša Malešević, 'The Act of Killing: Understanding the Emotional Dynamics of Violence in the Battlefield', *Critical Military Studies* (2019), 11–12.

[75] J. Glenn Gray, *The Warriors: Reflections on Men in Battle* (New York: Harcourt, Brace and Company, 1959): 27.

Also, as has been noted: it is vital to differentiate between the will to *fight* and the will to *serve*,[76] a difference which the literature at large allegedly has tended to ignore.[77] The question of what drives soldiers to sign up for duty is of course different from the question of why soldiers actively fight (and fight well) in battle. But reasonably, the will to fight involves more than simply deciding to serve on the one hand, and finding oneself in the midst of battle on the other; the moral component of fighting is arguably broader than the issue of what drives soldiers to fight (effectively) in a specific combat situation.[78] Military campaigns and operations are not characterized by an incessant set of combat situations; there is always something *before* combat, *after* combat, *between* combat: instances of patrolling, returning to base, or of endlessly waiting.[79] The immediate danger of combat could convert into periods of inactivity and stalemate, which could call into question the relevance and purpose of military action.[80] Soldiers are perhaps even 'stunned with boredom',[81] which could be reinforced by widespread apathy in society at large and a lack of support from back home.[82] In those 'in-between moments'—once combat is over, or has not yet begun—questions of 'meaning' may very well arise.[83]

> Contrary to the perception that many people have about wars, they're not places of excitement and glamour, but places of boredom, long periods of time alone and thinking, places of fear, of bone weary tiredness, thirst, hunger, frustration, living in rain, mud, dirt, heat, sweat and most of all, wonder at why you're there at all.[84]

Chodoff has distinguished between precombat and in-combat motivation, between placing oneself in danger versus actively participating once finding oneself at risk;[85] Kellett, in turn, discusses motivation *before* and *after* battle, as opposed to motivation *in* battle.[86] To reduce the moral component of fighting and the experience of war at large to the specific combat situation

---

[76] Henriksen, 'Warriors in Combat', 188.

[77] Newsome, 'The Myth of Intrinsic Combat Motivation'.

[78] Cf. Keegan, *The Face of Battle*, 272.

[79] E.g. Bård Maeland and Paul Otto Brunstad, *Enduring Military Boredom: From 1750 to the Present* (Basingstoke: Palgrave Macmillan, 2009).

[80] Kellett, 'Combat Motivation', 320.

[81] Sebastian Junger, 'Why veterans miss war', TED Talk, 2014, https://www.ted.com/talks/sebastian_junger_why_veterans_miss_war [accessed 20,200,828]

[82] Maeland and Brunstad, *Enduring Military Boredom*, 40–41.

[83] Cf. Christine Sylvester, 'Experiencing War: A Challenge to International Relations', *Cambridge Review of International Affairs* 26, 4 (2013): 669–674.

[84] David Pye, cited in Maeland and Brunstad, *Enduring Military Boredom*, 37. Excerpt from his memoirs called *Tour of Duty '71*, originally published on his personal home page in 1998. David Pye served as an Australian combat soldier in the Vietnam War.

[85] Chodoff, 'Ideology and Primary Groups'.

[86] Kellett, 'Combat Motivation', 327.

delimits the field of perception; if we merely focus on the *here and now* of combat, we omit perspectives that potentially would inform our understanding of what influences soldiers in the field. The idea of camaraderie as the key driving force behind the will to fight restricts the issue of will to the sphere of the local, and obscures the broader socio-political context and discourses that soldiers are inevitably embedded in.

Third, without discrediting soldiers' own experiences of comradeship and in-group solidarity and loyalty in times of combat, one may also question the notion of the group as an isolated entity, which is often reflected and reiterated in empirical and theoretical accounts. In short, the literature often reflects an idea of the small group, the military unit, as detached from the social and from society. Consider, for instance, how Collins regards nationalism and ideology as a feature restricted to the home front, largely irrelevant for combat soldiers,[87] or Henriksen's claim that once soldiers join their unit, their social world shrinks 'from being that of the state, to that of the unit'.[88] This notion is also reflected in discussions on intrinsic and extrinsic combat motivation,[89] where it is projected that once becoming a soldier, one is merely subjected to the pressures of the military institution, as opposed to societal norms and expectations at large. The idea here is that intrinsic motivations (such as nationalism, militarism, morality) are those which soldiers bring *into* military life, whereas extrinsic motivations are those developed and nourished *in* military life. In other words, the distinction builds on an idea that motivations are cultivated either *before* or *after* joining the military. Fundamentally, this suggests that intrinsic motivations are static and 'essential' in the sense that once in the military, these convictions or beliefs are not changeable. Coker even suggests that intrinsic motivations are 'genetic or culturally constructed in childhood'.[90] Similarly, Fennell differentiates between factors foregrounded in the literature that could be considered endogenous and exogenous, that is: originating from *within* the military organization, or from the *outside*, as if the inside and outside are distinct spheres.[91]

The differentiation between distinct and separate levels of influence in explaining why soldiers fight (as the micro–macro dichotomy suggests), the focus on the *here and now* of combat and the idea of the unit as an isolated entity potentially delimit the possibility of a holistic understanding of the moral dimension of fighting and how soldiers gain or lose motivation,

---

[87] Collins, 'Does Nationalist Sentiment Increase Fighting Efficacy?', 35–36.
[88] Henriksen, 'The Character of War and the Nature of Combat', 21.
[89] Coker, *The Warrior Ethos*; Henriksen, 'Warriors in Combat'; Newsome, 'The Myth of Intrinsic Combat Motivation'.
[90] Coker, *The Warrior Ethos*, 5.
[91] Fennell, 'In Search of the "X" Factor'.

potentially reconsider the whole enterprise, at different points throughout military operations. Dichotomies and the categorizing of different influencing factors tend to encourage us to set different set of explanations in opposition to each other; and it essentially makes ideational factors easy to disregard, as these simply become an 'abstraction', separated and removed from the *here and now* of the combat situation, the group, the army, even the military. With a strict focus on the *here and now* of combat, or the *within* dimension of military undertakings and developments, we might lose sight of societal influences which are always already present and which are subject to change. As Malešević argues: primary group solidarity, or more specifically solidarity within the military unit, has 'macrostructural origins'.[92] Group attachments are not given and inherent; however, the literature on social cohesion tends to neglect the 'macrohistorical contexts that make social cohesion possible in the first place'.[93] And perhaps this is to be expected: whereas ideational and societal influences are elusive and difficult for both researchers and interviewees to identify, micro-level or extrinsic sources of motivation—as in *I fight for my buddies*—are reasonably more tangible.[94] Recent scholarship has called for a broadening of the study of cohesion,[95] not just in terms of moving beyond the Eurocentrism which has characterized and defined the field[96] but also by 'rising above small groups' and further investigating so-called macro-level factors. It is important, Malešević argues, to steer attention to 'the social organisations that create, sustain and utilise the organisational and ideological means' that make 'microgroup bonds possible'.[97] In line with such a quest, to explore how the experience of combat and motivation is conditioned upon more societal phenomena such as shared beliefs and identifications, Nilsson, for instance, has studied how unit cohesion among the Peshmerga is influenced by ideas of Kurdish identity.[98] Connor et al., on their part, have explored how experiences of loyalty

---

[92] Malešević, *The Rise of Organised Brutality*, 290.

[93] Malešević, *The Rise of Organised Brutality*, 306.

[94] See also MacCoun et al., 'Does Social Cohesion Determine Motivation in Combat?', 250, and Peter Olsthoorn, 'Courage in the Military: Physical and Moral', *Journal of Military Ethics* 6, 4 (2007): 276, on how the importance of social cohesion has become common sense in the military, and how interviewees tend to draw on common sense.

[95] Ilmari Käihkö and Peter Haldén, 'Full-Spectrum Social Science for a Broader View on Cohesion', *Armed Forces & Society*, special issue: 'Broadening the Perspective on Military Cohesion' 46, 3 (2020): 517–522.

[96] See also Barkawi, 'Subaltern Soldiers'; Marco Nilsson, 'Primary Unit Cohesion among the Peshmerga and Hezbollah', *Armed Forces & Society* 44, 4 (2018): 647–665.

[97] Malešević, *The Rise of Organised Brutality*, 290.

[98] Nilsson, 'Primary Unit Cohesion among the Peshmerga and Hezbollah', 648.

at the micro-level are 'implicated within and conditioned by' social phenomena at the macro-level. To advance a thorough understanding of the forces influencing soldiers in the field, the literature would be well served by further studies critically interrogating how the experience of *here and now* connects to society and politics.[99]

Considering the Role of Society

As reflected above, much writing on combat motivation, morale, and cohesion reflects a notion of fighting, of violence, as a distinct phenomenon, with its own 'rhythm, dynamics and practices'.[100] But just as war and violence are not detached from the social,[101] soldiers are not detached from society.[102] The use of force is a socially situated practice, and so, reasonably, is the will to fight. In the context of the modern battlefield, it is difficult to imagine that 'the bond of battle'[103] is disconnected from the social and political processes that essentially serve to justify the use of force and give violence its power and meaning. For instance, societal expectations and conceptions of honourable soldiers affect soldiers' conduct in the field. As Kaspersen notes: 'Roles, socially constructed, mutually affected, and shaped by the interplay between expectations and experiences are influenced by the past, the present, and by societal and individual factors'.[104] Thus, it might be true that the battlefield is one of the loneliest places on earth,[105] but soldiers do not fight in a (moral) vacuum; their judgement and sense of morality are rather embedded in social and political discourse and reasonably influenced by (shifting) societal perceptions of what is right and feasible. In this respect, the *unit* and its practices are connected to society, to state, and to how the war at large is perceived and understood.[106]

---

[99] James Connor et al. ('Military Loyalty as a Moral Emotion', 13) similarly emphasize that the lived experience of loyalty among military personnel, and how it connects to political frames, is understudied.
[100] Sylvia Walby, 'Violence and Society: Introduction to an Emerging Field of Sociology', *Current Sociology* 61, 2 (2012): 96.
[101] Tua Sandman, *The Dis/appearances of Violence: When a 'peace-loving' State uses Force* (Stockholm University, Thesis (PhD), Stockholm Studies in Politics 180, 2019): 8.
[102] Kaspersen, 'New Societies, New Soldiers? A Soldier Typology', 2.
[103] Henriksen, 'The Character of War and the Nature of Combat'.
[104] Kaspersen, 'New Societies, New Soldiers? A Soldier Typology', 10.
[105] Bourke, *An Intimate History of Killing*, 65.
[106] The notion of the unit as *detached* from society brings to mind the discussion in military sociology on the relation between the military and civilian spheres of life, and whether 'warriors' are perceived to undergo a total, partial, or minimum transformation from their civilians selves. See Peter Haldén and Peter Jackson, 'Introduction: Symbolic and Mythological Perspectives on War and Peace Join the Archaic with the Modern', in *Transforming Warriors: the Ritual Organization of Military Force*, edited by Peter Haldén and Peter Jackson (Abingdon, Oxon: Routledge, 2016), 1–18.

The critical war studies literature,[107] or the cultural study of war and combat,[108] inspire us to appreciate and take note of the interlinkages between the act of using force, on the one hand, and social and political structures, on the other. For instance, just consider how 'societal culture impact upon the military' in terms of, say, norms on masculinity,[109] and how war time masculinities in turn feed into post-war discourse.[110] By acknowledging these interconnections, between warfare and society, one draws attention to the discursive structuration of war and violence, and its constitutive effects on subject formation and transformation.[111] War and combat are generative and constitutive forces, in the sense that experiences of war and violence generate reconfigurations of social identities. Subjects involved are 'cast into motion', and so are social and political orders.[112] As emphasized in the critical war studies literature, a key dimension of war is the making, remaking, and destroying of truth.[113]

Yet, in the literature on combat motivation, morale, and cohesion, social and political structures tend to be translated into concepts such as *ideology*,[114] *nationalism*,[115] *patriotism*,[116] or *propaganda*.[117] So-called macro-level forces are thus often reduced to something static and stable, seemingly unaffected by experiences on the ground. Ideology, for example, essentially refers to a fixed set of beliefs or principles and is largely presented as a fixed variable, which appears as a constant throughout the war or operation, perhaps even across decades or centuries. Barkawi also identifies how military scholarship often reproduces essentialism by 'attributing primary and enduring causal powers to national culture'.[118] What often slips from view is the *contingent*

---

[107] Tarak Barkawi and Shane Brighton, 'Powers of War: Fighting, Knowledge, and Critique', *International Political Sociology* 5, 2 (2011): 126–143; Marc von Boemcken, 'Unknowing the Unknowable. From "critical war studies" to a Critique of War', *Critical Military Studies* 2, 3 (2016): 226–241.

[108] E.g. John A. Lynn, *Battle: A History of Combat and Culture* (Cambridge, MA: Westview Press, 2004).

[109] Lynn, *Battle: A History of Combat and Culture*, xx; see also King, *The Combat Soldier*, 63.

[110] Jiří Hutečka, *Men under Fire: Motivation, Morale and Masculinity among Czech Soldiers in the Great War, 1914–1918* (New York: Berghahn, 2020), 12.

[111] Tarak Barkawi, 'Of Camps and Critiques: A Reply to "Security, War, Violence"', *Millenium: Journal of International Studies* 41, 1 (2012): 127.

[112] Barkawi and Brighton, 'Powers of War: Fighting, Knowledge, and Critique', 136.

[113] See e.g. Barkawi and Brighton, 'Powers of War: Fighting, Knowledge, and Critique'; von Boemcken, 'Unknowing the unknowable'.

[114] E.g. Chodoff, 'Ideology and Primary Groups'; Malešević, *The Rise of Organised Brutality*; Moskos, 'The American Combat Soldier in Vietnam'; Newsome, 'The Myth of Intrinsic Combat Motivation'; Strachan, 'Training, Morale and Modern War'; Wessely, 'Twentieth-century Theories on Combat Motivation and Breakdown'.

[115] E.g. Collins, 'Does Nationalist Sentiment Increase Fighting Efficacy?'; Newsome, 'The Myth of Intrinsic Combat Motivation'.

[116] E.g. Chodoff, 'Ideology and Primary Groups'; King, 'Discipline and Punish'; Moskos, 'The American Combat Soldier in Vietnam'.

[117] E.g. Barber and Miller, 'Propaganda and Combat Motivation'.

[118] Barkawi, 'Subaltern Soldiers', 25.

character of shared beliefs and ideas—how the shared notion of what is right or feasible always is susceptible to reconsideration. To be open to the shifts and changes in societal forces with regard to combat motivation among soldiers, and to better capture and account for it, one may suggest that concepts such as *perceived legitimacy* and *just cause* are worth taking into account and exploring further. For it is not 'the cause' as such that necessarily is of interest to the field of war studies,[119] but projected causes' perceived legitimacy among soldiers and its significance on moral power and fighting will. Along these lines, Fuller talks of the 'legitimate demand' theory, which holds that soldiers are motivated by 'the legitimacy of the ends in view, and to the validity of the military hierarchy's methods in pursuit of those ends'.[120] Moskos, on his part (although he contends that soldiers have little idea of political implications once the combat engagement is over) also refers to the 'shared beliefs' of soldiers and, specifically, the belief that one is fighting for a just cause.[121] As Friedman puts it: 'the troops must at least believe they are fighting for a moral purpose'.[122]

In democratic societies in particular, war is only possible if considered and perceived as legitimate, if *felt* legitimate and righteous. And, naturally, these perceptions and feelings may very well change. Motivation may change when exposed to the realities of combat,[123] and representations of experiences on the ground may upend societal perceptions of legitimacy and righteousness. The concepts of perceived legitimacy and just cause explicitly point to the contingency of social relations; the notion of legitimacy is always unstable, and our sense of justness or righteousness is open for constant reconsideration. And just as motivation and the will to fight on the 'macro' level, in society, has the potential of continuously shifting, so does motivation and will on the 'micro' level, among military units. For military personnel, one's sense of purpose and meaning may shift with changes in circumstances beyond one's control.[124] The will to fight thus ought to be considered uncertain and constantly *in the making*—not simply as in '*in* training' but as fundamentally contingent, always open for change. Just consider Sherman's seemingly obvious comment that one's cause for engaging in fighting is not necessarily what eventually *becomes* the cause,[125] or when Moskos depicts how the enthusiasm

---

[119] Cf. Shils and Janowitz, 'Cohesion and Disintegration in the Wehrmacht in World War II', 284.
[120] Fuller, *Troop Morale and Popular Culture in the British and Dominion Armies 1914–1918*, 21.
[121] Moskos, 'The American Combat Soldier in Vietnam', 26–29.
[122] Friedman, *On Tactics*, 94.
[123] Newsome, 'The Myth of Intrinsic Combat Motivation', 32.
[124] Chodoff, 'Ideology and Primary Groups', 581.
[125] Nancy Sherman, *The Untold War: Inside the Hearts, Minds, and Souls of Our Soldiers* (New York: W.W. Norton & Company, 2010).

and combat motivation among US soldiers varied throughout their rotation in Vietnam.[126] In her analysis of the Russian involvement in the Chechen wars and their lack of perceived legitimacy, Eichler notes that 'societal uncertainty about the true reasons for the war translated into moral uncertainty for the soldiers fighting'.[127] Welland also points to the interlinkages between the construction of legitimacy among soldiers in battle and in society at large; in her discussion on the framing of the war in Afghanistan, she notes how the idea of 'compassion' permeated the public's understanding of why the troops were fighting, but also helped the soldiers themselves to 'make sense of who they thought they were and what they thought they were doing'.[128] To believe that one's actions and sacrifices are promoting the common good is crucial for the motivation to fight, but also that these actions and sacrifices are recognized by the home front, by society at large.[129] It is, after all, essentially society, and not the military, that decides whether a particular war and its practices are to be considered legitimate.[130]

The wars and battlefields of today also differ significantly from those of the early twentieth century. Today, we have got used to military operations with a low level of intensity, which proceed over a long number of years.[131] The Western experience of war in recent decades has also been defined by 'wars of choice' rather than 'wars of necessity', that is: military engagements beyond national defence, which certainly require acts of justification to be considered acceptable. This, one may presume, makes the perceived legitimacy of combat operations more rather than less susceptible to reconsideration, to truth undone, both for home 'audiences' and for soldiers who are sent to battle in the name of the public. As a case in point, one may consider Ritchie's elaboration on *war failure* and the visual representation of war as quagmire, which disturbs and challenges national and governmental narratives.[132] Specifically referring to the messy representation of the US war in Afghanistan in the now infamous 'COIN slide' from an ISAF Joint Command briefing in 2009, she asks: What happens when war 'reaches the home front in the form of spaghetti?'.[133] The PowerPoint slide truly resembles spaghetti; it presents a

---

[126] Moskos, 'The American Combat Soldier in Vietnam', 31.

[127] Eichler, *Militarizing Men*, 114.

[128] Julia Welland, 'Compassionate Soldiering and Comfort', in *Emotions, Politics and War*, edited by Linda Åhäll and Thomas Gregory (London: Routledge, 2015): 124.

[129] Fernando Rodrigues Goulart, 'Combat Motivation', *Military Review* 86, 6 (2006): 96; Henriksen, 'The Character of War and the Nature of Combat', 23–24; King, *The Combat Soldier*, 74.

[130] Goulart, 'Combat Motivation', 96.

[131] Walby, 'Violence and Society', 99.

[132] Marnie Ritchie, 'War Misguidance: Visualizing Quagmire in the US War in Afghanistan', *Media, War & Conflict* (2021): 12.

[133] The slide covered the front page of the *New York Times* on 27 April 2010.

mind map with an infinite number of arrows that chaotically link together various actors and key obstacles to solve the conflict. Essentially, *spaghetti* potentially disturbs society's will to fight, as it reveals war as unmanageable and leaves its spectators with an impression of infinite and endless chaos, of non-progress, unfeasibility, and essentially illegitimacy. *Spaghetti* as a public perception, as well as a lived experience on the ground, also reasonably impacts the fighting will of troops who themselves are caught up in the ongoing mess of war. Delving into the case of the war in Afghanistan, it becomes clear that the home/front relationship has a bearing on the question of why soldiers fight and the moral component of fighting more generally, and that the contingency of perceived legitimacy may have practical implications for military operations. It is now common knowledge how the 20-year long Western military intervention in Afghanistan eventually turned out. As Stavridis puts it: by 2021, 'political patience' in the USA vis-à-vis the war 'had expired'.[134] In light of the failure of the military withdrawal, Western officers and soldiers now seem to find themselves struggling to make sense of the last two decades' military undertakings and sacrifices. The same kind of questions trouble Western publics as a whole. It is reasonable to assume that these post-war reflections, and the media representations of the immensely disruptive moment of the military withdrawal from Afghanistan, will have an impact on the moral component of fighting in the years to come. To move beyond the issue of military, social, and unit cohesion and reflect on the role of society vis-à-vis the moral component of fighting is arguably, therefore, critical not only for military scholars, but also for military practitioners themselves.

Conclusion

Combat is typically portrayed as an individual and fragmented experience, and as an experience that for each soldier, or warrior, is unique. Thus, it is vital to avoid simplistic understandings of what motivates soldiers to fight and carry on. As Lynn has phrased it: 'Soldiers bring different motivations, attitudes, and values to the field, just as they bear different arms and serve different masters'.[135] Soldiers engage in fighting, or abstain from fighting, for different reasons, and, one may presume, for various reasons all at once.

---

[134] James Stavridis, 'I Was Deeply Involved in War in Afghanistan for More Than a Decade. Here's What We Must Learn', TIME Magazine, 16 August 2021, https://time.com/6090623/afghanistan-us-military-lessons/ [accessed 07 September 2021].

[135] Lynn, *Battle: A History of Combat and Culture*, xvi.

And these motivations are always in the making, susceptible to change. This chapter has critically engaged with the literature on combat motivation, morale and cohesion, and has called for a reconsideration of the role of society in regard to the moral component of fighting and the question of why soldiers fight. It has uncovered how the role of society has largely slipped from view in historical and contemporary writings on why soldiers fight. The impact of 'macro-level' forces has generally been discarded as an 'abstraction' whereas the so-called 'micro-level' has been brought to the fore. Accordingly, debates have tended to predominantly focus on specific combat situations rather than considering the war and battlefield experience as a whole. Debates have also reproduced a notion of the military unit as an isolated entity, separated from society at large—a notion which does not seem to correspond to the circumstances of modern day battlefields.[136] Taken together, the question of why soldiers fight has generally been reduced to a matter of local circumstances. As a consequence, the broader socio-political context and discourses that soldiers inevitably are embedded in have largely, or often, been obscured. The chapter therefore argues that we should pay further attention to the role of society. Yet, the field would be well served by moving away from concepts such as ideology or nationalism, which have dominated the discussions on 'macro-level' forces, as these indicate something seemingly fixed and static. Of importance for the field of war studies is rather the *contingent* character of society's and soldiers' construction of meaning and sense of legitimacy, and how these (shifting) notions interlink and interact.

Moving forward, the field would benefit from studies bringing to light and further exploring—theoretically as well as empirically—how societal discourse and ever-shifting notions of legitimacy and righteousness influence and, in turn, are influenced by experiences on the ground. How and when do motivations and convictions shift and change? And what are the implications for military operations at large? First of all, we need more empirical work accounting for contemporary dynamics of war and battle, which might shed new light on the conclusions drawn from studies of historical cases such as the Second World War and the Vietnam War. Furthermore, if taking the association between battlefront and home front seriously, if inspired to take note of the interconnections between combat and society, we need to acknowledge that it matters how war and violence come into view. The portrayals of war and violence essentially shape our perception and truths about violent engagements and our sense of reality itself, our notions of legitimacy

---

[136] Cf. e.g. Kellett, 'Combat Motivation', 320, on the individual soldier's relative isolation on the battlefield during the Second World War.

and our sense of righteousness, of what is feasible, reasonable, and meaningful. And these notions of the home front impact and concern soldiers on the ground.[137] The representation of war has broader implications and a constitutive function; how war activities 'come home', as it were, thus essentially has a bearing also on the question of what enhances, maintains, or disrupts soldiers' willingness to engage in combat and war. Essentially, in writing about war, it is critical to take into account the social and political dimension of that which motivates soldiers to fight, and how the will to fight—as social structure at large—is radically uncertain and unstable. To acknowledge and critically explore the will to fight as a nonlinear process, constantly in the making, always already socially and historically situated, would—as here proposed—push the field forward.

---

[137] Already in the early 1980s, Anthony Kellett ('Combat Motivation', 328) noted that modern communications technologies meant that 'home front morale and beliefs have increasingly been transmitted to combat soldiers' and impacted their morale.

11

Military Health Services Supporting the Land Component in the Twenty-first Century

Martin C. M. Bricknell

> To sum up, the doctors were prepared to lay 15 to 1 that once a man got into their hands, whatever his injury, they would save his life and restore him to health. It's a fine thing that these odds were achieved with a handsome margin.
>
> **Field Marshall Montgomery. Commander 21st Army Group. Despatch. *The London Gazette*. 3 September 1946**

Introduction

The purpose of this chapter is to provide an overview of military health systems for a non-technical audience, explaining how they support the land component (army) of a country's armed forces with particular emphasis on deployed military operations. It will start by describing a 'military health system', its key capabilities and its relationships with both wider military forces and the wider civilian health system. It will then consider the twenty-first-century context covering both the implications of the changing character of the land battlefield and the lessons learned from coalition and NATO military experiences in Iraq and Afghanistan, and other military operations such as UN peacekeeping, and the response to the Ebola outbreak in 2014. A specific section will examine the current COVID pandemic and the role of military health services in the overall response to this crisis. The final section will integrate these observations into a view of the future requirements for health services support to land operations. The conclusion will

Martin C. M. Bricknell, *Military Health Services Supporting the Land Component in the Twenty-first Century*. In: *Advanced Land Warfare*. Edited by Mikael Weissmann and Niklas Nilsson, Oxford University Press. © Martin C. M. Bricknell (2023).
DOI: 10.1093/oso/9780192857422.003.0011

place these requirements into the wider context of the adaptation of armies to the twenty-first century.

A military health service (MHS) is a critical enabler of combat power alongside the other support services of logistics, field engineering, and personnel support. A MHS provides both a 'medically ready force' that is fit to fight and a 'ready medical force' that can both clear the battlefield of casualties and provide them with the best chance of survival and return to duty. The presence of credible and effective health services support (HSS) on the battlefield is an essential element of the moral component of fighting power that both motivates soldiers to fight and maintains the support of their families back home. Whilst often considered alongside logistics, HSS has many fundamental differences. Its role in maximizing the physical and mental fitness of soldiers through preventive medicine (the identification and mitigation of risks to health, including vaccinations), health promotion (supporting soldiers to maintain their health), and garrison healthcare acts as a personnel function. On the battlefield, HSS is a non-combatant function, protected under the Geneva Convention, combining evacuation and treatment to counter the principal challenge of time in the care of casualties. Battlefield medicine is a tactical activity delivered during the battle, contrasting with logistics and engineering support that are delivered before and after the battle.

In most armed forces, the Army (or land component) is the largest service by numbers of personnel and most casualties occur in the land domain (or environment). The medical services that support armies are both integrated within combat units and also function as independent medical units (e.g. field hospitals) that are commanded and operate as discrete entities. They may care for air force and naval personnel operating on land. They may rely on air forces to move casualties in the air (helicopters and aircraft) and may rely on navies to move casualties by sea. Most navies have a medical service to care for their personnel and operate medical facilities within ships and submarines. Except for designated hospital ships, Navy medical facilities are integral to warships and Navy medical services personnel do not command ships. Air forces have medical services to care for their personnel and to support the care of patients moved by air. In many countries, the civilian health services are an integral component of healthcare for armed forces personnel in the home base and a vital source of reserve medical manpower in the event of national mobilization. HSS is among the most inter-operable of military capabilities with many nations willing for their casualties to be treated by the military health services of other nations and to allow their military medical personnel to work within integrated medical units.[1] This

---

[1] Robin F. Cordell, 'Multinational Medical Support to Operations: Challenges, Benefits and Recommendations for the Future', *BMJ Military Health* 158, 1 (2012): 22–28.

contrasts with more substantial limitations in interoperability for combat, communications, engineering, or logistic support functions.

Military medicine has a long and proud history of innovation and technical advancement both in clinical practice and organizational design.[2] Armies have always been supported by a medical service, although it was not until the beginning of the twentieth century that combat wounds overtook disease as the principal cause of death in military service. Military doctors have often been at the forefront of developments in medical practice in both surgery and preventive medicine. Examples include Ambrose Pare (French), John Hunter (British), Sir Thomas Longmore (British), William Leishman (British), and Walter Reed (American). Wars have also led to advances in medical organization as championed by Baron Dominique Jean Larrey (French), George James Guthrie (British), and Jonathan Letterman (American). The organization of military medical services has not always been successful and the role of Florence Nightingale in transforming the British Army Medical Services and wider nursing is well known.[3] The substantial developments in military medicine during the First World War and the Second World War have been recorded in the official histories of these wars by many nations such as the USA, UK, Canada, Australia, and India. More recently, the campaigns in Iraq and Afghanistan have led to new advances in medical sciences across the care pathway for military casualties from battlefield first aid, resuscitation, surgery and intensive care, through to rehabilitation and recovery. Military medicine continues to influence civilian medical practice and there are many examples of clinical lessons from the care of battlefield casualties being adopted in civilian medicine.[4] More recently, military health services have been an integral component of the response to the COVID-19 pandemic both in ensuring the maintenance of military capability and in supporting the overall response to the crisis.[5] This crisis has also revealed the vulnerability of countries' health systems as a risk to overall national security.

This chapter will use NATO terminology, and particularly use Allied Joint Doctrine for Medical Support AJP 4.10(C)[6] as the capstone reference for the fundamental principles and agreed standards by which NATO nations

[2] Charles Van Way, 'War and Trauma: A History of Military Medicine', *Mo Med* 113, 4 (2016): 260–263; Van Way, 'War and Trauma', 336–340.
[3] I. Bernard Cohen, 'Florence Nightingale', Scientific American 250, 3 (1984): 128–137.
[4] Tom Woolley, Jonathan A. Round, Marylou Ingram, 'Global Lessons: Developing Military Trauma Care and Lessons for Civilian Practice', *British Journal of Anaesthesia* 119, 1 (2017): 135–142.
[5] Mohamed Gad, Joseph Kazibwe, Emily Quirk, Adrian Gheorghe, Zenobia Homan, and Martin Bricknell, 'Civil–Military Cooperation in the Early Response to the COVID-19 Pandemic in Six European Countries', *BMJ Military Health* 167, 4 (2021): 234–243.
[6] *Allied Joint Doctrine for Medical Support AJP 4.10(C)—UK Version* (Swindon: MoD, Defence Concepts and Doctrine Centre, 2019).

use their military health services to support national and multi-national operations. Specific definitions are shown in italics. This document has also been used by the United Nations Division of Healthcare Management and Occupational Safety and Health as a framework for the medical initiatives being undertaken under the Action for Peacekeeping initiative to improve the safety and security of peacekeepers.[7]

What is a Military Health System?

The term 'health services support (HSS)' is used to describe *all services provided directly or indirectly that contribute to the health and well-being of patients or a population.* The term 'military health system' is used to describe the whole organization that delivers HSS through the provision of 'military healthcare' that consists of the *measures and activities to sustain or restore the health and the fighting strength of all military personnel from enlistment to retirement through the full spectrum of military duties in garrison and on deployment.* The design for the operational structure of an MHS is based on the flow of a casualty through the NATO Roles of Medical Care from Point of Injury (PoI) to rehabilitation. The capabilities of an MHS are covered by the functions listed in the ten Instruments of Military Medical Care.[8] These are summarized at Figure 11.1.

Figure 11.1 shows the care pathway for a casualty from point of injury (POI) through Roles 1 to 4. 'Role 1' encompasses a set of primary health care (PHC) capabilities which includes but is not limited to triage, pre-hospital emergency care, and essential diagnostics. This also covers the role of Pre-Hospital Emergency Care (PHEC) during 'care under fire' and 'tactical field care' and the contribution of pre-hospital treatment teams (doctors, nurses, and paramedics) to deliver resuscitation under 'enhanced field care'. Medical treatment facilities (MTFs) will function as 'casualty collection points' and 'casualty clearing stations' prior to the evacuation of casualties to a Deployed Hospital Care (DHC) capability. PHC also includes general medical care, dentistry, mental health, and rehabilitation. DHC, or field hospitals, comprises Role 2 units that enhance the resuscitative spectrum of the Role 1 by treatment capabilities essential to preserve life, limb, and function and stabilize the patients' condition for further transport and treatment—this

---

[7] *Declaration of Shared Commitments on UN Peacekeeping Operations* (New York: United Nations, 2018).

[8] Michael Connolly, Martin Bricknell, and Timothy Hodgetts, 'United Kingdom Military Health Service Support to Operations', *International Review of the Armed Forces Medical Services* 88, 2 (2015): 5–14.

- Instruments of Care
  - Medical Command, Control, Communications, Computers (MedC4I)
  - Force Health Protection (FHP)
  - Pre-hospital Emergecy Care (PHEC)
    - Care under Fire (CUF)
    - Tactical Field Care
    - Enhanced Field Care
  - Medical Evacuation (MEDEVAC)
    - Forward MEDEVAC – forward evacuation
    - TACEVAC – tactical evacuation
    - STRATEVAC – strategic evacuation
  - Primary Health Care (PHC)
  - Deployed Hosital Care (DHC)
    - Damage Control Surgery (DCS)
  - Role 4/Fim Base
  - Medical Logistics
  - Medical Contribution to Security and Stabilisation
  - Research and Innovation

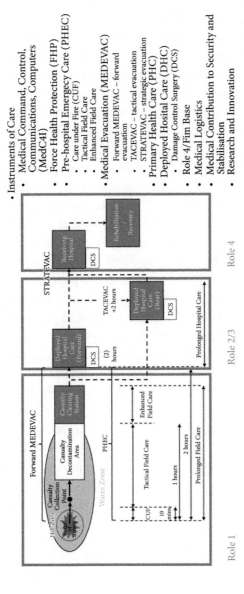

Fig. 11.1 Overall capabilities of a military health system

normally includes surgery and intensive care supported by imaging, laboratory diagnostics, and blood. Role 2 units can be further divided into 'forward', 'basic', and 'enhanced' depending on size, mobility, and sophistication of their support services. Role 3 units comprise a set of deployable specialist and hospital care capabilities which at least includes computed tomography (CT) and oxygen production in addition to all the R2 capabilities. R3 capabilities may reduce the need for the repatriation of patients and enable a higher standard of care prior to strategic evacuation. Finally, Role 4 comprises the full spectrum of military healthcare including highly specialized capabilities (such as reconstructive surgery, prosthetics, and rehabilitation) that cannot be deployed or will be too time consuming to be conducted in theatre. Role 4 medical support is a national responsibility and normally provided by (military or military contracted civilian) hospitals in the casualty's country of origin or at a regional hub (Firm Base). Role 4 is a component of the wider MHS that supports armed forces personnel and other beneficiaries in garrison and base activities. Individual MTFs in the care pathway are linked by the medical evacuation (MEDEVAC) capability that comprises: forward MEDEVAC which is the movement of casualties conducted from the point of injury/insult or a casualty collection point to the initial MTF, tactical MEDEVAC (TACEVAC) which is the movement of patients from one medical treatment facility to another within the area of operations, and strategic MEDEVAC (STRATEVAC) which is the movement of patients from intra-theatre MTFs, to an MTF outside the area of operations (usually Role 4). A large part of the role that the MHS undertakes to support the generation of medically ready forces is contained in the definition of Force Health Protection (FHP), *all medical efforts promote or conserve physical and mental well-being, reduce or eliminate the incidence and impact of disease, injury, and death, and enhance operational readiness and combat effectiveness of the forces.* Medical command, control, communication, computers, and information (MedC4I) provides the authority, processes, communications architecture, and information management resources employed in managing the MHS. The MHS may also provide non-clinical advice and training in humanitarian and disaster relief operations, security sector reform, and global health engagement. This is covered under the term 'medical contribution to security and stabilisation'. Finally, many MHSs have research and innovation capabilities to maintain and develop knowledge in topics such as aviation medicine, environmental medicine, underwater medicine, infectious disease, chemical, biological, nuclear, or radiological (CBRN) medicine, and mental health.

'Medical planning timelines' are overlaid upon the care pathway to provide guidance on the location of MTFs by time in MEDEVAC. These are most fully articulated in the UK version of the NATO timelines as '10-1-2(2)+2' also shown at Figure 11.1. This is ten minutes to first aid, 1 hour to advanced resuscitation, 2 hours to damage control resuscitation and surgery (which is assumed to take 2 hours) and an additional 2 hours for TACEVAC for further resuscitative, diagnostic, surgical, and specialist care capabilities necessary to stabilize the patient for strategic evacuation. Where these timelines cannot be met, it is anticipated that medical personnel may have to use techniques in 'prolonged field care' to minimize further deterioration of the casualty in PHEC and 'prolonged hospital care' in DHC. In spite of these measures the casualty fatality rate might be higher than expected from previous military operations if medical planning timelines are not met.

In addition to the operational role of an MHS, there might also be a substantial component that delivers healthcare in garrison clinics and military hospitals to members of the armed forces and wider beneficiaries such as families and veterans. This might represent the largest source of expenditure for the MHS. As well as general medical care, armed forces personnel have a requirement for occupationally-focused health services such as medical fitness evaluation, rehabilitation, dentistry, and mental health support. This component of an MHS might be delivered through a mix of military personnel, defence civilians, and contractors. Thus, military medical personnel are often committed to both the operational and garrison components of an MHS, which contrasts with combat personnel, logisticians, communicators, and other military functions who might be solely focused on training when not deployed.

Whilst most MHS comprise the generic functions described above, there are often significant differences between countries attributable to the overall size of the armed forces, institutional history, defence policy, and arrangements for public health services. The most significant difference is in the provision of military hospitals with many large armed forces having a complete military health system (including hospitals) to care for substantial numbers of personnel and wider beneficiaries (e.g. United States, China, India, Russia, Jordan). However, there may be no dedicated military hospitals in countries with small militaries and equitable access to health services for all citizens (United Kingdom, Canada, Australia, Norway, Sweden). Other countries, especially those with employer-derived social security systems, might have a small number of military hospitals supporting central garrisons with more isolated detachments receiving support from local civilian services (France, Germany, Belgium, Netherlands, Italy, Spain). These variations influence the arrangements for the reception and care of military casualties in the Role 4 component of the operational care pathway and have implications for the timeliness and recovery arrangements for army personnel medically evacuated from military operations. The total numbers of military medical personnel required to support the army during large-scale combat operations is likely to be far higher than that required to maintain the health of the armed forces in routine duties. It would be inefficient and unaffordable to employ such large numbers of medical personnel in full-time service and so most MHSs have arrangements for volunteers to be held in medical reserve units for mobilization in times of crisis. Thus, there is a synergy between military and civilian health services that is often deeper than other functional military services such as communications and logistics.

Each country will have its own structures for their military medical units and their relationships with the wider support units of logistics and engineering. Most countries have small Role 1 units that are integral to combat units (infantry, armour, and artillery) and might also have these within other large, battalion sized units (communications, logistics, etc.). Larger Role 1 units such as ambulance companies and casualty clearing stations may be held within composite logistic units or may operate as independent medical battalions. DHC is usually organized within field hospitals of around battalion size and held at divisional level, although they might be designed to detach smaller Role 2 units to be assigned to support combat brigades and battalions. There are multiple terms for such small DHC units[9] such as; forward surgical teams (US Army), forward resuscitative surgical system (US Marine Corps), ground manoeuvre surgical groups (UK), Antenne Chirurgicale (French).[10]

The Twenty-first-century Context

Advances in military medicine during US operations in Iraq and Afghanistan from 2001 to 2014 have led to the lowest case fatality rates in the history of warfare.[11] This has been attributed to improved battlefield first aid (primarily the use of tourniquets), better resuscitation (including the use of massive blood transfusions), and more rapid pre-hospital transport by

[9] Yi-Ling Cai, Jin-Tao Ju, Wen-Bao Liu, and Jian Zhang, 'Military Trauma and Surgical Procedures in Conflict Area: A Review for the Utilization of Forward Surgical Team', *Military Medicine* 183, 3–4 (2018): e97–e106.

[10] Ghislain Pauleau, Tristan Monchal, Yvain Goudard, Stéphane Bourgouin, and Paul Balandraud, 'Surgical Facilities on the Field: Update about the French Military Medical Service on Operations', *International Review of Armed Forces Medical Services* 91, 1 (2018): 5–9.

[11] Jeffrey T. Howard, Russ S. Kotwal, Caryn A. Stern, Jud C. Janak, Edward L. Mazuchowski, Frank K. Butler, Zsolt T. Stockinger, Barbara R. Holcomb, Raquel C. Bono, and David J. Smith, 'Use of Combat Casualty Care Data to Assess the US Military Trauma System during the Afghanistan and Iraq Conflicts, 2001–2017', *JAMA Surgery* 154, 7 (2019): 600–608.

MEDEVAC helicopters. There was further transformation in the treatment provided to casualties across the remainder of the care pathway including the use of computed tomography (CT) scanning for diagnosis, development of intensive care units for aeromedical evacuation, advances in wound care, and developments of new prosthetics to improve recovery. The full story of these achievements has been captured by many nations,[12] and indeed recorded as 'lessons learned'.[13] The duration of the conflicts allowed organizational and clinical learning through the establishment of clinical registries to record all trauma patients and the translation of innovations in clinical practice and research into new clinical protocols. Whilst an undoubted success, this was achieved using relatively unlimited medical resources (people, money, and equipment), a limited threat from indirect fire, complete control of the air, no substantial attacks on MTFs, and low numbers of casualties.

High risk national and multi-national military operations (NATO, European Union, African Union, United Nations) have continued since 2014 in countries such as Afghanistan, Iraq, Somalia, Mali, Central African Republic, and South Sudan. MHSs have continued to care for trauma casualties, with time, distance, and smaller medical footprints contributing to make health service support very challenging, especially if medical planning timelines are exceeded. The medical support arrangements for United Nations peacekeeping operations have been undergoing a programme of transformation similar to that undertaken by NATO forces in Afghanistan. This has included the development of common processes and procedures,[14] the introduction of the UN Buddy First Aid Course,[15] and a new policy for casualty evacuation in the field.[16] Military medical services have also contributed to international humanitarian and disaster relief operations in response to

[12] Arthur L. Kellermann and Eric Elster, E. eds. *Out of the Crucible: How the US Military Transformed Combat Casualty Care in Iraq and Afghanistan* (Fort Sam Houston, TX: Borden Institute, 2017); Ian Greaves, ed., *Military Medicine in Iraq and Afghanistan: A Comprehensive Review* (Boca Raton: CRC Press, 2018).

[13] Martin Bricknell and Martin Nadin, 'Lessons from the Organisation of the UK Medical Services Deployed in Support of Operation TELIC (Iraq) and Operation HERRICK (Afghanistan)', *BMJ Military Health* 163, 4 (2017): 273–279; Thijs van Dongen, *Military Medical Support Organization: Lessons Learned from the Dutch Deployment in Afghanistan* (Universiteit Utrecht, 2017).

[14] Min Yu, R. Li, L. Qiu, 'Overcoming new challenges in medical support for UN peacekeeping operations', *International Review of Armed Forces Medical Services* 91, 1 (2018): 20–28.

[15] Martin Bricknell, Claire Booker, Adarsh Tiwathia, Jillan Farmer, 'The Development and Introduction of the United Nations Buddy First Aid Course', *International Review of the Armed Forces Medical Services* 93, 2 (2020): 20–24.

[16] *Casualty Evacuation in the Field. United Nations Department of Operational Support* (New York: United Nations, 1 March 2020).

earthquakes (e.g. Nepal and Haiti), typhoons and hurricanes (the Philippines and Caribbean), and the response to the outbreak of Ebola in West Africa. These roles illustrate the potential role for cooperation between civil and military health services during natural disasters and other non-conflict emergencies.[17] MHSs may also be involved in providing medical care to isolated civilian populations as part of 'hearts and minds' projects or may be involved in capacity-building programmes with military or civilian health services of partner nations.[18]

Advances in military technology, as shown in recent conflict (Ukraine, Syria, Yemen, and Nagorno-Karabakh) has shown that the future battlefield may become even more lethal in the land environment, especially in peer-on-peer conflict. The development of unmanned aircraft and improved links between sensor technologies, missiles, and indirect fire has substantially increased the vulnerability of large, immobile military units such as field hospitals. There is also evidence that the neutrality of healthcare facilities under international humanitarian law is not being respected and these are even being directly targeted to undermine morale.[19] Alongside conventional weapons, the threat from CBRN weapons has not receded. Both may result in casualty rates that far exceed recent experience. However, many nations are reducing the size of their MHS by shrinking garrison health facilities and closing military hospitals. This is most apparent in the United States through the shift of responsibility for garrison healthcare from the Army, Navy and Air Force to the Defence Health Agency.[20] The majority of MHS rely upon the medical reserves from the civilian system to augment the active duty medical component. However, there is also a global shortage of health professionals resulting in many MHS having fewer personnel than required. In response to a shortfall in MHS there has been an expansion of commercial health services to manage MTFs and MEDEVAC on security and peace-keeping missions; with contracted medical facilities in Kosovo, Mali, and Somalia.

[17] Adam Kamradt-Scott, Sophie Harman, Clare Wenham, and Frank Smith III, 'Civil–Military Cooperation in Ebola and beyond', *The Lancet* 387 (2016): 104–105.

[18] Roberto N. Nang and Keith Martin, 'Global Health Diplomacy: A New Strategic Defense Pillar', *Military Medicine* 182, 1–2 (2017): 1456–1460.

[19] Preeti Patel, Fawzia Gibson-Fall, Richard Sullivan, and Rachel Irwin, 'Documenting Attacks on Health Workers and Facilities in Armed Conflicts', *Bulletin of the World Health Organization* 95, 1 (2017): 79–81; Mohammed H. Afzal and Anisa J. N. Jafar, 'A Scoping Review of the Wider and Long-term Impacts of Attacks on Healthcare in conflict zones', *Medicine, Conflict and Survival* 35, 1 (2019): 43–64.

[20] Terry Adirim, 'A Military Health System for the Twenty-First Century', *Health Affairs* 38, 8 (2019): 1268–1273.

COVID-19—A Game Changer?

The impact of the COVID pandemic on military health services merits discrete analysis. The priority of MHSs has been to protect the health of their beneficiaries by communicating health advice, segregating military personnel into isolation cohorts, introducing pre-deployment and post-deployment quarantine, and adapting arrangements for the provision of healthcare including remote teleconsultation and treating COVID patients. Despite these measures, there have been significant outbreaks of COVID in military units, especially ships. All armed forces have tried to maintain military outputs despite the constraints of COVID protection measures so as to prevent the health crisis becoming a security crisis. Most nations' armed forces have also been heavily committed to supporting the civilian response to the crisis with the MHS providing augmentation to most components of the health economy. Military medical personnel have provided COVID testing centres, case-tracing, care to civilians in military hospitals, military augmentation to the civilian health services (ambulances, hospital, care homes), and vaccination centres. Military biomedical manufacturing and research has also supported civilian industries.[21] This crisis has shown the vulnerability of most nations' health systems to pandemic threats and thus the MHS could be considered as a national strategic reserve. This crisis is also a reminder of the challenges of caring for CBRN casualties, particular those from biological agents.

The experience of the COVID response and lessons for military medicine are slowly emerging in the academic literature. The second- and third-order impact of the COVID crisis may increase the risk of conflict through exacerbating existing fault lines or creating new sources of tension.[22] In addition to the impact on health services, it has exposed vulnerabilities in strategic communications, supply chains, cyber protection, and societal consent to government. Countries will have to balance the costs of enhancing security and defence capabilities with the need to invest in economic recovery and protection from the continuing health and social costs of the pandemic. All these factors will determine how much can continue to be spent on MHSs

[21] Gad et. al., 'Civil–Military Cooperation'; Jori Kalkman, 'Military Crisis Responses to COVID-19', *Journal of Contingencies and Crisis Management* 29, 1 (2021): 99–103; Fawzia Gibson-Fall, 'Military Responses to COVID-19, Emerging Trends in Global Civil–Military Engagements', *Review of International Studies* 47, 2 (2021): 155–170.

[22] Christoph O. Meyer, Martin Bricknell, and Ramon Pacheco Pardo, *How the COVID-19 Crisis has Affected Security and Defence-related Aspects of the EU: Part II—In Depth Analysis* (Brussels: European Parliament, 2021).

within Defence, alongside how much the MHS remains a national strategic reserve to support the response to any future health crisis.

The Future of Health Services Support to the Land Component in Military Operations

Several authors have examined the implications of peer-on-peer conflict, or 'large-scale contingent operations' (LSCO) on HSS in the land environment.[23] In spite of technological developments in the other environments, it remains likely that the outcome of war will continue to be decided in the land environment. Armies will continue to require the ability to fight and win against their country's enemies. Conflict by its very nature causes casualties to one's own forces, the enemy, and non-combatants. Armies will continue to need to have HSS to care for casualties from combat, disease, and non-battle injury. The increased volume, precision, and reach of indirect fire weapons is likely to result in casualty rates that approach those of the Second World War; more likely replicating the scale and intensity of the German/Russian front rather than the Western European front. This threat will be compounded by an increasing lack of respect for International Humanitarian Law and the protections for medical units under the Geneva Conventions. Both the Healthcare in Danger project of the International Committee of the Red Cross[24] and the World Health Organisation's Attacks on Healthcare Initiative[25] were established as a result of evidence that health facilities were being directly targeted by government security forces and non-state armed groups. Military medical personnel will need to be competent and equipped to care for casualties from all forms of weapons (including potential weapons using new technologies such as lasers and bio-engineered biological agents). Land-based MHSs will need to align with the military plan, to conduct MEDEVAC whilst at risk of being targeted by the enemy, and to locate MTFs according to medical planning timelines even if they are at risk from enemy

---

[23] Martin Bricknell, Antony Finn, and Joanne Palmer, 'For debate: Health Service Support Planning for Large-scale Defensive Land Operations (Part 1)', *BMJ Military Health* 165, 3 (2019): 173–175; Martin Bricknell, Antony Finn, and Joanne Palmer, 'For Debate: Health Service Support Planning for Large-scale Defensive Land Operations (Part 2)', BMJ Military Health 165, 3 (2019): 176–179; Brent Thomas, *Preparing for the Future of Combat Casualty Care* (Santa Monica, CA: RAND, 2021); Matthew Fandre, 'Medical Changes Needed for Large-Scale Combat Operations: Observations from Mission Command Training Program Warfighter Exercises', Military Review, May–June 2020: 37–45.

[24] 'HCID Initiative', Healthcare in Danger Project, International Committee of the Red Cross, accessed 10 November 2021, https://healthcareindanger.org/hcid-project/.

[25] 'Stopping attacks on healthcare', Attacks on Healthcare Initiative, World Health Organisation, accessed 10 November 2021, https://www.who.int/activities/stopping-attacks-on-health-care.

fire. There will also be a threat to medical communications and equipment from electronic and cyber warfare.

Military medical planners will need to think more strategically and be able to adapt to greater uncertainties than in the recent past. HSS to armies will continue to be a 'joint' endeavour in the treatment and evacuation of casualties to a safe location in their home nations, extending into the national civilian health system to provide long-term care including after completion of military service as a veteran. Like many other land capabilities, the Army MHS will need to be interoperable with their sister medical services in the maritime and air environments so that patients can be treated and moved between environments. Given the likely nature of multi-national and coalition operations, this interoperability will need to extend to partners and allies. The frameworks for medical standardization within UN, NATO, EU, ABCA (Australia, UK, Canada, USA, and New Zealand) will need to widen and deepen to maximize the efficiency and effectiveness of multi-national cooperation in the care of casualties. MHS personnel will also need to conduct medical planning with local security partners, civilian authorities, and non-government organizations in order to meet their duties under humanitarian law.[26] Many of the coordination arrangements that were established to support the response to the COVID crisis, especially mutual support in the provision of medical material and the movement of patients, are likely to be relevant in the event of significant casualties from conflict. This includes the maintenance of the single 'European' military medical headquarters, the Multinational Medical Coordination Centre/European Medical Command (MMCC/EMC) that is designed to operate at the operational level to support both NATO and EU missions.[27] This is complemented by the NATO Military Medical Centre of Excellence which supports the transformation of military medical capability across NATO and partner nations.[28]

Whilst there may be a public expectation that the survival rates for military casualties will be matched in the next conflict, the future character of land warfare may render this impossible. The creation of the concepts of prolonged field care and prolonged hospital care (see Figure 11.1) reflect the likelihood of delays in the evacuation of casualties against the medical planning

[26] *Protecting Health Care: Guidance for Armed Forces* (Geneva: International Committee of the Red Cross, 2020).

[27] Ronnie Michel, 'Challenges for Medical Support in National and Collective Defence', Worldwide Military-Medicine.com, 30 July 2020, https://military-medicine.com/article/4131-challenges-for-medical-support-in-national-collective-defence.html.

[28] Tomáš Vašek, Jaroslav Žďára, Petr Král, Milan Růžička, Michal Potáč, Petr Smola, and John Quinn, 'Evidence Based Medicine: Lessons Learned from the NATO Military Medical Center of Excellence', *Romanian Journal of Military Medicine* 123, 3 (2020): 153.

timelines because of dispersed forces or enemy action. Although originating from the Special Operations environment, they also apply across the land environment. These concepts identify the need to mitigate the impact of this delay on the probability of survival for casualties by the development of clinical capabilities that can be used further forward in the care pathway.[29] This will involve the administration of blood closer to the point of injury, development of non-surgical methods to reduce bleeding from chest and abdominal wounds, and diagnostic tools supported by communications that allow senior clinicians to support junior personnel. The larger number of casualties will make difficult triage decisions more likely as there will be fewer medical personnel and less equipment available to treat each one. Medical personnel will have to accept that during mass casualty events (MASCAL) patients who might have been saved with the resources of previous campaigns may die, or may not even have life-saving treatment started. This will be psychologically challenging for healthcare professionals and will pose policy questions regarding triage and wider aspects of medical ethics that will be similar to those faced during the COVID pandemic.[30]

Concurrently new technologies may change aspects of field medical services. Advances in automated clinical decision-making through big data and artificial intelligence, supported by robust military communication systems may enable healthcare professionals to practise at a higher skill level than their current qualification by extending the reach of senior medical advisers. These communication systems could also support the movement of medical data across the care pathway separate from the patient so that receiving MTFs are informed of the medical condition of patients before they arrive. This same data, with the clinical information removed, could be used to manage the regulation of casualties by providing a common operating picture of their location across the care pathway to inform medical commanders in medical units and in medical staff branches in military headquarters. These same advances will support the provision of remote and distributed medical education programmes including the use of physical and virtual simulation of patients so that military healthcare personnel can maintain and develop their professional knowledge wherever they are deployed. This may offset concerns about skill fade for medical personnel who might be deployed for long periods to support military operations with

[29] Sean Keenan and Jamie C. Riesberg, 'Prolonged Field Care: Beyond the "golden hour"', *Wilderness & Environmental Medicine* 28, 2 (2017): S135–S139; Mike Smith and Richard Withnall, 'Developing Prolonged Field Care for Contingency Operations', *Trauma* 20, 2 (2018): 108–112.

[30] Christoph Jänig, Jennifer M. Gurney, Roger Froklage, Robin Groth, Christine Wirth, Hendrik van de Krol, Willi Schmidbauer, and Christoph Güsgen, 'Facing COVID-19: Early Recognition and Triage Tool for Medical Treatment Facilities with Limited Resources', *Military Medicine* 186, 1–2 (2021): e44–e51.

low casualty rates. Most of these applications of information technology to clinical practice and healthcare management are already in use in civilian health services and can be adapted for use in the military environment once issues such as cyber protection, information security, medical confidentiality, and common data architectures between national systems have been addressed.

Many of the opportunities for change in military supply chains may also apply to HSS including the use of drones for resupply of medical materiel, especially low density, high value commodities such as blood. It is even possible that unmanned aerial and ground vehicles could be used for MEDE-VAC of stabilized casualties through the care pathway. Additive manufacture could be used to provide spare parts for medical equipment, and distributed communications could be used to provide diagnostic and repair services for electrical medical equipment directly by manufacturers.

Advances in military technologies and biotechnology may also change the nature of the people who serve in the armed forces. Physical capability may not be a barrier to military service as the human control of weapons shifts away from the battlefield. Software engineers and drone operators may be able to undertake their duties in a sedentary position within a warm environment without the physical fitness requirements for ground combat. These possibilities may be extended through the development of technologies that enhance human performance. Human power may be augmented by exoskeletons or other external machines. Psychoactive drugs may reduce the need for sleep or improve concentration. Brain–nerve–machine interfaces may enable humans to be directly connected to machines and to speed up the reaction to external events or to control weapons in completely new ways. Military medical personnel will need to consider the ethics of research in these fields, the implications of the adoption of these technologies in the military environment, and the management of any physical and psychological harm that might result from their use.

At the tactical level, the land battlefield will be much more challenging for military medical services. Medical units will need to be dispersed and camouflaged (across all forms of the electromagnetic spectrum) to avoid being targeted. It will be necessary to have dedicated 'reserve' medical capacity both to be deployed to reinforce HSS in areas of high casualties but also to replace medical units that have been damaged or destroyed by enemy action. Whilst much of the MHS will be a reserve capability, it must be held at the same readiness as the remainder of the Army in order to be mobilized, equipped, and trained at the same pace as the forces that they support. This will require a significant holding of medical stockpiles.

The preceding paragraphs have considered the implications of HSS to LSCO, which is akin to the scale and complexity that NATO had anticipated during the Cold War. At the same time, armed forces are expanding their ambition for global deployments in support of peacekeeping, military diplomacy, and training to counter threats in the Asia-Pacific region. All of these activities are likely to require HSS support and a higher level of persistent commitment of medical units and personnel than the last decade after the ending of combat operations in Afghanistan.

Conclusions

In conclusion, the addition of this chapter within this book ensures that military health services are not overlooked as a military capability alongside all other military capabilities that combine to deliver land power. Changes in the character of combat in the land environment will impact military health services by making the likelihood of casualties greater alongside an increase in the threat to land medical units. MHS make a significant contribution to the military instrument of power, although increasingly more joint and defence rather than 'army'. MHS organizations and people will endeavour to maintain the substantial advances in healthcare that have improved the survival of military casualties during first two decades of the twenty-first century by exploiting emerging and developing technologies particularly in IT. However, this may be more difficult in a less permissive land environment. The COVID-19 crisis has also shown the importance of the armed forces (especially the mass provided by armies) as a source of assistance to the civilian response to non-conflict emergencies, with the MHS acting as a strategic, flexible medical reserve. Thus, whatever the economic pressures on defence budgets, MHSs will still be required to support armies, both to 'generate the medically ready force' and to generate the 'ready medical force'.

PART II
CASE STUDIES

12

The Operational Cultures of American Ground Forces

Bruce I. Gudmundsson

Over the course of the past century, the operations carried out by the ground forces of the USA have been shaped by the interplay of two very different cultures.[1] In some instances, one or the other of these two operational cultures has played the dominant role in the conception, coordination, and conduct of martial undertakings.[2] At other times, the two cultures combine to create chimeras of various kinds, enterprises in which the actions of some participants accord with one of these two cultures whilst the deeds of others reflect the prejudices, practices, predispositions, or precepts of the other.

The two competing operational cultures of American ground forces share a common origin in the work of Eben Swift. Born in 1854 at Fort Chadburne, Texas, where his father was serving as a military surgeon, Swift graduated from the United States Military Academy at West Point in 1876.[3] In the two decades of military service that followed his graduation from West Point, much of which was spent in frontier forts of the type so often seen in popular depictions of the 'Wild West', Swift devoted his leisure hours to the study of European military literature.[4] In particular, he spent a great deal of time with

---

[1] As used in this chapter, the term 'operations' refers to the things that a military organization does with respect to an enemy rather than the employment of formations for strategic purposes at the 'operational level of war'.

[2] The concept of 'operational culture' at the heart of this chapter should not be confused with the very different concept of the same name featured in Paula Holmes-Eber and Barak A. Salmoni, *Operational Culture for the Warfighter* (Quantico: Marine Corps University Press, 2011) or Paula Holmes-Eber, Patrice M. Scanlon, and Andrea L. Hamlen in *Applications in Operational Culture* (Quantico: Marine Corps University Press, 2009).

[3] The most complete biography in print of Eben Swift can be found in the pages of the old alumni magazine of United States Military Academy: 'Eben Swift and the Five-Paragraph Order', *Assembly* 38, 1 (1979), 9, 24, 111. For an overview of the role played by Swift at the schools at Fort Leavenworth, see Timothy Nenninger, *The Leavenworth Schools and the Old Army* (Westport: Greenwood Press, 1978), 43–48, 73.

[4] Swift was able to read both French, which had been a required subject at West Point, and Spanish, which served as a *lingua franca* in many parts of the American frontier in the late nineteenth century. For evidence of his command of these languages, see his review of three French books (only one of which had been translated into English) in *The North Carolina Historical Review* 2, 2 (April 1925): 255–259, and

Bruce I. Gudmundsson, *The Operational Cultures of American Ground Forces*. In: *Advanced Land Warfare*.
Edited by Mikael Weissmann and Niklas Nilsson, Oxford University Press. © Bruce I. Gudmundsson (2023).
DOI: 10.1093/oso/9780192857422.003.0012

the works of Julius von Verdy du Vernois, an officer of the German Army who, in the course of writing several dozen volumes, advocated an approach to the study of the military art that he called the 'applicatory method'.[5]

In the middle years of the 1890s, whilst he was teaching at a school for junior officers at Fort Leavenworth, Kansas, Swift composed a progressive programme of professional education for the officers of the garrison of an imaginary post located on the 'borderland' of the 'most distant possession' of the USA. Published as 'The Lyceum at Fort Agawam', this ideal curriculum made exclusive use of the aforementioned 'applicatory method'. Specially, it employed one-sided map problems, both fictional and historical; two-sided map manoeuvres; and outdoor excursions during which students composed orders, sometimes written and sometimes verbal, for units imagined to be engaged in warlike activity on the countryside in question.[6]

The programme of professional education laid out in 'The Lyceum at Fort Agawam' began with a series of eight single-step map problems in which students wrote out orders for imaginary units presumed to be facing specific problems upon real pieces of ground. (Swift emphasized the commission of such orders to paper by referring to these activities as 'written exercises'.) The next eight classroom exercises in the line-up were two-sided contests in which the imaginary forces in question were depicted upon a map (or maps) by blocks cut to scale and the results of engagements were adjudicated by an umpire.[7] (Swift referred to these by two names, sometimes calling them 'map maneuvers' and sometimes using the German term '*Kriegsspiel*'.[8]) The last eight of the indoor exercises in Swift's curriculum bore some resemblance to the one-sided map problems engaged in the first part of the course. However, rather than being works of fiction designed to draw attention to

the bibliography to his lecture on 'The Military Geography of Chili', in Arthur L. Wagner et al., *Military Geography* (Fort Leavenworth: United States Infantry and Cavalry School, 1895), 66.

[5] For a brief biography of Julius von Verdy du Vernois (1832–1910), see the lengthy obituary serialized in the *Militärwochenblatt* (Numbers 130 through 134) in 1910. For an attempt to trace the deeper roots of the applicatory method, see Bruce I. Gudmundsson, 'The Education of the Enlightened Soldier', *MCU Journal* 9, 1 (Spring 2018): 33–44.

[6] Eben Swift, 'The Lyceum at Fort Agawam', *Journal of the Military Service Institution of the United States* XX, LXXXVI (March 1897): 233–277.

[7] The way Swift imagined the conduct of 'map maneuvers' in 'The Lyceum at Fort Agawam' was in keeping with his adaptation of a French translation of a book by Verdy du Vernois on the subject of wargaming. For Swift's adaptation, see Julius von Verdy du Vernois (translated by Eben Swift), *A Simplified War Game* (Kansas City: Hudson-Kimberly, 1897). For the French translation from which Swift worked, see Julius von Verdy du Vernois (translated by Matthieu Morhange), *Essai de Simplification du Jeu de Guerre* (Brussels: C. Muquardt, 1877). For the German original, see Julius von Verdy du Vernois, *Beitrag zum Kriegsspiel* (Berlin: E.S. Mittler, 1876). For the context of the innovations in wargaming introduced by Verdy du Vernois, see Werner Knoll, 'Die Entwicklung des Kriegsspiels in Deutschland bis 1945', *Militärgeschichte* XX (1981): 180–182.

[8] Swift employed the singular form of the word '*Kriegspiel*' to designate both single war games and multiple exercises of that sort. This may stem from a lack of familiarity with the plural form of the original German word ('*Kriegsspiele*'). Alternatively, this practice may reflect a desire to coin an abstract expression comparable to 'the study of military history' or 'the sham battle'.

commonplace conundrums, these one-sided map problems, which together made up what Swift called the 'study of military history', were drawn from the annals of the march on Atlanta during the last year of the American Civil War.[9]

The last phase of Swift's 'Lyceum' consisted of what he called 'war rides'. The first of these resembled a grown-up version of the children's game of 'hide-and-seek', with four teams of horsemen tracking a group attempting to evade detection.[10] Subsequent 'war rides' bore a closer resemblance to the written exercises worked out at the start of the programme, but with real ground taking the place of paper maps. As was the case with the written exercises, these latter 'war rides' combined an open-ended search for custom-tailored solutions, what might be called the military analogue of academic freedom, with an insistence that students use a rigid format for the composition of orders. (This 'invariable model', as Swift called it, limited each order to five obligatory paragraphs. The first of these described the general situation and, in particular, the activity of the enemy. The second paragraph contained a succinct statement of the mission of the unit in question. The third paragraph promulgated a plan for fulfilling that mission. The last two paragraphs dealt, respectively, with arrangements for logistics and the transmission of information.)[11]

An empathetic reading of 'The Lyceum at Fort Agawam' reveals the work of a thoughtful mind attempting to simultaneously promote both predictability in small things and liberty of action in larger matters. Indeed, a particularly perceptive reader might even conclude that the system set down by Swift used his 'invariable model' to provide a familiar framework that reduced, for both leaders and the led, the psychological price exacted by encounters with necessarily novel notions. Such subtlety, however, proved hard to transmit from one mind to another. Thus, whilst some officers embraced the custom-tailoring of solutions at the heart of Swift's system, others found comfort in the predictability of his five-paragraph format.

In the years that followed the publication of his ideal curriculum, Swift himself seems to have become fonder of the formulaic aspects of his ideal curriculum and, at the same time, less enthusiastic about opportunities it offered

---

[9] The method Swift called the 'study of military history' corresponds closely to the one described in Julius von Verdy du Vernois, *Kriegsgeschichtliche Studien nach der applikatorische Methode, I Heft, Taktische Details Aus der Schlacht von Custozza* (Berlin: E.S. Mittler, 1876). This work was translated into English by G. F. R. Henderson as *A Tactical Study of the Battle of Custozza* (London: Gale and Polden, 1884) and French by Léonce Grandin, as *Études d'Histoire Militaire d'après la Méthode Appliqué* (Paris: J. Dumaine, 1877).

[10] 'The Lyceum at Fort Agawam', 270–271.

[11] 'The Lyceum at Fort Agawam', 243–244.

for creative problem solving. In 'The Lyceum at Fort Agawam', for example, he had raised the possibility that an officer who had mastered the art of composing orders might dispense with the five-paragraph format. Nine years later, in a pamphlet devoted to the elaboration of his five-paragraph order, he argued that 'it is also found that officers who have once been instructed in this way will, even after long experience, closely follow the accepted model'.[12] During Swift's first tour of duty at Fort Leavenworth (1894–1897), he informed his students that, when solving map problems, 'any idea that is not manifestly wrong will usually be considered right, if it be developed in a logical way'.[13] During his second tour of duty (1904–1906), he devoted a great deal of time and trouble to both the creation of 'approved solutions' and the introduction of various measures (such as preliminary 'recitations') that predisposed students toward them.[14] 'To propose problems to a class of officers without giving information as to the character of errors committed or as to the kind of solution which is considered right, and without having come to a conclusion as to what would be a proper answer, is not a satisfactory method of instruction'.[15]

The ossification of the teaching methods used at Fort Leavenworth took place at a time when Verdy du Vernois, the author who had introduced Swift to the applicatory method had come to reject even the modest relics of formal frameworks that could be found in his earlier writings.[16] This change reflected a growing tendency within the German Army of the last decade of the nineteenth century to condemn 'schemes', 'patent solutions', and any other practices that served to limit the freedom of officers to address the peculiarities of the situations that they encountered.[17] Paradoxically, as American officers following the trail blazed by Swift translated newer German works about the applicatory method, they made arguments in favour of this philosophy available to their colleagues.[18] In the case of a collection of map problems

---

[12] Eben Swift, *Field Orders, Messages, and Reports* (Washington, DC: War Department, 1906), 15.

[13] For a detailed description of how these methods were used, see Arthur L. Wagner, 'Department of Military Art', *Appendix B to H. S. Hawkins, Annual Report, US Infantry and Cavalry School*, 1 August 1896, 1–23. For the quotation, see page 19 of the same document.

[14] Used extensively at West Point as well as in many civilian schools of the day, a 'recitation' was a short speech, made without notes, in which a student provided a précis of a reading assignment.

[15] Eben Swift, 'Department of Military Art, Infantry and Cavalry School', Appendix B to J. F. Bell, *Annual Report of the Commandant of the Infantry and Cavalry School and Staff College, for the School Year Ending*, 31 August 1905, 6–7.

[16] For examples of this trend away from fixed formats, compare the editions of Julius von Verdy du Vernois, *Studien über Felddienst* published in 1887 and 1895 with those published in 1900 and 1908.

[17] For more on the movement away from forms and formats within the German Army in the late nineteenth and early twentieth centuries, see, among many others, Dirk W. Oetting, *Auftragstaktik: Geschichte und Gegenwart einer Führungs–konception* (Frankfurt am Main: Report Verlag, 1993).

[18] During his second tour of duty at Fort Leavenworth, Swift seems to have been more interested in the earlier works of Verdy du Vernois than the more recent products of that author's pen. Thus, when he found

translated for use at Fort Leavenworth, all illustrative directives were cast in the mould of the five-paragraph order.[19] Nonetheless, the translation retained a passage that reminded readers that the form of orders was secondary to their essence and that the sample solutions provided were 'aids to the memory, nothing more'.[20] In the instance of an official booklet issued to students at Fort Leavenworth, more than a third of the text consisted of lengthy quotations from the works of German officers who had advocated the custom tailoring of solutions to tactical problems. Indeed, of the fifty-one paragraphs borrowed from other publications, only one, which had been provided by an officer of the US Navy, had been written by someone other than a contemporary German foe of form and format.[21]

The new German military literature found some friends at Fort Leavenworth, the most senior whom was John F. Morrison. Fresh from observing the battles of the Russo-Japanese War (1904–1905), Morrison served for six continuous years (1906–1912) at the Fort Leavenworth schools, first as an instructor and then as administrator. Thanks to the habit, recommended by Verdy du Vernois, of working through map exercises of his own design, Morrison arrived at Fort Leavenworth with an uncommonly open-minded attitude toward the applicatory method. The chief task of the student engaging in an applicatory exercise, he believed, began with the discovery of the essence of the problem at hand. Once the student figured this out, the resulting solution would be so robust that minor mistakes in the realm of technique would have little effect upon the outcome.[22] Notwithstanding his long tenure at Fort Leavenworth, Morrison managed to convert few of his colleagues to his philosophy. What little progress he may have made, moreover, was quickly undone by the entry of the USA into the First World War.

---

time to adapt a second translation of a book by his favourite German author, Swift chose to Americanize a 1877 French translation of a slim volume that first emerged from the press in 1876. See Julius von Verdy du Vernois (translated by Eben Swift), *A Tactical Ride* (Fort Leavenworth: Staff College Press, 1906); Julius von Verdy du Vernois (translated by F.G.A. Peloux), *Un Voyage-Manoeuvre de Cavalerie* (Paris: Berger-Levrault, 1877); and Julius von Verdy du Vernois, *Beitrag zu den Kavallerie-Übungs-Reisen* (Berlin: E.S. Mittler, 1876).

[19] J. Franklin Bell, *Annual Report of the Commandant of the US Infantry and Cavalry School, US Signal School, and Staff College, for the School Year Ending, 31 August 1906*, 17.

[20] Otto Griepenkerl (translated by C. H. Barth), *Letters on Applied Tactics* (Kansas City: Franklin Hudson, 1908), 5.

[21] Harold B. Fiske, *Some Notes on the Solution of Tactical Problems* (Fort Leavenworth: Press of the Army Service Schools, 1916).

[22] For a lively description of the way that Morrison taught, see George C. Marshall, 'Letter to Colonel Bernhard Lentz, 2 October 1935', reproduced in Larry I. Bland and Sharon Ritenour Stevens, *The Papers of George Catlett Marshall, Volume 1, The Soldierly Spirit, December 1880–June 1939* (Baltimore and London: The Johns Hopkins University Press, 1981), 45–47. For examples of Morrison's problems, introduced with an explanation of his approach to the applicatory method, see John F. Morrison, *Seventy Problems* (Fort Leavenworth: US Cavalry Association, 1914).

The wartime fashion for branding all things German as inherently anti-American provided the partisans of form, format, and formula within the Army with a rhetorical advantage they had previously lacked.[23] At the same time, the most celebrated of America's alliances provided the champions of mechanical methods with a fresh source of inspiration, literature, and relationships. Moreover, whilst the rejection of recently imported German materials lasted until well after the end of the war, the explicit embrace of French manuals, methods, and models continued for more than two decades.[24]

Of the many items that the US Army borrowed from its French counterpart during the First World War, the most influential was the concept of 'doctrine'.[25] Conspicuously absent from American military culture of the years before 1917, this concept called for the development of a detailed description of the way that the units and formations fielded by an army ought to act.[26] As the sharing of such a script necessarily required a multitude of mutually compatible manuals, the adoption of this concept resulted in both the creation of a presumably consistent collection of official publications and the rejection of a heterogeneous body of texts read before the war.[27] Similarly, as written instructions rarely suffice to enforce conformity, the introduction of the concept of doctrine correlated with the establishment of a number of new schools for junior officers and the recasting of the schools at Fort Leavenworth as

[23] For a vivid illustration of the decline in dependence upon German ideas and examples on the eve of the USA's entry into the First World War, compare the first lecture, delivered on 29 January 1917, with the last lecture, given on 5 March 1917, of the collection published as Notes on Infantry, Cavalry, and Field Artillery (Washington: Government Printing Office, 1917).

[24] For an account of the influence of the First World War on the schools at Fort Leavenworth that makes no mention of the role played by French models, see Peter J. Schifferle, America's School for War: Fort Leavenworth, Officer Education, and Victory in World War II (Lawrence: University Press of Kansas, 2010), 9–17.

[25] Prior to 1918, the word 'doctrine' rarely appeared in American military literature. When it did, it usually referred to a specific teaching about a particular phenomenon, the most frequently mentioned of which was the 'Monroe doctrine'. For a notable exception, which described doctrine as a 'never-ending progressive' process utilizing 'the collective mind of the service', see Dudley W. Knox, 'The Role of Doctrine in Naval Warfare', Journal of the Military Service Institution of the United States LVII (July–September 1915), 70–90.

[26] For an early description of the new concept of doctrine, see Hugh A. Drum, 'Annual Report, 1919–1920, School of the Line', reproduced in Charles H. Muir, Annual Report, The General Service Schools (Fort Leavenworth: The General Service Schools Press, 1920), 17–24.

[27] In his report for the academic year that ended in the summer of 1920, the assistant commandant of the schools at Fort Leavenworth noted that 'none of the previous text-books could be used and new ones had to be written as they were required'. However, in an equally official document covering the same period, the director of one of the component schools, who was an otherwise enthusiastic proponent of the new concept of doctrine, reported the teaching of 'the tactical principles and methods enunciated in our FSR [Field Service Regulations], DR [Drill Regulations], Griepenkerl, Buddecke, von Alten, and [Morrison] Seventy Problems'. Leroy Eltinge, 'Annual Report, 1919–1920, Assistant Commandant', and Drum, 'Annual Report, 1919–1920, School of the Line', reproduced in Charles H. Muir, Annual Report, The General Service Schools (Fort Leavenworth: The General Service Schools Press, 1920), 7 and 21.

institutions for the teaching of doctrine related to divisions, army corps, and armies to mid-career professionals.[28]

The doctrine enthusiasts in the post-war US Army borrowed many elements from the French phenomenon that inspired their enterprise.[29] Thus, for example, they adopted the French method, codified in the last year of the First World War, of organizing the executive staffs of formations into four sections, as well as techniques for the organization and employment of field artillery, light tanks, and infantry heavy weapons. At the same time, they took pains to explain that the edifice they were building both reflected the peculiarities of American society and suited the needs of the rapidly raised armies the USA was likely to mobilize in the future.[30] '... American traits and characteristics', wrote one of the leaders of the doctrine movement, 'are too distinctive, too enduring, too decisive and too valuable to be sacrificed or to be subordinated to the teachings and methods of races not so blessed'.[31]

Notwithstanding the great pains taken to create a doctrine that was both national and prescriptive, many American military officers continued to display interest in, and, indeed, enthusiasm for, the German tradition of *ad hoc* problem solving. Thus, in 1923, a new edition of the senior field manual of the US Army, the *Field Service Regulations*, began with an introduction that included obvious, but uncredited, borrowings from its German counterpart. These passages stressed the uselessness of 'set rules', the occasional need to depart from prescribed methods, the importance of allowing subordinates 'a certain independence in the execution of tasks', and the importance of initiative and the seizure of opportunities, even at the cost of 'an error in the choice of means'.[32] Similarly, the desire to replace works of German origin

---

[28] For the change in the mission of the schools at Fort Leavenworth, see William K. Naylor, 'Annual Report, 1919–1920, General Staff School', reproduced in Charles H. Muir, *Annual Report, The General Service Schools* (Fort Leavenworth: The General Service Schools Press, 1920): 14. For a list of the new schools created in the year following the end of the First World War, see Peyton C. March, *Report of the Chief of Staff, United States Army, to the Secretary of War, 1920* (Washington, DC: Government Printing Office, 1920), 44–45.

[29] For a brief overview of the development of the French model of detailed doctrine as it applied to infantry units, see P.A. Cour, 'L'Évolution des Doctrines et Reglements Avant la Guerre et la Valeur Technique de Notre Infanterie', Revue Militaire Générale XVIII, 3 and 4 (March and April 1921). For a much longer treatment of the evolution of French doctrine as a whole, see Lucius (pseudonym), 'La Refonte des Règlements et Notre Doctrine de Guerre', Revue Militaire Générale, serialized in Volumes XVII through XX (1920 through 1923).

[30] For an early manifesto of the partisans of an 'American doctrine', see the pamphlet issued to students at the start of the 1919–1920 school year at Fort Leavenworth: *Explanation of Course and Other Pertinent Comments, 12 August 1919* (Fort Leavenworth: The Army Service Schools, 1919).

[31] Hugh A. Drum, 'Annual Report for the School Year 1921–1922 (Assistant Commandant)', reproduced in Hanson E. Ely, *Annual Report, The General Service Schools* (Fort Leavenworth: The General Service Schools Press, 1922), 24–25.

[32] Compare, for example, paragraph 38 of Felddienst Ordnung (Berlin: E. S. Mittler, 1908), 16 with the sixth paragraph of the introduction to *Field Service Regulations, United States Army, 1923* (Washington, DC: Government Printing Office, 1924), III.

with manuals written by, and for, Americans failed to prevent the translation of contemporary German military writings, whether by officers studying
in various schools, officers detailed to such duty, or, in a few instances, by
civilians hired for that purpose.[33]

In the Army school system, the proliferation of doctrinal manuals both
facilitated the composition of approved solutions and enhanced the authority
of such documents. Thus, the student who chose to solve a map problem in a
way that differed from the method previously provided him found himself at
odds, not merely with his instructor, but with a formulation of doctrine that
had been blessed by the highest authority. This, for many, altered the meaning ascribed to the term 'applicatory method'. Before the First World War,
English-speaking students of the German Army described the applicatory
methods as the application of 'knowledge' or 'theory' to 'concrete cases'.[34]
After the First World War, many American soldiers came to believe that the
applicatory method was a matter of applying doctrinal templates to specific
situations.[35]

In the schools at Fort Leavenworth, and the institutions that imitated
them, instructors quickly adopted the custom of marking solutions as if they
were grammar school compositions, with points deducted for each deviation,
whether of style or of substance, from the approved solution. This proliferation of arbitrary standards led some students to submit solutions that had
little to do with genuine beliefs and others to embrace fatalism of a kind that
discouraged serious study. ('Reading an approved solution', said a character
in a musical satire written by officers at Fort Leavenworth, 'is like playing
bridge with your wife. Everything you did was wrong'.)[36] At the same time,
it imbued approved solutions, and the doctrinal manuals upon which they
were based, with an unwarranted air of infallibility. ('There is always the tendency to look at military art as an exact science', wrote the aforementioned
satirist in a more serious venue, 'for it facilitates marking'.[37])

The Rococo quality of approved solutions also owed much to the definitive
experience of most of the Americans who fought in France during the First
World War, the Meuse-Argonne campaign of the last 47 days of that conflict.

[33] Hugh A. Drum, 'Annual Report for the School Year 1920–1921 (Commandant)', reproduced in Hugh
A. Drum, *Annual Report, The General Service Schools* (Fort Leavenworth: The General Service Schools
Press, 1921), 9.
[34] See, among others, Spenser Wilkinson, *The Brain of an Army* (London: A. Constable, 1895): 160 and
'The Lyceum at Fort Agawam', 239.
[35] See, for an example of this change in attitude, the definition of the 'applicatory system' provided
in Herbert J. Brees, *Methods of Training (Provisional)* (Fort Leavenworth: General Service Schools Press,
1925), 6.
[36] Bernard Lentz, *At Kickapoo* (Fort Leavenworth: Privately Published, 1922), 8.
[37] Bernard Lentz, 'The Applicatory Method', *The Infantry Journal* XX, 6 (June, 1922): 606.

Carried out by the largest army yet fielded by the USA, this operation took place at a time when the enemy was rarely able to offer much in the way of sustained resistance. As a result, there were many occasions when it appeared that traffic jams, straggling, and the 'stumblings, blunderings, failures, appeals for help, and hopeless confusion' of higher headquarters did more to hinder forward movement than anything that the Germans did.[38] Thus, it was not surprising that many veterans of the American Expeditionary Force came to the conclusion that success in war was largely a matter of attending to the details of internal organization.

In August 1920, the US Marine Corps founded the Field Officers' School, an institution for the education of mid-career officers that borrowed much from the recently reconstituted schools at Fort Leavenworth. This similarity was, to a large extent, a matter of convenience. Borrowing problems, publications, and policies from comparable courses preserved instructors at the schools for Marine Officers from the time, trouble, and expense involved in *ex nihilo* creation.[39] For some Marines, however, the texts, techniques, and teaching methods developed by the Army, were not only the products of the 'prolonged and exhaustive study of the best military minds in the country', but would also prepare Marines to work with, and for, their sister service counterparts.[40] At the same time, the Marines serving at the new school, which was located at Quantico, Virginia, made allowance for the sort of work the Marine Corps was likely to be called upon to do in the near future, and, indeed, in places like Haiti and Santo Domingo, was already doing. Such missions required that the school employ 'problems requiring independent thought and decision' in order to 'develop initiative, correct thinking and ready decision on the part of subordinate officers'.[41]

It was not until the academic year that began in 1926 that the Field Officers' School devoted a substantial part of its curriculum to the task of preparing Marines to lead units doing things other than operating as part of Army formations. In that year, it introduced a five-week course in 'overseas operations'

<hr />

[38] For a dispassionate catalogue of the self-inflicted difficulties suffered in the first few days of the Meuse-Argonne campaign of 1918, see General Headquarters, American Expeditionary Force, *Notes on Recent Operations, No. 3* (Chaumont: American Expeditionary Force, 1918). The colourful characterization of the deeds of higher headquarters comes from George C. Marshall, 'From the Chief's Office', *Infantry Journal* (March–April 1940): 185–193, quoted in Paul F. Gorman, *The Secret of Future Victories* (Fort Leavenworth: USACGC Press, 1994), 36. For more on the same subject, see Schifferle, *America's School for War*, 14–17.
[39] For an account of the first ten years of the Marine Corps Schools, see Randolph C. Berkeley, 'The Marine Corps Schools', *The Marine Corps Gazette*, May, 1931, 14–15.
[40] Robert Dunlap, 'Recommendations Based on Report of Critique on Joint Army-Navy Problem Number 3, by Officers of Marine Corps Schools, June 1 to 5, 1925', typescript found in Folder 756, Historical Amphibious File, Marine Corps Archives.
[41] 'Professional Notes', *The Marine Corps Gazette*, December, 1920, 409–410.

that dealt with both the design of the defences for improvised naval bases and the landing of Marines on hostile shores.[42] Rather than using the sort of minutely-marked problems that were then being used to teach the portion of the programme of instruction imported from Fort Leavenworth, this 'course within a course' made a much greater use of less formal problems that were discussed in small 'conference groups'.[43]

In December of 1929, the *Marine Corps Gazette* published a remarkable article on the subject of military education. Written by James Carson Breckinridge, 'Some Thoughts on Service Schools' argued for the replacement of arbitrary methods of teaching with 'open forums for the discussion and dissection of special episodes'. These, he argued, would result in the 'habit of thinking and analyzing (but not of fulfilling a ritual) that will be suitable to every situation encountered in military life'.[44] In other words, Breckinridge was calling for a revitalization of the pre-war applicatory method, one that involved both a return to the open-ended spirit of the original technique and its extension, beyond the realm of the tactics of conventional warfare on land, to all of the problems that a Marine might encounter in the course of his varied service.

Tragically, Breckinridge does not seem to have been aware of the existence of open-ended alternatives, whether German or American, of the ossified version of the applicatory method borrowed from Fort Leavenworth.[45] Rather, he recommended that Marine Corps schools for officers draw upon the spirit, if not the precise teaching methods, of the Experimental College at the University of Wisconsin, thereby committing the fatal rhetorical mistake of suggesting that Marines emulate an institution that was best known for the scruffy appearance, poor manners, and rowdy behaviour of its students.[46] Thus, although he served as commandant of the Marine Corps Schools, and thus the direct superior of the director of the Field Officers' School, for a combined total of more than four years, he proved unable to implement a thorough-going reform of the teaching methods used there.

---

[42] For the formation of this course on 'overseas operations', see Dion Williams, 'The Education of a Marine Officer', *Marine Corps Gazette,* August 1933, 19.

[43] For more on the evolution of the curriculum at the Field Officers School, see Bruce I. Gudmundsson, 'Ambiguous Application: The Study of Amphibious Warfare at the Marine Corps Schools, 1920–1933', in *On Contested Shores,* edited by Timothy Heck and B.A. Friedman (Quantico: MCU Press, 2018), 174–179.

[44] J. C. Breckinridge, 'Some Thoughts on Service Schools', *Marine Corps Gazette,* December 1929, 230–238.

[45] See, for the assumptions about the applicatory method held by Breckinridge, 'Tactical Problems', an unpublished essay found in the papers of James Carson Breckinridge (Box 19, Folder 4) on file at the Marine Corps Archives.

[46] Alexander Meiklejohn, *The Experimental College* (New York: Harper and Brothers, 1932) and Erin Abler, 'The Experimental College: Remembering Alexander Meiklejohn and an Era of Ideas', *Archive: A Journal of Undergraduate History,* 5 (2002): 50–75.

Nonetheless, Breckinridge did manage, in an indirect way, to reduce the influence of the Fort Leavenworth version of the applicatory method upon the operational culture of the Marine Corps. Thanks to his emphasis on the study of naval matters, landing operations, and small wars, many classes that had been borrowed from the Army were displaced by work on subjects for which neither approved solutions nor doctrinal manuals had been written.[47]

The pathos of the failure of Breckinridge to implement his vision is amplified by the proximity of resources for the engagement of the 'special episodes' he seems to have had in mind. With respect to materials, the book used by Marines at Quantico in the early 1930s to study the most successful amphibious operation of the First World War, *The Army and Navy during the Conquest of the Baltic Islands*, had been written in such a way that each chapter ended at the point where a leader made an important decision. In other words, it was written as a series of 'special episodes'.[48] With respect to method, Breckinridge seems to have entirely missed the reform of the applicatory method that had been taking place at the Army Infantry School at Fort Benning, Georgia.

Between 1927 and 1932, George C. Marshall, who had studied under John F. Morrison at Fort Leavenworth, wrought a series of remarkable changes at the Infantry School.[49] Like Morrison, Marshall was convinced that the 'bunk, complication, and ponderosities' that played such a large role in instruction at Fort Leavenworth needed to be replaced by teaching that helped students to develop the ability to uncover the 'essentials' of a given situation.[50] In contrast to Morrison, Marshall took great pains to recruit a group of talented instructors who were capable, not only of employing his approach, but of sustaining it after his inevitable departure.[51] Thus, throughout the

---

[47] Breckinridge served two tours as commandant of the Marine Corps Schools: July 1928 through December 1929 and April 1932 through January 1935. For details of his accomplishments during his second period of service, see Gudmundsson, 'Ambiguous Application', 181–184.

[48] For the original work, see Erich von Tschischwitz, *Armee und Marine bei der Eroberung der Baltischen Inseln im Oktober 1917* (Berlin: Eisenschmidt, 1931). For the translation by an officer of the US Army, see Erich von Tschischwitz (translated by Henry Hossfeld), *The Army and Navy during the Conquest of the Baltic Islands* (Fort Leavenworth: Command and General Staff School Press, 1933). For the translation by an officer of the Marine Corps, see Erich von Tschischwitz (translated by Samuel Cumming) Translation of *Army [Armee] und Marine bei der Eroberung der Baltischen Inseln im Oktober 1917*, (typescript on file at the Library of the Marine Corps, Quantico).

[49] For a short and sympathetic account of Marshall's military service, see Larry I. Bland, 'George C. Marshall and the Education of Army Leaders', *Military Review* 68 (October 1988): 27–37.

[50] For a description of Marshall's reforms, see the letter he wrote to Stuart Heinzelman on 4 December 1933 in *The Papers of George Catlett Marshall: 'The Soldierly Spirit', December 1880–June 1939 (Volume 1)*, edited by Larry I. Bland and Sharon R. Ritenour (Baltimore and London: The Johns Hopkins University Press, 1981), 409–413.

[51] For an account of the experience of an instructor at the Infantry School during Marshall's tenure there, see Leslie Anders, *Gentle Knight: The Life and Times of Edwin Forrest Harding*, (Kent, OH: Kent State University Press, 1985), 118–134.

fourth decade of the twentieth century, the Infantry School provided the Army with an alternative to what Marshall called the 'scholasticism' of Fort Leavenworth.[52]

In the course of transplanting the mindset of Morrison to its new home, Marshall and his collaborators made use of teaching methods from 'The Lyceum of Agawam' that had fallen by the wayside over the course of the past three decades.[53] The best documented of these were the 'historical map problems' that placed students in the role of a commander who, at some point in the past, found himself faced with a mission to fulfil, an obstacle to overcome, or a dilemma to resolve. As students could compare their solutions to those made in real life by the historical decision maker, historical map problems allowed instructors to dispense with approved solutions. (Indeed, they served to remind students that 'the schematic solution will rarely fit a definite case.[54]) Better yet, historical map problems allowed students to see the real-world results of a given decision and thus reflect upon the possible impact of the courses of action that they had proposed. Best of all, problems based on real events invariably contained elements of 'friction', whether in the form of unfavourable weather, casualties, shortages, poor intelligence, or badly composed orders, that rarely, if ever, appeared in exercises based upon imaginary scenarios.[55]

Between 1931 and 1939, historical map problems were a regular feature in the *Infantry School Mailing List*. (Not to be confused with the *Infantry Journal*, which was published by the Infantry Association, the *Mailing List* was the official journal of the Infantry School and, as such, was supported by public funds.) In the same period, the Infantry School also published short accounts of tactical engagements that could easily be converted into historical map problems. (Some of these were published in the *Mailing List*, others in the form of a book called *Infantry in Battle*.) Thanks, in part, to these methods of dissemination, as well as the efforts of officers who had served at the Infantry School, historical map problems found their way into many of the infantry units of the Army and Marine Corps.[56] There were even a few instances where

---

[52] For a highly sympathetic account of the reforms wrought by Marshall at the Infantry School, see Jörg Muth, *Command Culture: Officer Education in the US Army and German Armed Forces, 1901–1940 and the Consequences for World War II* (Denton: University of North Texas Press, 2011), 137–147.

[53] 'Editorial Note on Infantry School Teaching, 1927–1932', *The Papers of George Catlett Marshall, Volume 1*, 319–321.

[54] 'Infantry Problems', *The Infantry School Mailing List* 2 (1930–1931), 41.

[55] Many examples of historical map problems used at Fort Benning can be found in the volumes of *The Infantry School Mailing List* published between 1931 and 1939.

[56] For an example of a historical map problem from the Infantry School that was made available to Marines, see 'A Skirmish in Nicaragua', *The Leatherneck*, November 1938, 7–8 and 60.

instructors at Fort Leavenworth used the technique.[57] By the end of the 1930s, however, as the officers trained by Marshall and his disciples left Fort Benning, the popularity of historical map problems had waned. Thus, in 1939, the anonymous author of the last historical map problem to appear in the pages of the *Mailing List*, found it necessary to begin his article with an apology for the method he was using.[58]

Whilst the old applicatory method enjoyed its brief renaissance at Fort Benning, the formalism that had taken root at the start of the twentieth century, and blossomed soon after the end of First World War, continued to dominate instruction at Fort Leavenworth.[59] Thus, during the decade leading up the start of the Second World War, the Army possessed two very different operational cultures, each of which had its own champions, its own literature, and its own traditions. Marvellous to say, each of those cultures assumed that the chief task of the peacetime Army was to set the stage for the mobilization of a much larger force of the type raised in the First World War. However, whilst Marshall argued that wartime citizen-soldiers were best led by officers who knew how to quickly devise simple solutions to a wide variety of problems, the champions of the approach developed at Fort Leavenworth believed that hastily-trained fighting men needed to be provided with written instructions drawn up at leisure, both in the form of detailed doctrinal manuals and finely-formatted field orders.[60]

Throughout the interwar period, the foreign military organization of greatest interest to officers of the ground forces of the USA was the Army of the French Republic. Nurtured by the attendance of American officers at French military schools and the translation of French manuals, this relationship sometimes took the form of attempts at unequivocal imitation.[61] (This was particularly true in the case of the field artillery of the US Army, which was both armed with weapons of French design and enamoured of French methods.)[62] As a rule, however, the chief product of this relationship

[57] Arthur R. Walk, 'A Critical Analysis of the Military History Course at the Command and General Staff School during the Years 1931–1933, with Some Comparisons and Suggested Changes' (Student Paper, Command and General Staff School, Fort Leavenworth, 1933).

[58] Anonymous, 'The Battle at Rocourt', *The Infantry School Mailing List* XVII (January 1939), 1.

[59] For a detailed description of the teaching methods used at Fort Leavenworth between 1934 and 1936, see J. P. Cromwell, 'Are the Methods of Instruction Used at this School Practical and Modern?' (Student Paper, Command and General Staff School, Fort Leavenworth, 1936).

[60] For a detailed account, rich in anecdote, of instruction at the Fort Leavenworth schools in the 1930s, see Muth, *Command Culture*, 115–137.

[61] For the controversy over the American adoption of what was, to a large extent, a direct translation of its French counterpart, see William Odom, *After the Trenches: The Transformation of Army Doctrine, 1918–1939* (College Station: Texas A&M University Press, 1999), 118–131.

[62] Bruce I. Gudmundsson, *On Artillery* (Westport: Praeger, 1993), 109–110.

seems to have been reassurance. That is, the existence of the French Army as the paragon of a doctrinaire military organization provided psychological comfort to American devotés of a similar operational culture.

As might be imagined, the fall of France in June of 1940 put an immediate end to any and all American calls for the adoption of artefacts of the French military establishment. (In the relatively rare instances where such imitation took place, such as the original design for tank destroyer units, the American officers involved refrained from mentioning the provenance of the models in question.)[63] At the same time, there was no attempt to purge American military culture of features that had been borrowed from its French counterpart during the First World War. If anything, the rapid expansion of both the Army and the Marine Corps in preparation for the entry of the USA into the Second World War buttressed demand for a system that provided each soldier, from the recently drafted private through to the recently commissioned junior officer, with dependable directions to follow.

In the Marine Corps, the demand for doctrine was tempered by the possibility that, rather than taking part in the Second World War, the USA would find itself faced with a long cold war against Germany or Japan. In such a scenario, Marines might well find themselves returning to the Caribbean, there to fight 'small wars' against the proxies of a power intent upon control of the approaches to the Panama Canal. This possibility led to the republication, in the form of a single book, of pamphlets written by veterans of Marine operations in Haiti, Santo Domingo, and Nicaragua in the years between 1912 and 1934. The resulting *Small Wars Manual*, whilst containing some of the sort of prescriptions that filled the pages of conventional doctrinal manuals, placed a great deal of emphasis on the custom-tailoring of solutions, not merely to specific military situations, but to the political problems that were invariably intertwined with them.[64]

In the Army, the power of doctrine had to contend with the influence of the many disciples that Marshall had made at the Infantry School as well as with Marshall himself, who had become chief of staff of the US Army on 1 September 1939, the very day that Germany had begun its invasion of Poland. In the realm of personnel, Marshall enjoyed an enormous degree of freedom when it came to the selection of leaders. In the realm of policy, he was able to

---

[63] For the French original, see Éric Denis et François Vauvillon, 'Le Chasseur de Chars Laffly W 15 TTC et les Batteries Anti-Chars Automotrices', *Histoire de Guerre, Blindés et Matériel* 85 (October, November, December 2008): 7–21. For the American copy, see Christopher Gabel, *Seek, Strike, and Destroy, US Tank Destroyer Doctrine in World War II* (Fort Leavenworth: Combat Studies Institute, 1985), 20–21.

[64] *Small Wars Manual*, Headquarters, US Marine Corps (Washington, DC: Government Printing Office, 1940).

encourage large-scale free-play exercises in which leaders who were fond of lengthy orders found themselves at a great disadvantage.[65]

In both the Army and the Marine Corps, the avatars of the approved solution found themselves at odds with a third aspect of operational culture, that of the officer who had studied tactics on his own time. A notable example of the benefits of this practice is provided by John S. Wood, who used the leisure afforded to him by nearly 10 years of service as a professor of military science at civilian schools to pursue a programme of self-education that challenged the approach taught at both the Staff College at Fort Leavenworth and its French equivalent, the *École Supérieure de Guerre*.[66] As was the case with his fellow autodidact (and friend) George S. Patton, the synthesis that resulted from this clash of thesis and anti-thesis bore a closer resemblance to the teachings of the Infantry School of the 1930s[67] than those of the schools that these officers had actually attended.

In the course of American participation in the Second World War, the existence of two separate operational cultures often resulted in conflict between commanders. A particularly stark incident of this sort took place during the battle for Saipan in the summer of 1944.[68] Whilst often presented as a product of inter-service rivalry, this 'battle of the Smiths' had more to do with the way each of the parties had studied the art of war during the long years of peace. Holland M. Smith, a Marine officer who commanded all of the American ground troops in that battle, had formed a low opinion of the instruction he had received at both the Naval War College and the Marine Corps Schools. (The instructors at the latter institution, he wrote, 'could not handle situations which refused to square with theory'.)[69] Thus, when Ralph C. Smith, a former instructor at Fort Leavenworth, persisted in employing his division in the methodical manner celebrated in approved solutions, the senior Smith relieved the junior Smith of his command.[70]

The Second World War presented American officers of both operational cultures with a large number of novelties. These included a myriad of new

---

[65] For a detailed description of these large-scale exercises, see Christopher R. Gabel, *US Army GHQ Maneuvers of 1941* (Fort Leavenworth: Combat Studies Institute, 1991).

[66] For a detailed report, by an American officer, of the programme of instruction at the French staff college, see Leon W. Hoyt, 'The École Supérieure de Guerre', *Marine Corps Gazette*, December, 1926, 219–225.

[67] Hanson W. Baldwin, *Tiger Jack: Major General John S. Wood* (Fort Collins: Old Army Press, 1979), 77–79.

[68] Extensive accounts of the 'battle of the Smiths' can be found in Harry A. Gailey, *Howlin' Mad versus the Army* (Novato: Presidio Press, 1986) and Norman Cooper, *Fighting General* (Quantico: Marine Corps Association, 1987).

[69] Holland M. Smith, *Coral and Brass* (New York: Charles Scribner's Sons, 1949), 57.

[70] Eric Pace, 'Gen. Ralph C. Smith, Honored for War Bravery, Dies at 104', New York Times, 26 January 1998, A–17.

weapons, new means of transport, and new realms of conflict. For those dependent upon doctrine and the approved solution, the integration of each of these innovations required a formal process of study, composition, approval, and promulgation. For those who celebrated the uniqueness of each situation, however, new devices, new modes of operation, and new spheres of struggle differed little from the specific incarnations of the familiar factors found in each problem solved in the course of service, schooling, or self-study. Thus, an artillery officer like John S. Wood or an engineer like Bruce C. Clark proved as capable of mastering the art of handling all-arms mechanized formations as an experienced tank officer like George S. Patton. At the same time, the doctrinaire commanding general of the American Tenth Army, Simon B. Buckner, Jr, proved unable to make decisive use of the powerful collection of tanks and other armoured vehicles at his disposal on the island of Okinawa.[71] (A distinguished graduate of the senior of the two schools then at Fort Leavenworth, Buckner had also taught at that institution for three years.)[72]

During the Second World War, the proliferation of new phenomena of importance to military leaders led American military schools to make extensive use of films, slide shows, demonstrations, and lectures. At the same time, a shortage of instructors resulted in the use of the same tools to cover many subjects previously taught by means of the applicatory method.[73] In order to judge the effectiveness of these passive forms of instruction, the schools made extensive use of written examination, many of which were developed with the help of civilians trained in the academic field of education. At the Infantry School, there were so many of these that the teaching staff resorted to the use of machine-graded multiple-choice 'bubble tests' of a type that would become familiar to many American students, both military and civilian, in the second half of the twentieth century.[74]

The combination of presentation and examination adopted by American military schools during the Second World War shared the same assumptions about military operations as the interwar combination of doctrine and the

---

[71] For the overall approach employed by Buckner on Okinawa, see Nicholas Evan Sarantakes, ed., *Seven Stars: The Okinawa Battle Diaries of Simon Bolivar Buckner, Jr. and Joseph Stilwell* (College Station: Texas A&M Press, 2004), particularly pages 30 and 51. For a critique of the handling of American armour in the battle for Okinawa, see Bruce I. Gudmundsson, 'Okinawa', in *No End Save Victory*, edited by Robert Cowley (New York: G.P. Putnam's Sons, 2001), 638.

[72] For Buckner's time at Fort Leavenworth, see the annual reports for the General Service Schools for the academic years ending in 1925 through 1928.

[73] For a characterization of the frequently theatrical lectures that replaced interactive exercises at Fort Leavenworth, see Michael D. Stewart, *Raising a Pragmatic Army: Officer Education at the US Army Command and General Staff College, 1946–1986* (University of Kansas, Doctoral Thesis, 2010), 34.

[74] Ronald R. Palmer, Bell I. Wiley, and William R. Keast, *The Procurement and Training of Ground Combat Troops* (Washington, DC: Center of Military History, 1991), 294.

approved solution. In both approaches, the definitive task of military education was the transfer of the martial analogue of a script for a film or a play, a set of instructions that, with a minimum of adjustment and improvisation, would provide the student with everything he needed to know to fulfil his assigned role on active service. Likewise, both systems equated mastery of this script with the ability to manipulate its most marginal aspects, whether those were the finer points of formatting or distinctions between arbitrary categories.

The chief difference between the doctrinaire curricula of the interwar period and its Cold War counterpart lay in the speed with which the military schools changed the material they presented. Before the Second World War, the nuts-and-bolts of American ground operations, whether physical or conceptual, had remained essentially the same for twenty years. ('From 1923 through 1944', explained one senior American general, 'the fundamentals of combat at division level ... did not change significantly'.)[75] After the Second World War, the Army acquired new equipment, undertook new missions, and adjusted to tectonic shifts in the strategic environment with such rapidity that doctrinal manuals, and the lessons based upon them, quickly became obsolete. Thus, in 1957, when Army Chief of Staff Maxwell Taylor told the graduating class of the Command and General Staff College, 'the Army is burning its old military textbooks, to clear away the old and make way for the new' he was telling them nothing that they did not already know.[76]

Over the course of the three decades that followed the end of the Second World War, openly radical transformations in the realm of Army doctrine alternated with self-consciously conservative 'retromorphoses'.[77] The first of the futuristic transformations was based on the short-lived presumption that, as the atomic bomb had created an era of 'push-button warfare', ground forces would serve chiefly as constabulary organizations, concerned less with fighting other armies than with occupation, administration, and police work.[78] This 'constabulary era' ended sharply in June of 1950 when Communist ground forces of an entirely conventional type invaded South Korea. The resulting return to traditional ways of doing business lasted for seven years

---

[75] James S. Wheeler, *Jacob L. Devers, A General's Life* (Lexington: University Press of Kentucky, 2006), 79.

[76] Andrew J. Bacevich, *The Pentomic Era: The US Army Between Korea and Vietnam* (Washington, DC: NDU Press, 1986), 73.

[77] For an accessible account of doctrinal turnover during the first three decades of the Cold War, see Robert A. Doughty, *The Evolution of US Army Tactical Doctrine* (Fort Leavenworth: Combat Studies Institute, 1979).

[78] For the transformation of Army forces in Europe into a 'constabulary' force, see *European Command, Reorganization of Tactical Forces, VE Day to 1 January 1949* (Karlsruhe: Historical Division, European Command, 1949).

before giving way to the 'Pentomic era', a period defined by a highly original doctrine for ground combat operations that assumed the promiscuous
use of atomic munitions of types designed for use on the battlefield.[79] In the
early 1960s, this fanciful approach to fighting, well supplied with futuristic
prototypes and freshly-minted jargon, yielded to a second resumption of a
vision for ground combat operations based heavily upon the experience of
the last year of the Second World War.[80]

Whilst the Army of the first half of the Cold War engaged in its game of
doctrinal musical chairs, the Marine Corps managed to combine substantial cultural stability with considerable change in both equipment and modes
of operation. This happy situation stemmed, in part, from a law, passed in
1952, that gave the Marine Corps definitive responsibility for the development and maintenance of expertise in the realm of amphibious operations.[81]
Ownership of this unequivocal mission preserved Marines from the peren-
nial identity crisis that plagued their Army counterparts. At the same time,
the close relationship between the Marine Corps and amphibious warfare
allowed Marines to experiment with a variety of techniques and technologies,
whether for various types of landings or for 'subsequent operations ashore',
without endangering their sense of who they were.[82]

The experience of the first 20 years of the Cold War shaped the very different ways in which the Army, on the one hand, and the Marine Corps, on
the other, embraced the challenges posed by the war in Vietnam. Formations
fielded by both services made extensive use of helicopters, which had originally been adopted for use on a battlefield rich in tactical nuclear weapons,
to enhance conventional operations against formed bodies of Communist
fighters.[83] However, where the leadership of the Marine Corps embraced the
Combined Action Program, which married Marine rifle squads to units of
part-time soldiers defending their home villages, the Army leadership argued

[79] The definitive study of the 'Pentomic Army' of the late 1950s and early 1960s is Bacevich, *The Pentomic Era*. For the one aspect of this revolution that survived its demise, see Christopher C. S. Cheng, *Air
Mobility: The Development of a Doctrine* (Westport: Praeger, 1994).
[80] For an account of the transformation of the Army that took place in the early 1960s, see Peter Camp-
bell, *Military Realism: The Logic and Limits of Force and Innovation in the US Army* (Columbia: University
of Missouri Press, 2019), 62–74.
[81] For background on Public Law 416 of the 82nd Congress, the Douglas–Mansfield Bill of 1952, see
Alan Rems, 'A Propaganda Machine Like Stalin's', *Naval History Magazine* 33, 3 (June 2019).
[82] For an account of the changes that took place in the Marine Corps in the first three decades of the Cold
War, see Kenneth J. Clifford, *Progress and Purpose: A Developmental History of the United States Marine
Corps, 1900-1970* (Washington, DC: Government Printing Office, 1978), 71–112. For an exploration of the
relationship between the changes that took place in the 1950s and the amphibious identity of the Marine
Corps, see G. F. Cribb, Jr, 'Embarkation Ready', *The Marine Corps Gazette*, August, 1959, 20–26.
[83] For a nuanced introduction to the helibourne operations conducted by the Army in Vietnam at the
start of the 'main force war', see J. Paul Harris, *Vietnam's High Ground: Armed Struggle for the Central
Highlands, 1954-1965* (Lawrence: University Press of Kansas, 2016), 220–400.

that all ground combat units be exclusively employed to prosecute the 'main force' war.[84]

Viewed from the point of view of the operational culture, the war in Vietnam wrought the same sorts of effects as the Pentomic reforms of the late 1950s and the constabulary transformation of the second half of the 1940s. That is, it created a period rich in peculiarities that both followed and preceded a return to normalcy. Indeed, what might be called the 'neo-classical revival' of the 1970s, which took place at a time when the Soviet Union had achieved parity in the realm of nuclear weapons, was even more conventional than the retromorphoses of the two preceding decades. In particular, it assumed that the ground forces of the North Atlantic Treaty Organization (NATO) and those of the Warsaw Pact could fight each other on German soil without necessarily resorting to atomic weapons of any kind.

The presumption that the chief mission of the Army was the waging of conventional war in Central Europe coincided with an increasingly favourable view of the German military tradition. One contributor to this phenomenon was frequent contact with members of the Army of the Federal Republic of Germany, many of whom had been trained by veterans of the Second World War who had fought against the forces of the Soviet Union. Another was an artefact of the post-war publishing industry, which had discovered that books about German soldiers were both easier to write and more likely to sell than works that told tales of their American counterparts.[85] A third reason for the increased willingness of American military men to learn from the German experience rested upon the still-fresh memory of defeat in the Vietnam war, which greatly reduced the power of the argument that, after losing two world wars, the German military tradition had nothing of value to study, let alone imitate.

As had been the case with the first wave of American enthusiasm for German military culture, admiration and understanding were two very different things. William E. DePuy, who had fought in Europe as a young infantry officer in 1944 and 1945, attributed the prowess of his erstwhile foes to 'doctrine', 'battle drill', and 'standard operating procedures', all of which were concepts alien to the German military tradition. Thus, in 1973, when DePuy

[84] For a sympathetic account of the Combined Action Program, see Ronald E. Hays II, *Combined Action: US Marines Fighting a Different War, August 1965 to September 1970* (Quantico: Marine Corps University Press, 2019).

[85] The ability of English-speaking authors to write about the German experience of the Second World War was greatly enhanced by microform publication, by agencies of the US government, of scores of millions of pages of German documents and hundreds of retrospective studies written by former German officers. For guides to these products, see Robert Wolfe, *Captured German and Related Records* (Athens: Ohio University Press, 1975) and *Catalog German Studies, 1945–1952* (Karlsruhe: Historical Division, Headquarters, European Command, 1952).

took charge of the newly formed Training and Doctrine Command, he set in motion an ambitious programme to provide the Army with a comprehensive library of prescriptive doctrinal manuals, each of which would explain to soldiers of a particular specialty exactly what they were expected to do on a Central European battlefield.[86]

In 1976, DePuy published, as the cornerstone of his 'system of field manuals', a completely reworked edition of the Army's field service regulations. In sharp contrast to previous versions of *Operations*, which had dealt largely with matters peculiar to senior commanders and their staffs, the new book was 'intended for use by commanders and trainers at all echelons'. Similarly, where previous editions of *Operations* had dealt largely in definitions, axioms, and platitudes, DePuy's *magnum opus* made an internally-consistent argument, not merely for a particular approach to the defence of West German territory against the ground forces of the Warsaw Pact, but for a specific set of combat techniques. (Some of these were products of his own experience. Others were drawn from the annals of the war between Israel and its Arab neighbours that had taken place in October of 1973.)[87]

In March of 1977, *Military Review*, the official journal of the Army Command and General Staff College at Fort Leavenworth, published a comprehensive critique of DePuy's edition of *Operations*. Written by William S. Lind, a civilian serving on the staff of Senator Gary Hart, this article took the new manual to task for, among many other things, its promotion of a tactical system that resembled that of the French Army of 1940. Instead of these 'firepower/attrition' tactics, Lind argued, the Army would be better off adopting the 'maneuver' tactics employed by the German Army in the Second World War and the Israeli Defense Forces in the recent war in the Middle East.[88]

The debate over the 1976 edition of *Operations* took place at a time when the Marine Corps was suffering from the martial equivalent of a crisis of faith. As the Inchon landings of 1950 faded into an increasingly distant memory, many Marines were hard pressed to imagine scenarios in which a full

---

[86] For a concise treatment of the connection between DePuy's experience of combat against German opponents and the reforms he implemented in the Army of the 1970s, see Paul H. Herbert, *Deciding What Has to Be Done: General William E. DePuy and the 1976 Edition of 100-5, Operations* (Fort Leavenworth: Combat Studies Institute, 1988), 15–18 and 75–96. For a biography that pays a great deal of attention to this relationship, see Henry G. Golem, *General William E. DePuy: Preparing the Army for Modern War* (Lexington: University Press of Kentucky, 2008).

[87] *Field Manual 100–5 Operations*, United States Army (Fort Monroe: US Army Training and Doctrine Command, 1976), cover page.

[88] William S. Lind, 'Some Doctrinal Questions for the US Army', *Military Review* 57, 3 (March 1977): 54–65.

regiment, let alone a division or more, would land upon a hostile shore.[89] When combined with the positive examples provided by recent wars in the Middle East and the 'neo-classical revival' taking place in the Army, this phenomenon led many Marines to explore the possibility of a different sort of landing operation, a 'Blitzkrieg from the Sea' carried out by sea-borne mechanized forces.[90] This, in turn, led many forward-thinking Marines to the study of the German military tradition and, in particular, the experience of the German Army of the Second World War.[91]

Over the course of the late 1970s, what had begun as an interest in the mechanization of the Marine Corps landing forces became something more cerebral and, as such, independent of any particular items of equipment. This 'maneuver warfare' movement drew inspiration from the theories of John Boyd, a retired fighter pilot who began to share his thoughts, in the form of an evolving series of briefings, in 1976. It also owed much to the work of William S. Lind, who did much, by writing, speaking, and hosting informal gatherings, to make Boyd's work accessible to Marines. Lind, an enthusiastic student of German military history, also introduced many Marines to relevant aspects of the German military tradition.[92]

In the Marine Corps of the 1980s, the influence of the manoeuvre warfare movement grew considerably. One reason for this was a marked improvement in the quality of people volunteering for service and the consequent increase in the number of people attracted to a philosophy that emphasized the importance of the initiative, creativity, and professionalism of junior leaders. Another was the involvement of Marine Corps units on the 'northern flank' of NATO, which led, among other things, to the study of the Russo-Finnish Winter War of 1939–1940 and contemplation of the possibility that Marines would have to fight Soviet ground forces.[93] A third contributor to the influence of the manoeuvre warfare movement within the Marine Corps of the 1980s was the absence of any attractive alternative. Thus, Marines

[89] For an influential expression of the scepticism about the viability of large-scale amphibious operations in the 1970s, see Jeffrey Record and Martin Binkin, *Where Does the Marine Corps Go from Here?* (Washington, DC: Brookings Institution, 1976).

[90] For an application of the term 'Blitzkrieg from the Sea' to landing operations conducted by mechanized forces, see Richard S. Moore, 'Blitzkrieg from the Sea: Maneuver Warfare and Amphibious Operations', *Naval War College Review* 36, 6 (November–December 1983): 37–48.

[91] For an extensive discussion of the role of German examples in the early days of the 'Maneuver Warfare movement' in the Marine Corps, see Marinus, 'Learning from the Germans: Part I', *The Marine Corps Gazette*, December, 2020, 52–55.

[92] For an account of the role played by William S. Lind in the manoeuvre warfare movement, see Fideleon Damian, *The Road to FMFM 1: The United States Marine Corps and Maneuver Warfare Doctrine, 1979–1989* (Kansas State University, Masters Thesis, 2008), 29–37.

[93] For the connection between the manoeuvre warfare movement and study of the Winter War, see Michael D. Wyly, 'Fighting the Russians in Winter: Three Case Studies, Leavenworth Papers no. 5', *Marine Corps Gazette*, April, 1983, 76–77.

opposed to the new philosophy could imitate the methods learned in Army schools, indulge in nostalgia, or wax rhapsodic about the power of particular weapons. They failed, however, to come up with an overall approach capable of competing with manoeuvre warfare.[94]

In March of 1989, Alfred M. Gray, Jr, then serving as commandant of the Marine Corps, promulgated a formal explanation of manoeuvre warfare. Called *Warfighting*, this little book might well be described as the anti-thesis to the 1976 edition of *Operations*. Where *Operations* dealt in specific techniques, *Warfighting* provided a philosophy. (The first of the four chapters of the work was called 'The Nature of War'.) Where *Operations* prepared the Army to fight in a particular location, *Warfighting* presumed that Marines needed to be ready to fight 'in every clime and place'. Where *Operations* was the harbinger of a 'system of field manuals', *Warfighting* reduced most other manuals, whether published by the Army or the Marine Corps, to the status of 'reference publications'.[95] In other words, the publication of *Warfighting* was, among many other things, an explicit repudiation of the prescriptive approach to doctrine championed by William E. DePuy.

In May of 1989, at a conference convened at the Marine Corps base at Quantico to discuss manoeuvre warfare, Hasso von Uslar, a military officer serving at the embassy of the German Federal Republic in Washington, demonstrated a simple map problem. This encouraged John F. Schmitt, the Marine who wrote most of *Warfighting*, to start a working group dedicated to the revival of such 'tactical decision games'. Soon thereafter, the *Marine Corps Gazette* made exercises of this sort, many of which were composed by Schmitt, a regular feature.[96] (Marvellous to say, whilst some members of the working group based their tactical decision games on historical events, no one made any attempt to revive historical map problems of the type championed by George S. Marshall.)[97]

Whilst the Marine Corps embraced 'a new conception of war', the Army continued to follow the course laid out by DePuy in the 1970s.[98] Thus, whilst

---

[94] The most articulate opponent of manoeuvre warfare in the Marine Corps of the 1980s was Gordon D. Batcheller. For his critique, see the articles on the subject that he wrote for the *Marine Corps Gazette* in this decade: 'Let's Watch Where We're Going!', June, 1981, 18–19; 'Reexamining Maneuver Warfare', April, 1982, 22–23; and 'Sorting Out Maneuver and Attrition', January, 1987, 79.

[95] *Fleet Marine Force Manual 1: Warfighting*, United States Marine Corps (Washington, DC: United States Marine Corps, 1989), cover page.

[96] For a collection of these map problems, see John F. Schmitt, *Mastering Tactics: A Tactical Decision Games Workbook* (Quantico: Marine Corps Association, 1994).

[97] The author of this article, who was serving in the Marine Corps at this time, both attended the manoeuvre warfare conference in May of 1989 and participated in the tactical decision game working group.

[98] The phrase 'a new conception of war' is taken from the title of Ian T. Brown, *A New Conception of War: John Boyd, the US Marines, and Maneuver Warfare* (Quantico: Marine Corps University Press, 2018).

the two editions of *Operations* that were published in the 1980s described ideal battles that differed considerably from those depicted in the edition of 1976, they shared with their predecessor the presumption that the purpose of doctrine was the provision of a set of scripts which, whilst requiring 'judgement in application', minimized the custom tailoring that leaders in the field might be called upon to do.[99] Each also rested on a set of subsidiary publications that, as a rule, changed more slowly than the 'keystone' field manuals.[100]

The experience of the Gulf War of 1991 convinced many American soldiers of the essential soundness of the Army's operational culture. This victory coincided with a sea-change in the realm of academic military history. The 1990s saw the rise of a generation of scholars, many of whom were retired Army officers or civilian employees of the Army, who celebrated the triumph of American arms in the Second World War as the natural outcome of the Army's operational culture.[101] The resulting atmosphere of self-satisfaction was exacerbated by the collapse of the Soviet Union, which deprived the Army of the stimulus that had long been provided by the imminent possibility of war in Central Europe. Nonetheless, a number of army officers found much to like in the manoeuvre warfare movement and, over the course of the 1990s, formed a counter-culture comparable (in quality, if not in influence) to that of the Infantry School of the 1930s.[102]

The Marine Corps of the 1990s suffered from a relapse of the identity crisis from which it had suffered in the 1970s. In first half of the decade, the official response to this problem took the form of a renewed emphasis on the naval character of the Marine Corps and, in particular, the use of Marine units afloat to provide the USA with a global emergency response force.[103] In the second half of the decade, the Marine Corps entertained a fad worthy of Army's Pentomic era. Successively known as 'Green Dragon', 'Sea Dragon', and 'Hunter Warrior', this took the form of an attempt to replace traditional

---

[99] For a thorough account of the replacement of the 1976 edition of Operations with that of 1982, see John L. Romjue, *From Active Defense to AirLand Battle: The Development of Army Doctrine, 1973–1982* (Fort Monroe: Army Training and Doctrine Command, 1984).

[100] For a discussion of the problem of harmonizing subsidiary field manuals with keystone doctrinal publications, see Michael P. Coville, *Tactical Doctrine and FM 100-5* (Fort Leavenworth: School of Advanced Military Studies, 1991).

[101] For a paragon of this genre, see Keith E. Bonn, *When the Odds Were Even, An Operational History of the Vosges Campaign October 1944–January 1945* (Novato: Presidio Press, 1994).

[102] For several views of the manoeuvre warfare movement within the Army of the early 1990s, see Richard D. Hooker, Jr., *Maneuver Warfare: An Anthology* (Novato: Presidio Press, 1993).

[103] Carl E. Mundy, Jr., 'Reflections on the Corps: Some Thoughts on Expeditionary Warfare', *Marine Corps Gazette*, March, 1995, 26–29.

ground combat units with swarms of six-man reconnaissance teams, each of which was able to direct the fire of long-range missiles of various kinds.[104]

Throughout the 1990s, many Marines looked to manoeuvre warfare as a means of mitigating the identity crisis of that decade and, in particular, of distinguishing the Marine Corps from the Army. However, public embrace of the artefacts of a philosophy, the most salient of which was *Warfighting*, did not require a full understanding of its tenets, let alone its implications. Thus, throughout the decade, many Marines continued to confuse manoeuvre with movement, to refer to reference publications as 'doctrine', and to embrace formal planning processes that aped those of the Army.[105] The close association between the Marine Corps and manoeuvre warfare, moreover, led easily to the assumption that everything that Marines did, or, indeed, had done in the past, was *ipso facto* a reflection of that philosophy. In 1996, for example, the Marine Corps published a concept paper that, among other things, described the painfully slow exploitation of the Inchon landing of 1950, in which American forces required eleven days to advance less than 18 km, as a paragon of 'operational maneuver from the sea'.[106]

At the end of the twentieth century, few serving in the American ground forces remembered Eben Swift. Fewer still had any knowledge of 'The Lyceum at Fort Agawam'. The pattern established by that article, however, persisted for a century. Like Swift, American soldiers and Marines struggled with the fundamental paradox of war on land, the coexistence of the organizational need for order, and the inherently chaotic nature of armed conflict. They thus attempted to strike a balance between predictability and creativity, prefabrication and custom-tailoring. At times, this resulted in moments of brilliant improvisation, adaptation, and boldness, and even periods when such virtues were fostered in a systematic way. On the whole, however, Americans who fought on land, like Swift himself, preferred method to manoeuvre.

---

[104] For an official view of this enterprise, see Charles C. Krulak, 'Innovation, the Warfighting Laboratory, Sea Dragon, and the Fleet Marine Force', *Marine Corps Gazette*, December, 1996, 12–17. For a very different view, see, John F. Schmitt, 'A Critique of the Hunter Warrior Concept' *Marine Corps Gazette*, June, 1998, 13–19.

[105] For an example of these tendencies, see Paul A. Hand, 'Planning the Battalion Attack: A New Paradigm for an Old Process', *Marine Corps Gazette*, December, 1995, 22–28.

[106] United States Marine Corps, *Marine Corps Concept Paper 1: Operational Maneuver from the Sea* (Washington, DC: Headquarters, United States Marine Corps, 1996), 15 and Russell H. S. Stolfi, 'A Critique of Pure Success: Inchon Revisited, Revised, and Contrasted', *The Journal of Military History* 68, 2 (April 2004): 505–525.

13

People's Liberation Army Operations and Tactics in the Land Domain

Informationized to Intelligentized Warfare

Brad Marvel

Introduction

China's People's Liberation Army (PLA) is perhaps the most carefully observed and studied military in the world today. Forty years of near-constant reform radically altered the composition and capabilities of the PLA, transforming it from a poorly-equipped and poorly-trained revolutionary mob to a modernized and professionalized military. The modern PLA presents a true multi-domain capability set, an emerging joint backbone, and a unique operational structure built upon decades of relentless study and experimentation. Indeed, the PLA's modernization efforts are not yet complete: new operational concepts and new systems are under development and being integrated on a seemingly daily basis. As China moves toward its lofty national goals of the mid-twenty-first century, the PLA will seek to assert itself not only as a world-class military, but as a showpiece emblematic of Chinese resurgence as an international power.[1]

This chapter begins by discussing the historical background and the impetus for change that shaped Chinese military thinking, along with the strategic and political dynamics that influenced the PLA's era of modernization. It then moves into a detailed discussion of the PLA's current and future operational concepts, describing the modern Chinese way of war.

---

[1] Views and opinions presented in this chapter are those of the author alone, and do not necessarily represent those of the United States Government or Department of Defense.

Brad Marvel, *People's Liberation Army Operations and Tactics in the Land Domain*. In: *Advanced Land Warfare*. Edited by Mikael Weissmann and Niklas Nilsson, Oxford University Press. © Brad Marvel (2023). DOI: 10.1093/oso/9780192857422.003.0013

Part I: Building the Modern People's Liberation Army

The Era of Reform

The PLA of the late 1970s was a catastrophe. The tumultuous years of civil war and subsequent land reform, the brutality of the Cultural Revolution, and the chaos that followed Mao's death in 1976 left China with a bloated, dated, and unwieldy[2] military unprepared for the rigours of modern warfare. The PLA had far too many generals with far too much political power, a long-obsolete force structure that was a vestige of the Chinese Civil War and Korean Wars of previous generations, and mountains of ageing and increasingly useless obsolete equipment.[3] PLA Army (PLAA) tactics still centred on the mass infantry assault conducted at the corps echelon: combined arms operations—let alone joint operations—were virtually non-existent.[4]

As Deng Xiaoping consolidated his power through the late 1970s, his 'Four Modernizations' effort triggered what would become a decades-long era of reform for the PLA. Despite noting the relatively dire state of China's military, Deng deliberately rank-ordered his four main modernization efforts with the PLA at the bottom, behind the more economically-focused areas of agriculture, industry, and technology.[5] The reason for this decision was simple: Deng believed that without a renovated and liberalized economic backbone, China would be unable to pay for a new military. Interestingly, this bifurcation between economic development and national security would end up as one of the major features of Chinese national policy for a generation or more in the years that followed.

China's need for military modernization was demonstrated dramatically during the Sino-Vietnamese war in the late winter of 1979. Although PLA forces ostensibly achieved their very limited campaign objectives, performance virtually across the board was dire. Facing a Vietnamese opponent whose local forces consisted largely of militia and irregular troops, the PLA struggled to mount and sustain operations only a few dozen kilometres from the Chinese border.[6] Chinese casualties were extremely high for both soldiers and vehicles, whilst severe capability limitations in command and control,

---

[2] Benjamin Lai, *The Dragon's Teeth: The Chinese People's Liberation Army—Its History, Traditions, and Air Sea and Land Capability in the 21st Century* (Havertown, PA: Casemate, 2016).

[3] Defense Intelligence Agency, 'Military and Security Developments Involving the People's Republic of China', *Defense Intelligence Agency*, 2020.

[4] Lai, *The Dragon's Teeth*.

[5] Andrew Chuter, '30 Years: Deng Xiaoping—Enabling China's Rise', Defense News, 25 October 2016.

[6] Xiaoming Zhang, *Deng Xiaoping's Long War: The Military Conflict Between China and Vietnam, 1979–1991* (Chapel Hill: University of North Carolina Press, 2015).

firepower, joint integration, and combined arms manoeuvre were all laid bare for PLA and Communist Party of China (CPC) leaders to ponder.[7]

Deng remained steadfast in his rank-order of national priorities despite the PLA's dubious performance in the Sino-Vietnamese War, and thus, the PLA remained buried beneath the needs of the civil economy for virtually the entirety of the 1980s.[8] Development of the PLA's operational concept—now called *People's War in Modern Conditions* (现代条件下的人民战争)—limited professionalization, stop-and-start system modernizations, and the early stages of reduced politicization among PLA officers characterized the decade.[9] Deng's national strategy was otherwise quite successful—China's economic rise through the 1980s was one of the most dramatic in human history—but the PLA stagnated. As the USA and the Soviet Union went through their final modernization cycle of the Cold War, the PLA saw only modest improvements.

Two nearly back-to-back major global events kicked the PLA's modernization effort into overdrive. The first event was the military response to the Tiananmen Square demonstrations in the spring of 1989. In front of a huge global audience, the PLA demonstrated major fundamental shortcomings in basic military competencies: thousands of officers were openly insubordinate and PLA units proved completely unprepared for the task of breaking up the mostly-unarmed protestors. PLA units were eventually forced to resort to deadly force to suppress the uprising; this mix of incompetence and brutality playing out in front of the world's media severely damaged both PLA morale and prestige.[10]

The second major event was the Persian Gulf War, or Operation Desert Storm. With the wounds of their abysmal performance at Tiananmen still bleeding, China watched as a US-led coalition seemingly effortlessly decimated an Iraqi military that in many ways resembled the PLA. This set off serious questions about the PLA's ability to resist a similar campaign and drove a top-to-bottom reexamination of PLA doctrine, equipment, and training.[11] The lessons taken from careful study of both these events became the basis for today's modernized and professionalized PLA.

---

[7] Charlie Gao, 'The National Interest', 25 April 2021, https://nationalinterest.org/blog/reboot/how-1979-sino-vietnamese-war-made-china-superpower-183484.

[8] Chuter, '30 Years: Deng Xiaoping—Enabling China's Rise'.

[9] C. Fred Bergsten, Charles Freeman, Nicholas R. Lardy, and Derek J. Mitchell, eds, *China's Rise: Challenges and Opportunities* (Peterson Institute for International Economics, Washington, DC, 2008), ch. 9.

[10] Dennis J. Blasko, *The Chinese Army Today: Tradition and Transformation for the 21st Century*, 2nd ed. (London: Routledge, 2012).

[11] Michael Dahm, 'China's Desert Storm Education', *Proceedings*, U.S. Naval Institute, Vol 147/3/1,417, March 2021, https://www.usni.org/magazines/proceedings/2021/march/chinas-desert-storm-education.

The Chinese Strategic Construct and the PLAA's Modern Role

The PLA got its start as a revolutionary army, and many of these sensibilities still influence Chinese military thinking today. Mao's *People's War* (人民战争) outlined his vision for China's military as essentially an extension of Marxist-Leninist and Maoist political theory.[12] The early Chinese Red Army and PLA adopted *People's War* with enthusiasm, and went on to win great victories against both the Imperial Japanese Army and Republican Chinese forces in the Sino-Japanese War and Chinese Civil War respectively, using Mao's philosophies as their primary guidepost.

Mao's greater vision for China imagined a largely insular and self-sufficient nation with little need for any meaningful power projection away from Chinese shores.[13] The Mao-era PLA embodied this vision: a massive land force built to grind down opponents whilst fighting on Chinese territory. In many ways, this approach drew on the past: China faced serious and persistent existential threats to its territory throughout its history, and typically, Chinese forces found themselves fighting in their own backyard. Whilst this approach simplified strategic matters and granted significant tactical advantages to Chinese defenders, the economic, political, and human costs of centuries of fighting in their own territory was eye-watering.

As Chinese thinking evolved in the aftermath of Mao's death, the obvious shortcoming of *People's War*—that it limited major military actions to within Chinese borders—became more acute. The Gulf War drove this point home to the CPC and PLA leaders: it was now clear to all that passively waiting inside one's own territory as a powerful opponent massed combat power nearby was no longer a viable national defence strategy. The PLA's immediate response—*People's War in Modern Conditions*—introduced a radical departure from Mao's original *People's War*: *active defence* (积极防御).[14] Essentially, active defence enabled the PLA to pre-emptively engage threatening enemy forces away from Chinese borders in certain situations. The active defence concept initially manifested in the development of a litany of powerful, long-range interdiction and strike platforms largely designed to influence tactics and strategy in the Western Pacific.

Active defence soon evolved into a much broader expansion of power projection. As China's internal economy exploded, global export markets and the commerce lanes supporting them became critically important to China's

---

[12] Blasko, *The Chinese Army Today*.
[13] Edward Friedman, 'Reconstructing China's National Identity: A Southern Alternative to Mao-Era Anti-Imperialist Nationalism', *The Journal of Asian Studies* 53, 1 (1994).
[14] Alexander Chieh-cheng Huang, 'Transformation and Refinement of Chinese Military Doctrine: Reflection and Critique on the PLA's View', in *Seeking Truth From Facts: A Retrospective on Chinese Military Studies in the Post-Mao Era*, edited by James C. Mulvenon and Andrew N. D. Yang (Santa Monica, CA: RAND, 2012).

vision for its future. At the same time, China began asserting itself much more prominently on the global stage, laying down a sweeping plan outlining the CPC's vision for China's re-ascension to the status of a global power. This plan was neither vague nor farfetched: the CPC laid out a specific date (2049, the PRC's centennial anniversary) and specific goals to guide the process. Whilst most of these goals are economic, political, or diplomatic in nature, the PLA received its own metric it was to achieve: the status of a 'world-class military'.[15]

It is this endstate—the 'world-class military'—that today informs PLA leaders and developers. The quaint days of early *People's War* and its pure notions self-defence are long gone, replaced by an expressed desire to control and protect global sea lanes, deploy brigade- or even division-sized forces abroad, the expansion of overseas basing, and an ever-growing integrated network of powerful combat systems capable of ranging far into the Western Pacific.[16] Building these capabilities required the PLA to accelerate its ongoing reforms and necessitated a comprehensive rework of its professional development and military education, equipment, and structure. These reforms are ongoing and are likely not complete, although it appears the PLA's current structure will be in place for the foreseeable future.

The PLA's current operational concept is *People's War in Conditions of Informationization* (信息化条件下的人民战争) or just *Informationized Warfare*.[17] This concept evolved from *People's War in Modern Conditions* during the most recent period of reform. It displays a more mature understanding of non-kinetic and multi-domain capabilities, along with advanced, comprehensive reconnaissance and intelligence on top of the venerable fire-and-manoeuvre model that dominated *Modern Conditions*.[18] It is thought of by the PLA as an evolution of Mao's original *People's War*: despite its significant differences, the PLA does not consider it a refutation or rejection of Mao's original construct. *Informationized Warfare* is characterized heavily by a focus on winning 'local wars', the PLA's term for smaller, limited conflicts with regional opponents. The PLA viewed this objective as a critical waypoint along the path to its world-class military: winning a local war requires true joint integration and the ability to deploy and sustain an expeditionary force.

Moving into the mid-twenty-first century, PLA thinkers envision an evolution to *Informationized Warfare*: a future concept called *Intelligentized*

---

[15] Defense Intelligence Agency, 'Military and Security Developments Involving the People's Republic of China'.
[16] Department of the Army, 'ATP 7-100.3—Chinese Tactics', 2021.
[17] People's Liberation Army, *Army Combined Operation Tactics under the Conditions of Informationization* (Beijing: Shijiazhuang Army Command Academy Press, 2012).
[18] Department of the Army, 'ATP 7-100.3—Chinese Tactics'.

Warfare (智能化作战).[19] The informationized construct seeks to develop an agile and interconnected multi-domain force. The intelligentized construct envisions that same force enabled by the widespread and streamlined use of emerging technologies such as artificial intelligence, data analysis, and robotics. Intelligentized leaders are presented decision trees informed by mountains of data, simplified and distilled to enhance their situational understanding. The intelligentized war is won through superior decision-making and better manipulation and exploitation of data, allowing the PLA to dominate the information domain and gain a decisive advantage long before any shots are fired in battle.[20] In that sense, *Intelligentized Warfare* can be thought of as the ultimate modern expression of Sun Tzu's 'highest form of generalship'.

The Structure of the Modernized PLAA

The evolution of China's ground forces through the era of reform saw them move from a clumsy and obsolete corps-based model dating from the Korean War to a light, agile, and fully modern brigade-based model in a roughly 25-year timespan. Whilst this is a remarkably tight timeline by military modernization standards, the PLAA's development is the byproduct of many years of conceptual thinking, experiments—both successful and failed—and stop-and-start initiatives: the path to the modern PLAA was anything but straight. The solution the PLAA arrived at should look very familiar to western observers in many ways, although there are some critical differences that characterize a uniquely Chinese solution.

The Theater Command (TC) is the PLA's joint operational headquarters for a theatre. The Theater Command Army is the PLAA headquarters within the TC. Each TC houses some number of Group Armies (GA), the PLAA's new corps-level headquarters. The GA, in turn, houses twelve brigade-sized organizations: six combined-arms brigades (CA-BDEs) and six support brigades. The CA-BDE is the PLAA's basic tactical building block, and will be the focus of much of the rest of this document. CA-BDEs themselves are comprised of a number of combined-arms battalions (CA-BNs), along with reconnaissance, artillery, air defence, and support battalions.[21]

---

[19] Michael Dahm, 'Chinese Debates on the Military Utility of Artificial Intelligence', *War on the Rocks*, 5 June 2020, https://warontherocks.com/2020/06/chinese-debates-on-the-military-utility-of-artificial-intelligence/.

[20] Jerry A. Smith, 'Intelligentization: China's Road to AI Warfare Algorithms', 5 February 2020, https://www.linkedin.com/pulse/intelligentization-chinas-road-ai-warfare-algorithms-smith/. [Accessed August 2021].

[21] Department of the Army, 'ATP 7-100.3—Chinese Tactics'.

The PLAA developed three distinct varieties of the CA-BDE: light, medium, and heavy. The light CA-BDE consists of motorized infantry CA-BNs enabled by a high-density of dismounted anti-tank weaponry and lightweight, high-mobility tactical transport. The medium CA-BDE is considered mechanized infantry, built around CA-BNs consisting of both wheeled and tracked infantry fighting vehicles (IFVs), enabled by assault guns and mobile firepower. The heavy CA-BDE mixes armour and mechanized infantry companies in its CA-BNs, and is enabled by a powerful mix of heavy tube and rocket artillery and advanced air defence systems.

Study of the new PLAA structure alongside the PLA's operational concept clearly demonstrates China's vision for its ground forces, one that will evolve as informationized warfare transitions to intelligentized warfare. The heart of this vision is tactical system warfare.

Part II: From Informationized to Intelligentized Warfare

Tactical System Warfare

System Warfare is one of the basic conceptual constructs informing the PLA's capability development. The term 'system warfare' is very wide ranging and varies somewhat in translation, but the basic idea is quite simple and can be explained through two key ideas: the *operational system* (作战体系) and the *node* (节点).

The *operational system*[22] is the centrepiece of a system warfare campaign. It can be thought of as an evolution of the basic concept of a task force: a task-organized military unit assembled to achieve a specific goal or conduct a specific mission. The operational system simply approaches this concept from a perspective of tasks or missions in lieu of units.

The *node* is a blanket term for any targetable entity whose destruction or suppression reduces or impacts the performance of multiple other enemy battlefield systems.[23] In the same way as 'system warfare', node has a number of different terms and translations, but the basic idea is the same regardless of differences in wording. The most classic example of a node is the enemy's communications network or the enemy command team, but there are many divergent examples of nodes that are less-traditional: the radar systems supporting counter-fire or integrated air defences; the reconnaissance

[22] Edmund J Burke., Kristen Gunness, Cortez A. Cooper III, and Mark Cozad, *People's Liberation Army Operational Concepts* (Santa Monica: RAND Corporation, 2020).
[23] Department of the Army, 'ATP 7-100.3—Chinese Tactics'.

platforms that feed intelligence across a number of different subordinate units or services; or electronic defence assets that protect critical electronic warfare systems across the theatre.

Taken together, the *operational system* seeks to target and either exploit or destroy *nodes*, usually through an asymmetric pathway that avoids enemy strengths and targets enemy weaknesses.[24] Examples of this approach at the strategic and operational levels are numerous and oft-written about: the use of anti-ship ballistic missiles to destroy or standoff enemy surface vessels (in lieu of a major naval surface confrontation); the use of submarines or anti-ship missiles to target enemy transport vessels (in lieu of allowing the enemy to transport, land, and mass land combat power), or the use of information warfare to undermine the enemy's political will domestically (in lieu of directly confronting powerful enemy ground forces in close combat).

The PLA's approach at tactical echelons leverages the same system warfare construct scaled down to inform tactical-level operations. One additional concept is introduced in tactical system warfare: the *group*.[25] Groups are also nothing more than task organized entities, the difference being that the group is assembled in order to perform a single tactical task in support of the operational system. Whilst the tactical system warfare construct seems to place no limits on the levels of task organization an operational system designer might adopt, in practice, the PLAA's experience with task organization largely mirrors that of western militaries: that there is a careful balance to be struck between the desired capabilities and organizational familiarity. Thus, wild reorganizations are largely avoided: instead, groups are constructed with an organic unit—usually a CA-BDE or CA-BN—at the centre. This unit is then supplemented with additional capabilities drawn from the theatre, the group army, or even adjacent brigades or battalions.

PLAA doctrine lists a seemingly endless number of group types—virtually every type of ground domain mission is represented, and the group naming convention is non-standardized and can vary wildly between different PLA publications and reports. Examples and descriptions of some of the most commonly seen group types include:

- *reconnaissance and intelligence group*: supports the operational group's ISR and forward security missions;
- *advance group*: serves as the reconnaissance and security element in support of an offensive group;

[24] Ibid.
[25] Ibid.

- *frontline attack group*: the main body of an offensive group, responsible for fixing and assaulting enemy positions;
- *depth attack group*: group that advances through the enemy's main defensive line once the frontline attack group achieves a breach;
- *Thrust manoeuvre group*: a mobile reserve that exploits any advantages in enemy rear areas created by the depth attack group;
- *cover group*: the reconnaissance and security element supporting a defensive group;
- *frontier defence group*: the group that establishes the main defensive line at the heart of a defensive group;
- *depth defence group*: a mobile reserve that serves as the main counterattacking force in a defensive group;
- *firepower group*: a collection of direct, tube artillery, and rocket artillery that provides mass fires in support of the operational system.

The composition of these groups tends to align with their mission: forward groups are usually built around scouts or dismounted infantry, main line groups are often built around motorized or mechanized infantry, whilst reserve or depth groups tend to be built around mechanized or armoured forces[26].

Conceptually, the idea of groups aligns well with the objectives of tactical system warfare. The commander can build the operational system's subordinate groups such that they are optimized to exploit enemy weaknesses and target enemy nodes, whilst simultaneously offsetting enemy advantages in combat power and ensuring the operational group does not have an exploitable vulnerability in any domain. The PLA views this model in much the same way as the US Army: multi-domain warfare.

The PLAA and Multi-domain Warfare

Throughout the modern era, military services tended to allocate the vast majority of their development resources toward their primary domain: land for armies, sea for navies, air for air forces, and so on. Although there were numerous examples of cross-domain assets—such as ground-based air defences and naval gunfire support—the vast majority of a service's interest focused on fighting and winning a single-domain fight.

[26] Department of the Army, 'ATP 7-100.3—Chinese Tactics'.

Contemporary military thinkers are taking the first steps to move away from this one-domain model. The reason is fairly simple: advanced modern systems now feature the range, precision, lethality, versatility, and targeting support to significantly influence multiple domains either sequentially or simultaneously. For example, long-range surface-to-surface fires can target ships at sea or distant airfields; long-range surface-to-air fires can effectively deny the use of vast areas of airspace to threat aircraft; and air-launched precision standoff munitions can penetrate dense air defences and target key assets in rear areas. In addition, new domains—space, cyber, and the electromagnetic spectrum—promise to influence future military operations at all echelons. In short, the modern battlefield features more dangers from more directions than ever before, whilst simultaneously offering greater opportunities for aggressive and creative leaders to leverage new systems, tactics, and techniques.

These two emerging trends—vulnerabilities seemingly from all directions coupled with more and better options for attack—were coupled together to form the multi-domain concept. In the USA, it was first called 'cross-domain', and eventually, 'Multi-Domain Operations'.[27] In China, it is referred to as all-domain, cross-domain, and multi-domain. The basic concept, however, is the same across all of these terms. It consists of two basic ideas:

1. your force must have few or no exploitable vulnerabilities in any domain; and
2. your force must be capable of synchronizing capabilities from across multiple domains to create windows of opportunity by exploiting enemy vulnerabilities.

It is a simple, even anachronistic approach to warfare in many ways, with a certain obviousness that prompts constant observations that 'this is nothing new'. Implementation of the concept, however, is enormously complex. It requires careful planning, a robust, protected, and agile command and control background, extensive sustainment, and extraordinarily well-trained leaders and planners.

Careful study of the US joint concept 'Air-Sea Battle' starting in the early 2010s seemed to kick PLA interest in multi-domain concepts into high gear.[28] Specifically, the ostensibly direct counters to developing Chinese capabilities in the Western Pacific not only alarmed Chinese military thinkers,

---

[27] Department of the Army, 'ATP 7-100.3—Chinese Tactics'.
[28] Dean Cheng, 'Chinese Lessons from the Gulf Wars', in *Chinese Lessons From Other People's Wars*, edited by Andrew Scobell, David Lai, and Roy Kamphausen (Carlisle, PA: Strategic Studies Institute, U.S. Army War College, 2011).

but prompted them to carefully examine their own approach to system warfare. Air-Sea Battle focused heavily on 'blinding' enemy sensors and suppressing enemy command and control whilst fighting from concealed or hardened positions. This approach largely mirrored the Chinese system warfare construct, and seemed to spur PLA thinking aimed more at reducing vulnerabilities in addition to asymmetrically attacking one's opponent.

The PLA's most recent large-scale reform effort in 2015 showed a great deal of influence from the nascent multi-domain concepts. First, the Chinese understanding of the definition of 'domain' crystallized. Two basic definitions emerged:

> Cross-domain operations refer to both cross-regional operations and cross-domain use of troops. The basic form of combat in future information warfare is integrated joint operations, and the realization of integration will center on cross-domain operations. Cross-domain operations are an important method to seize and maintain combat superiority, and it also makes joint operations present new changes different from the past.[29]

This understanding of 'domain'—as a cross-regional issue in addition to the traditional definition—made sense considering the PLA's emerging expeditionary mission. That China viewed domestic and regional/international areas of operations as distinct domains shows how significant the change was in the mindset between *People's War* and *People's War in Conditions of Informationization* regarding international power projection and military deterrence. It also showed that the PLA understood the severity of the challenges they faced in building an expeditionary force virtually from scratch.

Second, the PLA clearly articulated how the multi-domain concept dovetailed with their system warfare construct:

> Cross-domain operations transform the acquisition of operational advantage from capturing and maintaining symmetrical advantages to capturing and maintaining asymmetrical advantages. Symmetrical superiority contest refers to comparing the performance and scale of similar weapons and equipment in the same combat field, seeking combat superiority by the generational difference or quantity of weaponry and equipment, and using strength to defeat strength. The Asymmetric approach uses comparative advantages of different combat areas. Instead of launching a 'dignified battle' with the opponent, we focus on the opponent's weakness and use our strength to defeat the enemy's weakness. Cross-domain

[29] People's Liberation Army News, 'Cross-domain Operations Enhance Joint Operations Superiority', 22 August 2017, http://news.sina.com.cn/o/2017-08-22/doc-ifykcirz3693290.shtml.

operations fully reflect this: Domain superiority means leveraging the asymmetric advantages of other domains relative to the operational domain to achieve unexpected effects.[30]

Finally, the PLA reflected how this new concept would influence the organization of military forces as they arrayed for battle:

> Cross-domain operations have transformed the use of combat forces from 'distribution' to 'combination'. With the continuous development of cross-domain capabilities and the continuous growth of new quality combat forces, the previous model of using the military to distinguish between strategy, campaign, and tactics will be broken ... the former force grouping mode that allocates tasks based on capabilities will change to the combining combat forces based on tasks; the previous hierarchical command and deliberative collaborative 'quota'-style power control method will also change. It will be replaced by networked command and autonomous coordination of aggregated power control methods.[31]

Thus, the operational concept of the future PLAA emerged. The PLAA intends to fight as a multi-domain force called the *operational system*. This system is task-organized based on mission and enabled by advanced command networks and autonomous combat multipliers, and seeks to detect, target, and exploit enemy weaknesses whilst avoiding enemy strengths.

The PLAA's Operational Construct

All of the PLA's subordinate services are working to incorporate system warfare and multi-domain warfare ideas into their capability portfolios. The PLAA found itself in the midst of a substantial renovation right about the time these concepts began to gain traction, and so found itself somewhat conveniently able to implement and test various new methodologies with relative ease.[32] This sort of agility and adaptability was unheard of in the PLAA only a generation before and is one of the biggest success stories of the PLA's reformation efforts.

The PLAA's combat training centres (CTCs) are a mainstay of this era of learning. Conceptually, the PLAA's CTCs strongly resemble comparable US/NATO training facilities: vast training areas able to support a full

---

[30] Ibid.
[31] Ibid.
[32] Kevin McCauley, 'The People's Liberation Army Attempts to Jump Start Training Reforms', *The China Brief* 21, 3 (2021).

CA-BDE operation, substantial observer/trainer support, an opposing force (OPFOR) free to act as a thinking, adaptive enemy, and a comprehensive after-action review process designed to improve unit training and publish lessons learned from each rotation.[33] The PLAA enthusiastically adopted CTC rotations as flagship exercises, noting in particular how tough and realistic training exercises can help offset the PLAA's lack of real-world combat experience. This new approach to training and capability development is still immature and far from perfect—as discussed later in this document—but it is a fantastic illustration of the PLAA's modernized approach to land combat. Indeed, it is now very common for PLAA brigades to 'lose' the fight at their rotation—the kind of thing that was quasi-unthinkable just a generation before.[34] Amusingly, the PLAA's OPFOR is informally referred to as the 'Blue Brigade', whilst friendly forces are 'Red Force', an inversion of the US/NATO naming convention.

Although the PLA's classification of domains is not yet firmly codified, examination of the PLAA's operational construct shows recognition of three general domain types, each with its own specific challenges and its own specific solutions: the *physical domain*, the *electronic domain*, and the *cognitive domain*.[35] Whilst these domains are conceptually separate, cross-domain capabilities reinforce one another: strengths of capabilities in one domain offset weaknesses in another. In other words, multi-domain warfare is simply an extension of the ancient idea of combined-arms.

The Physical Domain

The physical domain encompasses those combat elements that directly, kinetically contest one another on land, sea, air, and in space. Confrontation in the physical domain is probably the oldest basic military concept, and remains the best-understood by both Western and Chinese leaders and planners. The PLAA's operational approach in the physical domain consists of building operational systems that have no obvious vulnerabilities and enjoy one or more major advantages over their opponent.[36] Firepower is the advantage most-often sought: the PLAA's density of tube and rocket artillery, ballistic missiles, and direct-fire weapons systems makes an assemblage of overwhelming firepower simple and straightforward. Groups must

---

[33] Gary Li, 'The Wolves of Zhurihe: China's OPFOR Comes of Age', *China Brief* 15, 4 (2015).

[34] People's Liberation Army Daily, 'Practical Admonitions from Zhurihe in the Smoke of Gunpowder', 29 November 2018, http://military.people.com.cn/n1/2018/1129/c1011-30431912.html.

[35] Various PLA writings also recognize a separate 'information' domain, or separate maritime, air, and space apart from the land domain. It is not yet clear if or how the PLA will finalize or codify its definitions of domains.

[36] Department of the Army, 'ATP 7-100.3—Chinese Tactics'.

be protected from enemy action by some combination of deception, armour, and manoeuvre, whilst enemy air power and firepower must be suppressed or offset by air defences and counter-fire, respectively.

In practical terms, this infers some basic practices. CA-BDEs will likely be reinforced by either attached or direct support from powerful GA artillery units in the form of an artillery or firepower group. Organic GA air defence assets—mostly guns and short-range air defence systems—are augmented by a mix of PLA Air Force long-range surface-to-air missile systems and manned aircraft. Reconnaissance and intelligence blends classic ground-based recon-naissance missions, such as scouting with advanced ISR platforms ranging from short-range drones all the way up to national-level intelligence assets. Manoeuvre groups seek to fix enemy defences with a mix of feints and demonstrations, then penetrate areas of weakness with powerful mechanized groups that can wreak havoc in rear areas.

All of this is accomplished through the building of task-oriented groups, built to bring a specific capability set to a specific mission, unified under a single command and control umbrella. These groups then target and exploit enemy weaknesses—asymmetrically wherever possible—in accordance with the basic principles of system warfare. In a recent interview, a PLA 'Blue Brigade' commander summed this approach up thus: 'the blue force brigade attaches exceptional importance to all new-type combat forces, including those established within its organization and those being put under its command on an ad hoc basis in wartime.'[37]

The Electronic Domain

The electronic domain encompasses the entirety of electronic warfare and cyber warfare: those military actions contested in the electromagnetic spec-trum (EMS) or over computer information networks. Whilst the military contest in the electronic domain is now well over a century old, modern reliance on information networks and advanced electronic emitters make the military contest in the EMS and cyberspace equal in importance to the phys-ical. The PLAA long sought advantages in electronic and cyber warfare as a way to offset threat advantages in weapons systems and soldier training; this approach continues to the present day.

The central idea governing PLAA operations in the electronic domain is *synthesis*.[38] Both EMS and cyber contests are viewed as essentially zero-sum: any advantages gained in information superiority or spectrum dominance

---

[37] Li, 'The Wolves of Zhurihe: China's OPFOR Comes of Age'.
[38] Department of the Army, 'ATP 7-100.3—Chinese Tactics'.

can be offset by vulnerabilities exploited by one's opponent. Thus, offensive and defensive electronic actions must be synthesized: one cannot achieve results without the other. Moreover, synthesis between disparate systems should be sought: the effects of a single cyber or electronic warfare platform may be easily offset through simple countermeasures, but the combined-arms effect of multiple platforms targeting multiple enemy vulnerabilities greatly increases the chances of a decisive outcome.

In practical terms, PLAA operational systems seek to integrate a variety of electronic domain systems—both organic and external—into a synergistic construct that both protects their formations and enables attacks on enemy systems. This blending of capabilities is referred to as *synthetic quality*.[39] The PLAA recognizes two basic actions in the electronic domain: *electromagnetic attack/protection* (actions on the EMS), and *network attack/protection* (actions on computer networks). PLAA units employ capabilities like wide-area jammers to suppress enemy over-the-air communications and emitters, and passive electronic reconnaissance platforms to collect and exploit enemy EMS targets. These systems were traditionally ground-based, but an increasing number are now deployed on manned and unmanned aircraft. The PLAA generally wants its actions in the electronic domain to be clandestine or invisible to the enemy: this precludes the enemy from employing countermeasures and enables PLAA reconnaissance and intelligence assets to collect from the enemy as long as possible. At tactical echelons, electronic domain capabilities will likely be centralized in the *Electronic and Network Warfare Group*, a task-organized unit attached to the operational system's command group.

The Cognitive Domain

The contest in the cognitive domain cuts to the heart of competition: what one's opponent believes is ultimately the most important factor determining the outcome of a contest. This idea is at the heart of Sun Tzu's writings and forms the basis for the PLAA's modern understanding of what they call 'cognitive domain operations' (认知域作战). Through most of human history, military effects in the cognitive domain were achieved by actions in the physical domain: military force was employed to impose one's will on the opponent. One achieved victory by convincing the opponent that additional resistance was pointless, which then allowed the victor to dictate the outcome of the competition. Although past generations had limited means for contesting each other in the cognitive domain, it was not until the information age that this domain became truly exploitable on its own.

---

[39] Ibid.

Chinese thinkers began studying Western-style psychological operations in the early 2000s, concerned mainly with the possibility of dissidents and competitors undermining CPC authority on the early internet and social media. As the information environment expanded exponentially through that decade, the idea of 'cognitive security' (国家认知空间安全) became a major point of emphasis for the CPC.[40] By 2018, the PLA had expanded its understanding of cognitive warfare to include strategic information operations targeting competitors' domestic information environments on a vast scale, all the way down to tactical actions on the battlefield seeking to undermine the opponent's will and pollute the opponent's situational understanding. The endstate for these actions is referred to as 'mind superiority', or more dramatically, 'brain control' (制脑权).[41]

PLAA tactical actions in the cognitive domain are governed by principles similar to those in the electronic domain: attack and defence are simultaneous, and the contest is zero-sum. The biggest difference in cognitive domain actions is efforts undertaken at much higher echelons and at much more distant ranges—those actions, for instance, targeting enemy political decision-makers and civilians—can have a direct effect on tactical confrontations: for instance, an enemy undermined by a lack of political will at home may suffer reduced morale or focus. This concept can be broken down even further: key enemy personnel may be targeted for psychological attack, using disinformation propagated through platforms like social media to influence their state of mind. The broad term for these actions is *psychological warfare* (心理战), a term very familiar to Western audiences.[42]

Direct psychological warfare operations seek to attack the enemy's *conviction* and *understanding*.[43] Attacks on conviction target enemy willpower and morale, whilst attacks on understanding target the enemy's decision-making process and situational understanding. The objective of tactical psychological warfare is to achieve an *information trap*, leading the enemy into make bad or inefficient decisions due to poor information or a compromised state of mind. Techniques for achieving an information trap range from simple deception efforts—concealment, demonstrations, feints—to advanced disinformation efforts propagated through media or data networks. As with information domain efforts, cognitive domain efforts are best employed such that the enemy is unaware of their presence.

---

[40] Nathan Beauchamp-Mustafaga, 'Cognitive Domain Operations: The PLA's New Holistic Concept for Influence Operations', *China Brief* 19, 16 (2019).

[41] Shen Shoulin Zhang Guoning, 'Recognize intelligent warfare', 1 March 2018, http://www.81.cn/jfjbmap/content/2018-03/01/content_200671.htm.

[42] Department of the Army, 'ATP 7-100.3—Chinese Tactics'.

[43] Ibid.

The PLA believes the importance of the cognitive domain will grow rapidly in the future, to the point where it may become warfare's new high ground. *PLA Daily* summed this idea up thus in 2017: 'With the development of brain science and technology, a new form of warfare with the brain as the center and the cognitive space as the domain is quietly starting. The issue of fighting for "brain control" has become a new strategic commanding height issue for the world's military powers to compete with each other.'[44]

Bringing it all Together: The *Stratagem*

Throughout the PLA's era of reform, the idea of the *stratagem* (计策) remained a more or less constant feature. The *stratagem* is one of the oldest and most closely-held concepts in Chinese military philosophy, one that influences Chinese thinking from the highest levels of politics all the way down to tactical military operations. The basic idea of the *stratagem* is winning a confrontation through a trick, a plan, or a scheme: simple enough in theory. Yet, it is this idea more than any other that differentiates the Chinese way of war from that of many of their competitors: whilst powerful militaries like the USA may include ideas like deception and trickery as one part of their doctrine, for the PLA, they are the centrepiece. Indeed, the PLA's understanding of multi-domain operations and their system warfare construct are all built around winning through the *stratagem*.[45]

For the PLAA, implementing the *stratagem* demands comprehensive integration of a wide variety of systems, tactics, and techniques. Physical, electronic, and cognitive domain capabilities must all be leveraged, employed through a carefully-constructed operational system. This effort begins long before any shots are fired on the plane of tactics: information and psychological warfare efforts work to undermine the enemy's home front and political support for the conflict, whilst a variety of long-range weapons systems attempt to deny access and dis-integrate enemy units before they can mass combat power against PLAA units.

Once the tactical fight commences, different elements of the PLAA's operational system are synchronized to present the enemy with multiple dilemmas, whilst an information attack manipulates the enemy's mindset, reduces morale and cohesion, and encourages the enemy to make the wrong decisions. Once an opportunity presents itself—ideally through an enemy

[44] People's Liberation Army Daily, 'Brain-making Warfare: A New Model of Future War Competition', 17 October 2017, http://military.people.com.cn/n1/2017/1017/c1011-29592326.html.
[45] Department of the Army, 'ATP 7-100.3—Chinese Tactics'.

mistake—decisive action is taken to exploit the breach and culminate in tactical action. Attack through the cognitive domain is relentless: once the PLAA commander is inside the enemy's decision cycle, that advantage must not be surrendered. The fusion of direct physical actions and threats, isolation and disorientation sowed by electronic attack, and confusion and hopelessness sowed by psychological attack all work together to convince the enemy unit they are defeated. In essence, the *stratagem* at the tactical level views the enemy's mindset as the centre of gravity, and the enemy unit's destruction as the objective.

Ongoing Challenges and Lessons

Any military professional or observer reading a document like this—or the PLA writings it was derived from—is sure to notice the high density of jargon and ornate language coupled with a very demanding and largely untested operational concept. Scepticism is certainly warranted: modern militaries have a long and colourful history of developing unworkable concepts veiled under buzzwords, and the PLA is certainly no exception to this practice.

It is a mistake, however, to think that the CPC and the PLA/PLAA are unaware of their current weaknesses, or that they are similarly unaware of the huge challenges that come along with implementing their operational concepts. In contrast to the CPC's well-deserved international reputation for secrecy—and sometimes duplicity—when it comes to internal Chinese matters, when it comes to the PLA, both the Chinese government and its military are remarkably willing to acknowledge shortcomings publicly. Xi Jinping himself acknowledged some of the PLA's most critical issues in a very public way with his 'Five Incapables', a mix of public speeches and articles published in 2015.[46] Xi called out the PLA as not building the military leaders capable of executing its operational construct; this public rebuke became enormously influential as the PLA worked to modernize its leader development.

The PLA/PLAA face numerous other challenges to fully realizing its operational concepts. This section discusses four in detail: lower echelon leadership; the C2 backbone; organizational inertia and corruption; and a lack of combat experience.

Lower Echelon Leadership and the NCO Corps

Fundamental leadership issues formed the basis for Xi's 'Five Incapables'. Although Xi's shotgun blast was directed at the entirety of the PLA, arguably

---

[46] Dennis J. Blasko, 'The Chinese Military Speaks to Itself, Revealing Doubts', *War on the Rocks*, 18 Febuary 2019, https://warontherocks.com/2019/02/the-chinese-military-speaks-to-itself-revealing-doubts/.

the most critical limitation in the entirety of the PLA is tactical-level unit leadership within the PLAA. *Informationized Warfare* envisions an agile, creative, and free-thinking force able to rapidly react to battlefield conditions and execute missions independently.[47] PLA junior officers, however, simply do not possess the sort of initiative to execute this concept as envisioned, and the PLAA's organizational culture is struggling to grow it. Or, as the 'Five Incapables' put it, 'Some cadre cannot (1) judge the situation, (2) understand the intention of the higher up authorities, (3) make operational decisions, (4) deploy troops, and (5) deal with unexpected situations.'[48]

So, too, lower-echelon leaders are hesitant to adopt the litany of changes in thinking and fighting brought about by years of constant reform. As one writer in 2015 described:

> during training or exercises, individual commanders did not study enough new combat forces and used them improperly. Some use new combat forces to 'support the facade', only for excitement, regardless of actual results ... some are immersed in traditional combat methods, so that they can only Run away from the system, or fight alone outside the system. Unless these phenomena are overcome, it will be impossible to give full play to the powerful role of the new combat force[49].

The PLA is addressing some of these shortcomings through the introduction of a professionalized NCO corps. The idea of the career NCO in the PLA is shockingly new: it was not until the very end of the twentieth century that the PLA started the programme. This first cohort of career enlisted professionals is now just reaching the end of their 30-year service term. Whilst NCOs successfully replaced officers at numerous technical billets and key staff positions, they have yet to firmly establish their positions as free-thinking combat leaders, mentors, and—when necessary—voices of experience and reason to the officer corps.[50]

In short, resolving the seemingly competing needs of an autocratic and hierarchical system of government with the PLA's need to recruit, train, and develop creative and independent-minded lower-level military leaders has not yet been truly resolved.

---

[47] People's Liberation Army, *Army Combined Operation Tactics under the Conditions of Informationization.*

[48] Dennis J. Blasko, 'The New PLA Joint Headquarters and Internal Assessments of PLA Capabilities', 21 June 2016, https://jamestown.org/program/the-new-pla-joint-headquarters-and-internal-assessments-of-pla-capabilities/.

[49] People's Liberation Army Daily, 'The New Combat Force of the People's Liberation Army Exercises Changed from "running a dragon" to the Protagonist of the Battlefield', 15 July 2015, http://military.people.com.cn/n/2015/0715/c172467-27305283.html.

[50] Marcus Clay and Dennis J. Blasko. 'People Win Wars: The PLA Enlisted Force, and Other Related Matters', *War on the Rocks*, 31 July 2020, https://warontherocks.com/2020/07/people-win-wars-the-pla-enlisted-force-and-other-related-matters/.

Command and Control Backbone

Perhaps no other technical capability undermines new operational concepts as much the command and control (C2) backbones built to support them. Combat developers envision vast, agile, interconnected forces and rapid, accurate sharing of information and intelligence, but in the field, network and communication backbones often fail to deliver the necessary speed, bandwidth, security, and resilience. So too do militaries conceptualize rapid and seamless task-organization, often overlooking the inherent challenges in blending units with very different capability sets and organizational culture.

Informationized Warfare seems to gloss over these challenges in large part, whilst *Intelligentized Warfare* presumes significant advances in networking, artificial intelligence, robotics, and data management will offset them.[51] Whilst huge efforts are being directed at developing the sort of robust and agile C2 structure these operational concepts demand, the PLAA struggles with both the technical aspects of modern C2 systems and with the free-flowing assemblage of forces envisioned. A brigade commander described this in 2015:

> The real difficulty in transformation of the armed forces is not just in acquiring new technology and building network systems, but rather in changing direction to develop new types of organizational structure capabilities having precision, flexibility, and integrated characteristics.

Organizational Inertia and Institutional Corruption

The most prominent pillar of Xi Jinping's domestic agenda is ending the culture of corruption that long-permeated Communist China (and, really, China long before that) and its government. This effort put the PLA squarely in his crosshairs: institutional corruption in China's enormous and politically active military was not only prevalent, but was, in many ways, a structural component of its daily operations. The defence industry and civil engineering projects were at the heart of this institutional corruption; the PLA was actually expected to supplement its national budget through graft and enterprise through things like military-run businesses and a robust military-backed real-estate industry.[52]

The general officer corps raised in this environment remained firmly entrenched as Xi took power and the PLA began its most recent round of reforms. Almost overnight, a low-grade civil war broke out between the

---

[51] Department of the Army, 'ATP 7-100.3—Chinese Tactics'.

[52] World Peace Foundation, 'China's Crackdown on Military Corruption', 2017, https://sites.tufts.edu/corruptarmsdeals/chinas-crackdown-on-military-corruption/.

old-school Chinese generals and the forces of reform. Historically it has often proved suicidal for the leader of an autocratic country to so directly take on the military, but Xi's ruthlessness and political savvy proved equal to the task: hundreds of PLA generals were dismissed, forcibly retired, and occasionally even imprisoned whilst Xi's position seemed completely unthreatened.[53] Whilst senior PLA personnel sometimes resisted these efforts—and usually lost their arguments—Xi's approach built a cadre of tremendously loyal mid-career officers. This cohort stood to gain much from the ongoing reforms: a better and more capable military, more opportunities for promotion, and greater national prominence for the PLA.

The outcome of Xi's war with his generals and the corresponding era of reform is still unclear. Ridding the force of ageing officers unable to embrace modernization seems superficially prudent, but a great deal of organizational knowledge and experience went with them. CPC control over the PLA has never been more assured, but, as discussed above, that brand of autocratic leadership seems directly at-odds with the PLA's operational concepts.

Lack of Combat Experience

The PLA has not conducted a major military operation in over 40 years. In this time, they have been through multiple major reform efforts and watched the rise of ongoing competition and confrontation with numerous regional and international powers. They have also watched as their pacing threat—the USA—fought its way through three major expeditionary military campaigns and numerous smaller excursions. Many US allies were also involved in these operations as well, creating one of the most combat-experienced military cohorts in human history.

The PLA is acutely aware of its lack of real-world combat experience. In fact, the PLA even coined a term for the worrying trends this lack of experience bred: the 'peace disease' (和平病).[54] This issue became a core talking point for Xi in the 2019 timeframe, and corresponds well with the rise in prominence of the PLAA's CTCs and other major military exercises. The PLA does not believe that training exercises can replace combat experience, but is firmly committed to creating the toughest and most realistic training environment possible in order to offset this disadvantage as much as possible.[55] So too must the litany of new concepts, ideas, systems, and personnel

---

[53] Dennis J. Blasko, 'Corruption in China's Military: One of Many Problems', 16 February 2015, *War on the Rocks*, https://warontherocks.com/2015/02/corruption-in-chinas-military-one-of-many-problems/.

[54] Liu Faqing, 'Rectify "peace sickness" and Consolidate Combat', 21 June 2018, http://www.81.cn/jfjbmap/content/2018-06/21/content_209078.htm.

[55] China Military Network Ministry of National Defense Network, 'Pay Attention to Actual Combat Military Training', 7 January 2019, http://www.81.cn/jfjbmap/content/2019-01/07/content_224687.htm.

be experimented upon and validated. The PLA is pushing forward with these efforts enthusiastically, but the spectre of failure in a real-world combat scenario is a constant dark cloud over the PLA's assessments of their readiness and combat power.

Conclusions

The PLA presents a sophisticated and well-constructed approach to developing a modern military force. Through a mix of meticulous study and years of experimentation, China's military thinkers arrived at a uniquely Chinese approach to modern warfare: an approach that is now, in turn, carefully studied by a litany of analysts worldwide. This approach is defined by a focus on multi-domain warfare, enabled by an agile force structure and the leveraging of a wide variety of different capability sets. This very modern vision is all constructed to support that most-Chinese of concepts: the *stratagem*.

Indeed, the *stratagem* helps describe much of how the PLA intends to approach conflict now and for the foreseeable future. By dominating the information or cognitive domain, Chinese forces believe they can influence, manipulate, and exploit the cognitive weaknesses of enemy forces, leading the enemy to make bad decisions, or even achieving victory without direct confrontation. When direct confrontation occurs, the PLA envisions a force capable of rapidly massing and decisively employing combat power to target and exploit enemy weaknesses in all domains. The PLA's focus on agility and adaptability underpins this operational concept at all echelons.

Assessing the true combat readiness of the PLA and PLAA remains a challenge both for external observers and for the Chinese themselves. Studying Chinese doctrine and military theory gives a clear understanding of where the PLA *wants* to go, but assessments of real-world units and particularly their performance in high-profile and realistic training scenarios send very mixed signals. As the PLA moves from *Informationized Warfare* to *Intelligentized Warfare*, observers and analysts must carefully assess how effectively the PLA implements their operational concepts and how effectively they deal with the technical, tactical, and social challenges they face as they try to realize their vision of a 'world-class military'.

14

A Strategy of Limited Actions

Russia's Ground-based Forces in Syria

Markus Balázs Göransson

Introduction

Russia's intervention in Syria (2015-) has marked a new direction in Russian military power projection abroad. It is Russia's first military operation outside of the former USSR since the Soviet-Afghan War (1979–1989), its first military campaign spearheaded by the aerospace forces and its first-ever expeditionary war. It is also Moscow's first long-range deployment of military forces since Nikita Khrushchev's dispatchment of Soviet troops to Cuba in Operation Anadyr in 1962.

Russia's military brass have stressed the importance of the Syrian intervention as a pivot in Russia's use of military force abroad. Russia's chief of the General Staff, Army General Valeriy Gerasimov said in 2019 that the intervention fits within 'a strategy of limited actions', developed to enable Russian forces to carry out 'tasks to defend and advance national interests outside the borders of Russian territory'.[1] Gerasimov noted that the strategy involves 'the creation of a self-sufficient grouping of forces based on force elements of one of the branches of the Russian Armed Forces possessing high mobility and the capability to make the greatest contribution to executing assigned missions', and that it relies on 'securing and retaining information superiority, advanced command-and-control and all-round support, and covert deployment of the necessary grouping'.[2] As observers have pointed out, this applies to a greater or lesser extent to the Syrian campaign, where Russia's

---

[1] Valeriy Gerasimov, 'The Development of Military Strategy under Contemporary Conditions. Tasks for Military Science', *Military Review*, November 2019, translated by Harold Orenstein and Timothy Thomas, https://www.armyupress.army.mil/journals/military-review/online-exclusive/2019-ole/november/orenstein-gerasimov/.

[2] Valeriy Gerasimov, 'V RF razrabotana strategiya ogranichennykh deystviy po zashchite ee interesov za predelami natsional'noy territorii—Gerasimov' [In the Russian Federation a strategy of limited action has been developed for the defence of its interests outside its national territory], *Interfax—AVN*, 2 March 2019, https://www.militarynews.ru/story.asp?rid=1&nid=503181&lang=RU.

Markus Balázs Göransson, *A Strategy of Limited Actions*. In: *Advanced Land Warfare*. Edited by Mikael Weissmann and Niklas Nilsson, Oxford University Press. © Markus Balázs Göransson (2023). DOI: 10.1093/oso/9780192857422.003.0014

aerospace forces have been assigned the lead role and where Russian military deployment has been marked by mobility and flexible decision-making.[3]

Russia's intervention in Syria has been described as an aerial or a non-contact operation. This is accurate to a degree, for the intervention has relied on the Russian aerospace forces, novel 4CISR technology and precision-guided missiles to minimize direct contact between Russian forces and the Syrian armed opposition. Yet the Syrian intervention is no Russian repeat of NATO's Operation Allied Force in Kosovo 1999, with which it is sometimes compared.[4] Rather, the operation has involved an important but overlooked ground-based contingent comprised of artillery troops, naval infantry, special operations forces (*spetsnaz*), military police, military advisors, and others. It is this land warfare component that sits at the centre of the present chapter and will be considered in terms of its role in the intervention. What part has this ground-based contingent played within overall Russian force employment in Syria? This is the question that will be addressed.

The chapter is divided into four parts. First it surveys other research about the Russian ground-based contingent, research that is illuminating though limited. Then it provides an overview of Russia's ground-based contingent in Syria. Third, it discusses six key strategic functions of the ground-based forces in Russia's overall force employment in Syria. Finally, it summarizes the findings, whilst reflecting on their implications for the development of Russian land warfare. The chapter covers the period between the start of Russia's military intervention in Syria in 2015 and 2021 and therefore does not address the changes wrought to the Russian military contingent in Syria after Russia's full-scale invasion of Ukraine in February 2022.

In examining the Russian ground-based contingent in Syria, the chapter draws on Russian- and English-language sources, including newspaper articles, think tank reports, and academic publications. Much of the material is, directly or indirectly, based on Russian or Syrian primary sources. This is important to note, for as Russia expert Timothy Thomas has pointed out, independent first-hand reports on Russian military actions are limited,[5] making much of the common understanding of Russian warfare in Syria contingent on information divulged by Russian and Syrian sources.

---

[3] Cf. Roger McDermott, 'Gerasimov Unveils Russia's "Strategy of Limited Actions"', *The Jamestown Foundation*, 6 March 2019, https://jamestown.org/program/gerasimov-unveils-russias-strategy-of-limited-actions/; Marina Miron and Rod Thornton, 'Emerging as the "Victor"(?): Syria and Russia's Grand and Military Strategies', *The Journal of Slavic Military Studies* 34, no. 1 (2021): 1–23; Dmitry Adamsky, 'Russian Lessons from the Syrian Operation and the Culture of Military Innovation', *Marshall Center Security Insights* 47, February 2020, https://www.marshallcenter.org/en/publications/security-insights/russian-lessons-syrian-operation-and-culture-military-innovation.

[4] Cf. Seth Jones, 'Russia's Battlefield Success in Syria: Will it Be a Pyrrhic Victory', *CTC Sentinel* 12, 9 (October 2019), https://ctc.usma.edu/russias-battlefield-success-syria-will-pyrrhic-victory/.

[5] Timothy Thomas, 'Russian Lessons Learned in Syria. An Assessment', *MITRE Center for Technology and National Security*, June 2020, 2, https://www.mitre.org/sites/default/files/publications/pr-19-3483-russian-lessons-learned-in-syria.pdf.

Undoubtedly, scholarly research of Russia's intervention will continue to evolve as more information comes to light.

Attentive readers will have noticed that the chapter uses the phrase 'ground-based forces', not ground forces. This follows on Russia researchers' Charles Bartles and Lester Grau's employment of that phrase in their study of the Russian ground-based contingent.[6] It captures the situation in which Russia's ground-based force constellation in Syria comprises also units not part of Russia's regular ground forces. The naval infantry, for example, belongs to the Russian navy, whilst private military companies (PMCs) that have been employed in Syria are nominally independent but often practically aligned with Russia's regular forces. Indeed, the PMCs are no stand-alone appendix to regular forces but operationally and strategically integrated with them. The widened vocabulary used in the chapter reflects a widened understanding of the ground military assets that Russia has used to project power in Syria.

Previous Research on Russia's Ground-based Forces in Syria

Western and Russian military thinkers have tended to foreground the actions of Russia's aerospace forces and play down the role of its ground-based forces. Russian airstrikes, supplemented by standoff weapons, have represented the most visible and powerful use of Russian kinetic power in Syria, whilst the initial thrust of the Russian operation was toward the use of aerial power, with the operation morphing into a more heavily ground-based one only later. Furthermore, early Russian public messaging signalled that there was no Russian ground war in Syria and that Russian special operations forces who were present on the ground were engaged not in warfare but in anti-terrorist operations.[7] Russian military and security scholar Anatoliy Tsyganok, writing in the first quarter of 2016, described the intervention as a joint operation between Russia's Aerospace Forces and Navy, with ground-based forces providing base security.[8] Others, too, have focused on the role of Russian

[6] Charles Bartles and Lester Grau, 'The Russian Ground-Based Contingent in Syria', *Foreign Policy Research Institute*, October 2020, https://www.fpri.org/wp-content/uploads/2020/10/report-4-bartles-grau-oct-2020.pdf.
[7] Vladimir Putin, 'Zasedanie kollegii Federal'noi sluzhby bezopasnosti' [Meeting of the board of the Federal Security Service], The President of Russia, 26 February 2016, http://kremlin.ru/events/president/news/51397.
[8] Anatoliy Tsyganok, 'Gruppirovka rossiiskikh voysk v Sirii v bor'be s IGIL (strategiiya i stsenarii)' [The grouping of Russian forces in Syria in the struggle with ISIS (strategy and scenarios)], *Vestnik Akademii Voyennykh Nauk* 1, 54 (2016): 11.

aerial power and long-range weaponry, neglecting the role played by Russian military advisers, *spetsnaz*, and artillery units.[9]

With time, Russian researchers have begun to acknowledge that the Russian ground-based assets in Syria have been engaged in aerial support alongside base security. As Dima Adamsky has pointed out, Russian scholars have described Syria as the first case where Russia has put into practice evolving operational and tactical ideas about *reconnaissance-strike* and *reconnaissance-fire complexes*. These concepts, which refer to systems of linking remote weaponry to real-time target intelligence, emerged out of Soviet military theoretician and chief of the general staff Nikolay Ogarkov's understanding that modern warfare will be based on the integration of ISR, C2, and fire systems, allowing the rapid identification and destruction of enemies from range.[10] Russia in Syria has relied on advance C2ISR systems, including the satellite-based Glonass navigation system, to identify targets and coordinate aerial and standoff military action. Yet the limitations of these technologies and the weak fighting strength of pro-Assad ground forces have required the use also of human intelligence and reconnaissance. This has been provided specifically by *spetsnaz* troops who have supplied forward reconnaissance, guided airstrikes, and assessed airstrike impact, whilst Russian military advisors embedded with Syrian units have also played an important role.[11]

Indeed, the role of Russian ground-based assets has gone beyond providing base security and supporting *reconnaissance-fire* and *reconnaissance-strike complexes*. Russia's military police units, military advisors and private military contractors have served other functions as well. As Bartles and Grau have argued, two major ways through which Russia's ground-based contingent has shaped the outcome of the Syrian conflict have been through (1) its deployment of a system of military advisors, who have helped to plan and coordinate tactical and operational action, and (2) its use of fire and artillery support, which has helped to give the pro-regime forces a technological edge over their adversaries.[12] Michael Kofman and Matthew Rojansky have further noted that Russian ground-based forces have fulfilled a variety of functions. For example, special operations forces have conducted diversionary operations, reconnaissance, and targeted killings, whilst demining units

---

[9] V.K. Novikov et al., 'Kontseptual'niy vzglyad na problemu ustoychivosti i bezopasnosti mira' [A conceptual glance at the issue of stability and security of the world], *Vestnik Akademii Voyennykh Nauk* 3, 56 (2016): 15.

[10] Dmitry (Dima) Adamsky, 'Russian Lessons Learned from the Operation in Syria: a Preliminary Assessment', in *Russia's Military Strategy and Doctrine*, edited by Glenn E. Howard and Matthew Czekaj (Washington, DC: The Jamestown Foundation, 2019), 384.

[11] Adamsky, 'Russian Lessons Learned from the Operation in Syria', 389.

[12] Bartles and Grau, 'The Russian Ground-Based Contingent in Syria', 2.

have cleared seized territories, with private military contractors spearhead-ing some high-risk engagements.[13] As Kofman and Rojansky put it, 'Syria continued to reveal the general Russian preference to use local forces first, mercenaries and other Russian proxies second, and its own forces last, only for decisive effect on the battlefield.'[14]

The following section gives an overview of the Russian ground-based contingent in Syria, detailing the structure and functions of various ground-based assets. It largely uses Bartles and Grau's categorization of the contingent as set out in their report *The Russian Ground-Based Contingent in Syria*, although it considers artillery as a separate force category and examines naval infantry units in isolation, rather than as a component of the broader cate-gory of coastal defence troops. In the subsequent section, the examination of individual force types will lead into a broader discussion of the operational and strategic functions of the ground-based contingent in Syria.

The Russian Ground-based Contingent

The Russian ground-based contingent in Syria counts a maximum of around 3,000 regular troops,[15] with an additional cohort of approximately 2,000 PMC fighters deployed to Syria at times.[16] (The latter number, however, appears to have dwindled after the Russian Wagner Group fatally clashed with US forces in February 2018 as will be discussed below.)[17] This means that the size of the Russian ground-based contingent in Syria has been only a frac-tion of those of previous Russian and Soviet deployments, including those in Georgia in 2008 (where an estimated 35–40,000 troops may have taken part[18]) and Afghanistan between 1979–1989 (80,000–120,000 troops). The small size of the Russian force is central to understanding its mode of deploy-ment. Kofman and Rojansky explain it as a result of both strategic necessity and strategic restraint. They note that Syrian military bases initially lacked the capacity to host a large number of Russian troops, forcing Moscow to adopt 'a more conservative and ultimately smarter approach to the battle

[13] Michael Kofman and Matthew Rojansky, 'What Kind of Victory for Russia in Syria?', *Military Review*, January 2018, https://www.armyupress.army.mil/Journals/Military-Review/Online-Exclusive/2018-OLE/Russia-in-Syria/.

[14] Kofman and Rojansky, 'What Kind of Victory for Russia in Syria?'.

[15] Kofman and Rojansky, 'What Kind of Victory for Russia in Syria?'.

[16] Kofman and Rojansky, 'What Kind of Victory for Russia in Syria?'.

[17] Aleksey Ramm and Nikolay Surkov, 'Chechenskiy spetsnaz budet okhranyat' aviabazu Kheimim' [Chechen spetsnaz will guard the Kheimim airbase], *Izvestiya*, 8 December 2016, https://iz.ru/news/650206.

[18] Ariel Cohen and Robert E. Hamilton, 'The Russian Military and the Georgia War: Lessons and Implications', *Strategic Studies Institute*, June 2011, 11, https://www.files.ethz.ch/isn/130048/pub1069.pdf.

space'.[19] However, even after bases were expanded at an early stage during the intervention,[20] the Kremlin resisted a large scaling-up of its force.[21] A likely rationale for this was to limit the risk of casualties and logistical and financial overstretch.

With that said, the Russian ground-based contingent did expand over time, if to a limited extent. As Russia's aerospace operations notched up successes during 2015 and 2016 and parts of Syria's armed opposition surrendered or retreated, the intervention shifted into a lower gear, moving from aerial bombardment and close-air support to what may be described as peace enforcement and stabilization. After the fall of eastern Aleppo in late 2016—a major success for the pro-regime forces—Moscow sent a detachment of Military Police to Syria. This detachment reportedly consisted of troops from the disbanded Chechen-dominated Vostok and Zapad *spetsnaz* battalions, which had experience of counterinsurgency and urban warfare in the Caucasus.[22] Strikingly, the troops were organized into military police battalions only on the eve of their deployment to Syria in December 2016,[23] which limited the degree of dedicated military police training they received. Further deployments of military police took place in subsequent years,[24] which helps to explain the overall increase in the Russian contingent in Syria. In October 2015, Russian media outlet RBC, referencing military experts, estimated that the contingent comprised only some 2,000 troops, including personnel from the aerospace forces;[25] later estimations have placed the number at around 5,000,[26] apparently excluding PMC contractors.

In his 2019 address referenced above, Valeriy Gerasimov spoke of 'self-sufficient' groupings of forces in relation to Russia's strategy of limited actions. But Russia's contingent in Syria has hardly been self-sufficient. Forces

---

[19] Kofman and Rojansky, 'What Kind of Victory for Russia in Syria?'.

[20] Cf. Matthew Bodner, 'Why Russia Is Expanding its Naval Base in Syria', *The Moscow Times*, 21 September 2015, https://www.themoscowtimes.com/2015/09/21/why-russia-is-expanding-its-naval-base-in-syria-a49697.

[21] Kofman and Rojansky, 'What Kind of Victory for Russia in Syria?'.

[22] Anon, 'SMI soobshchili ob otpravke chechenskikh batal'onov "Vostok" i "Zapad" v Siriyu' [Media reports of dispatch of Chechen battalions 'Vostok' and 'Zapad' to Syria], *Lenta.ru*, 8 December 2016, https://lenta.ru/news/2016/12/08/syria/; Ramm and Surkov, 'Chechenskii spetsnaz budet okhranyat' aviabazu Kheimim'.

[23] Anon, 'SMI soobshchili ob otpravke chechenskikh batal'onov "Vostok" i "Zapad" v Siriyu'.

[24] Anon, 'Batal'on voennoy politsii iz Ingushetii zavershit mirotvorcheskuyu missiyu v Sirii v mae' [Military Police battalion will end its peace-keeping mission in Syria in May], *Interfax*, 23 May 2017, https://www.interfax-russia.ru/south-and-north-caucasus/news/batalon-voennoy-policii-iz-ingushetii-zavershit-mirotvorcheskuyu-missiyu-v-sirii-v-mae.

[25] Maksim Solopov, 'Vezhliviy kontingent: skol'ko v Sirii rossiiskikh voyennykh' [The polite contingent: how many Russian soldiers there are in Syria], *RBC.ru*, 1 October 2015, https://www.rbc.ru/politics/01/10/2015/560d472d9a7947ed7fa0540d.

[26] Omar Lamrani, 'The Risks and Rewards of Moscow's Mission in Syria', *Stratfor*, 24 October 2019, https://worldview.stratfor.com/article/risks-and-rewards-moscows-mission-syria-putin-assad-erdogan.

have been dispatched to, and detached from, it over the course of the conflict in accordance with need and circumstance. If units of Military Police were sent to Syria on three-month tours at the end of 2016 and the start of 2017,[27] teams of sappers have, too, been deployed on short notice.[28] Further, as mentioned previously, PMC contractors were apparently removed from the country following the clash with US forces in February 2018. Stable maritime and air connections with Syria and short troop rotation times—often no more than three to six months—have supported continuing changes in the make-up of the Russian intervention force.

With that said, one may divide the Russian ground-based contingent in Syria into six main categories of troops: naval infantry, artillery, special operations forces (*spetsnaz*), military police, military advisors, and private military companies. This list is not exhaustive, as there are reports that airborne (VDV) troops have also served in Syria, providing security at Russian military bases, as well as groups of sappers, radio signallers, and radio intelligence troops.[29] Nevertheless, it covers the main deployed Russian ground-based force types. This section considers each of these force types in turn.

Naval Infantry

Highly mobile and highly trained, Russia's naval infantry are part of the country's rapid reaction capacity, forming a kind of naval sister force to the more famous airborne forces (VDV). The naval infantry units in Syria are drawn from Russia's Black Sea and Northern Fleets and were initially deployed at the start of the intervention to provide security at Russia's air base in Kheimim and naval base in Tartus, and have since remained.[30] The naval infantry's presence in Tartus has helped to ensure the smooth maritime supply of the Russian intervention force but may also be intended to function as a deterrent to a possible coastal incursion by Western forces, according to Charles

[27] Anon, 'Batal'on voennoy politsii vernulsya v Chechnyu iz Sirii bez poter', *Kavkazskii Uzel*, 26 June 2017, https://www.kavkaz-uzel.eu/articles/304999/.
[28] Anon, 'Rossiiskie sapery v Sirii', *Gazeta.ru*, 3 April 2016, https://www.gazeta.ru/army/photo/nashi_sapery_v_sirii.shtml.
[29] Anon, 'Na okhranu rossiiskikh baz v Sirii otpravili morpekhov i spetsnazovtsev', *Lenta.ru*, 1 October 2015, https://lenta.ru/news/2015/10/01/secure/.
[30] Thomas Gibbons-Neff, 'How Russian Special Forces are Shaping the Fight in Syria', *The Washington Post*, 29 March 2016, https://www.washingtonpost.com/news/checkpoint/wp/2016/03/29/how-russian-special-forces-are-shaping-the-fight-in-syria/; Bartles and Grau, 'The Russian Ground-Based Contingent in Syria', 8; Aleksey Ramm, Aleksey Kozachenko, and Bogdan Stepovoy, 'Vydadut bronyu: brigady morpekhov usilyat tankovymi podrazdeleniyami', *Izvestia*, 22 October 2019, https://iz.ru/923772/aleksei-ramm-aleksei-kozachenko-bogdan-stepovoi/vydadut-broniu-brigady-morpekhov-usiliat-tankovymi-podrazdeleniiami.

Bartles and Lester Grau.[31] With that said, their role in Syria reaches beyond that of base security and coastal defence. As elite units, naval infantry have been used in at least one rescue operation,[32] whilst the Russian defence ministry has said that Russian naval infantry carry out a 'wide spectrum of tasks' alongside base security in Syria, not specifying what those other tasks are.[33]

Artillery

Russia has supplied large quantities of artillery to the battlefields in Syria, including 152 mm Msta-B and 122 mm D-30 howitzers, the 300 mm Smerch, 120 mm Grad/Tornado-G, and 220 mm Uragan multiple launch rocket systems and the so-called TOS-1A Solntsepyok system.[34] This is advanced Russian weaponry that has helped to tilt the balance of firepower between pro- and anti-regime forces in Syra in favour of the former. In addition, Russia has deployed targeting technology, including Israeli-made drones, which has further increased the effectiveness of indirect fire.

Most of the artillery is operated by Syrian regime forces but some of it appears to be handled by Russian artillery units.[35] Where Syrian troops have conducted artillery strikes they have often done so under the supervision of Russian specialists or after being trained by the latter in the use of Russian artillery pieces. Russian expertise allows Syrian forces not only to wield more advanced Russian artillery pieces but also to implement new tactics of artillery deployment. One Russian colonel who served as a military advisor in Syria said he and other Russian advisors taught Syrian artillery troops tactical movement, as well as techniques of concealment and deception (*maskirovka*), in order to evade counter fire.[36] Another Russian officer said Syrian troops

---

[31] Bartles and Grau, 'The Russian Ground-Based Contingent in Syria', 8.

[32] Viktoriya Makarenko, 'V Novocherkasske pokhoronili morpekha, pogibeshego v Sirii vo vremya operatsii po spaseniyu letchikov' [In Novocherkassk they buried a navy infantryman who was killed during an operation to rescue airmen], *Novaya Gazeta*, 30 November 2015, https://novayagazeta.ru/articles/2015/11/28/66570-v-novocherkasske-pohoronili-morpeha-pogibshego-v-sirii-vo-vremya-operatsii-po-spaseniyu-letchikov.

[33] Anon, 'Minoborony RF soobshchilo o vypolnenii morskoy pekhotoy spetsial'nykh zadach v Sirii' [The Defense Ministry of the Russian Federation Tell of the Carrying Out of Special Tasks by the Naval Infantry in Syria], *Interfaks*, 27 November 2019, https://www.militarynews.ru/story.asp?rid=0&nid=522538&lang=RU.

[34] Bartles and Grau, 'The Russian Ground-Based Contingent in Syria', 4.

[35] See Tim Ripley, *Operation Aleppo. Russia's War in Syria* (Lancaster: Telic-Herrick Publications, 2018), 36–9.

[36] Ol'ga Grebenyuk, 'Esli imya tebe komandir' [If your name is commander], *Krasnaya Zvezda*, 10 August 2018, http://redstar.ru/esli-imya-tebe-komandir/?attempt=1.

lacking in professional skills were trained in 'how to position and target long-range guns, adjust fire, and equip and use shelters at battery positions'.[37]

Spetsnaz

Russian special operations forces, or *spetsnaz* (short for *spetsialnoe naz-nachenie*, special designation), had reportedly been present in Syria, much like Russian private military contractors, even before the official start of the Russian intervention in the autumn of 2015.[38] Whilst their activities have understandably been shrouded in secrecy, reports suggest their role has been threefold: providing ground reconnaissance, acting as forward air controllers, and carrying out high-value missions behind enemy lines.[39]

First, as British war reporter Tim Ripley has observed, Russian *spetsnaz* provided important ground reconnaissance for the VKS in the early stages of the intervention, compensating for unreliable intelligence from Syrian, Iranian, and other coalition partners.[40] Second, their small numbers and elite training have made them suitable as forward air controllers, assisting in the identification of targets for aerial attacks and guiding air power toward these. First commander of the Russian contingent in Syria, Colonel General Aleksandr Dvornikov, pointed out that the special operations forces conduct ground reconnaissance and help to lead aerospace forces to the targets.[41] Third, they carry out attacks on the armed opposition, both in concert with and independently of coalition forces, including assassinations and other special operations. One publicized incident occurred on 11 January 2021, when Russian *spetsnaz*, fighting in tandem with Syrian regime forces in Idlib Province, reportedly killed eleven armed fighters.[42]

---

[37] Yekaterina Vinogradova, 'Rossiiskie voyennye sovetniki sodeystvuyut rostu masterstva siriiskikh artilleristov' [Russian military advisors contribute to the growing mastery of Syrian artillery troops], *Krasnaya Zvezda*, 1 February 2021, http://redstar.ru/rossijskie-voennye-sovetniki-sodejstvuyut-rostu-masterstva-sirijskih-artilleristov/.

[38] Josh Cohen, 'Russia's Vested Interests in Supporting Assad', *The Moscow Times*, October 23 2014, https://www.themoscowtimes.com/2014/10/23/russias-vested-interests-in-supporting-assad-a40700; Kirit Radia and Rym Momtaz, 'Russian Anti-Terror Troops Arrive in Syria', *ABC News*, 19 March 2012, https://abcnews.go.com/Blotter/russian-anti-terror-troops-arrive-syria/story?id=15954363.

[39] Bartles and Grau, 'The Russian Ground-Based Contingent in Syria', 13.

[40] Ripley, *Operation Aleppo*.

[41] Anon, 'V Sovfede poyasnili status rossiiskogo spetsnaza v Sirii' [In the Federal Council on Defense and Security they clarified the status of Russian spetsnaz in Syria], *Interfax*, December 12 2016, https://www.interfax.ru/russia/541031. Chechen leader Ramzan Kadyrov said similarly that ethnic Chechen *spetsnaz* deployed to Syria gather information about the enemy, fix targets, and monitor results of air attacks: Anon, 'Shto mozhet delat' chechenskii spetsnaz v Sirii?' [What is Chechen spetsnaz doing in Syria?], *BBC*, 8 December 2016, https://www.bbc.com/russian/features-38239283.

[42] Adam Gur'ev, 'Rossiiskii spetsnaz unichtozhil 11 boevikov v Sirii' [Russian spetsnaz killed 11 militants in Syria], *Lenta.ru*, 13 January 2021, https://lenta.ru/news/2021/01/13/algab/.

Military Police

The establishment of a Russian military police force had been foreseen already in 2011 during the reformer Anatoliy Serdyukov's tenure as Defense Minister, yet, it was only in March 2015, in other words shortly before the start of Russia's intervention, that President Vladimir Putin confirmed its constitution, enabling it to come into being.[43]

The Russian military police battalions in Syria have had a far wider remit of tasks than is customary for Western military police forces. In Bartles and Grau's words, it may be more appropriate to view the role of the Russian battalions as one of 'expeditionary peacekeepers', tasked with promoting stability and security in post-violence contexts.[44] When the first Russian military police detachment was deployed in Aleppo after the fall of its eastern districts to coalition forces in late December 2016, reports suggest that its role was to escort humanitarian convoys, protect Russian and international personnel in the field (including during mine-clearing and humanitarian operations), and provide a more palatable face to the coalition forces to Aleppo's Sunni population than would Assad's troops and their Shia auxiliaries have done.[45] Such a focus on public outreach may explain the very high presence of Muslim Chechen and Ingush troops in the military police battalions.[46] It mirrors a similarly heavy use of Muslim troops during the Soviet Union's invasion of Afghanistan in 1979–1980.[47] Furthermore, according to Russian press reports, the Russian military police units in Syria may help to deter

---

[43] Anon, 'Putin utverdil ustav voennoy politsii' [Putin confirmed the service regulations of the military police], *TASS*, 27 March 2015, https://tass.ru/politika/1859938 (13 October 2021).

[44] Bartles and Grau, 'The Russian Ground-Based Contingent in Syria', 6.

[45] Evgenii Shestakov, 'Ne dopustit' provokatsii v Aleppo' [Don't allow provocations in Aleppo], *Rossiiskaya gazeta*, 30 January 2017, https://rg.ru/2017/01/30/minoborony-rf-zajmetsia-ohranoj-poriadka-v-sirii.html; see also Lyubov' Merenkova, '"Armiya islama" prosit Rossiyu razmestit' kavkaztsev u Damaska' ['The Army of Islam' asks Russia to quarter troops in Damascus], *Kavkaz Realii*, 7 April 2018, https://www.kavkazr.com/a/armiya-islama-prosit-rossiyu-razmestit-kavkaztsev-u-damaska/29149824.html; Maksim Solopov and Inna Sidorkova, 'Chechentsy zashchityat Aleppo ot maroderov' [Chechens defend Aleppo from marauders], *Gazeta.ru*, 8 December 2016, https://www.gazeta.ru/army/2016/12/08/10413125.shtml?updated.

[46] Anon, 'Voennaya politsiya iz Chechni nachala patrulirovanie na severe Sirii' [Military police from Chechnya started patrolling in northern Syria], *Interfax*, 25 October 2019, https://www.interfax.ru/world/681766; Anon, 'Chechenskie voennye prisutstvuyut v Sirii tol'ko v sostave voennoy politsii' [Chechen soldiers are present in Syria only in the ranks of the military police], *Interfax*, 19 October 2019, https://www.interfax-russia.ru/south-and-north-caucasus/main/chechenskie-voennye-prisutstvuyut-v-sirii-tolko-v-sostave-voennoy-policii-kadyrov; Anon, 'V Siriyu napravili noviy batal'on chechenskoy voennoy politsii' [A new battalion of Chechen military police was sent to Syria], *RBC*, 20 April 2017, https://www.rbc.ru/rbcfreenews/58f7d42c9a79473273877b3f; Bartles and Grau, 'The Russian Ground-Based Contingent in Syria', 7.

[47] Cf. Jiayi Zhou, 'The Muslim Battalions: Soviet Central Asians in the Soviet-Afghan War', *The Journal of Slavic Military Studies* 25, 3 (2012): 302–328.

abuse of Aleppo's Sunnis by other coalition troops.[48] According to this interpretation, the forces have been inserted into post-conflict zones partly as a buffer between conquerors and conquered in an attempt to reduce the risk of renewed violence. This may help to explain the fact that the Kurdish People's Defense Units (YPG) in December 2016 requested to hand over control over its section of Aleppo to the Russian military police rather than to Assad's forces, with whom it had a history of conflict.[49]

Moreover, military police units have been deployed to areas contested by Turkish-backed militias in Aleppo and Idlib Provinces. Again, their presence appears to have been intended to have a calming effect. The town of Saraqib in Idlib Province sits strategically on the M5 motorway that connects Aleppo and Homs, at the location where the M5 merges with the M4 running from the city of Latakia. Syrian regime forces seized Saraqib in February 2020 but were soon beaten back by Turkish-backed militias.[50] When Syrian forces retook the city, Russian military policemen quickly moved in. This inaugurated a period of relative calm in the area, with the Russian troops apparently acting as a cordon sanitaire, deterring Turkish-sponsored attacks.[51] Russian military police units were credited in the Russian press with a similar stabilizing effect after they took over the manning of guard posts from Syrian government forces in the Ayn-Isa district near Raqqa. The government forces had drawn repeated fire from Turkish and Turkish-backed units, but after the arrival of the Russian units, the situation was reported to become more stable.

Importantly, Russia's military police battalions in Syria consist largely of *spetsnaz* troops, who, as mentioned, in several cases were transferred to their military police units shortly before their deployment to the Middle East. At least one of the battalions was given additional training at the new centre for the training of *spetsnaz* forces in Gudermes by veterans of the elite Alfa

[48] Maksim Solopov and Inna Sidorkova, 'Chechentsy zashchityat Aleppo ot maroderov' [Chechens defend Aleppo from marauders], 8 December 2016, https://www.gazeta.ru/army/2016/12/08/10413125.shtml?updated.

[49] Aleksandr Rybin, 'Kurdskii vopros ostalsya bez otveta' [The Kurdish request remains unanswered], *Gazeta.ru*, 30 December 2016, https://www.gazeta.ru/army/2016/12/30/10457615.shtml.

[50] Lidiya Misnik, 'Srazheniya v Sirii: rossiyane zanyali Sarakib' [Clash in Syria: Russians took Saraqib], *Gazeta.ru*, 3 March 2020, https://www.gazeta.ru/army/2020/03/03/12986905.shtml; Aleksei Nikol'skii, 'Rossiiskaya voyennaya politsiya vvedena v siriiskii Serakab' [Russian military police is brought into Syrian Saraqib], *Vedomosti*, 2 March 2020, https://www.vedomosti.ru/politics/articles/2020/03/02/824243-siriiskii-serakab.

[51] It was however not a complete success, as at least one further attack, apparently carried out by pro-Turkish militia, wounded three Russian military policemen. Anon, 'Troe rossiiskikh voennykh poluchili lyegkie raneniya pri obstrele v Sirii' [Three Russian servicemen received light wounds from arms fire in Syria], *RT*, 29 December 2020, https://russian.rt.com/world/news/817834-voennye-siriya-obstrel. Anon, 'Rossiiskaya voennaya politsiya pribyla na nablyudatel'niy post v Ayn-Ise na severe Sirii' [Russian military police arrived at observation posts in Ayn-Isa in the north of Syria], *Kommersant*, 29 December 2020, https://www.kommersant.ru/doc/4637061.

and Vympel *spetsnaz* units.[52] It is likely that the *spetsnaz* troops were selected for the rigour of their training and their experience of counterinsurgency operations in the Caucasus, granting them a certain stature and authority. Presumably, they were also better prepared for the demands and dangers of operating in war-torn Syria.

Military Advisors

In an interview in the Russian government newspaper *Rossiskaya Gazeta* on 23 March 2016, Lieutenant-General Aleksandr Dvornikov, who was the first commander of the Russian forces in Syria, said Russia had established a system of military advisors at 'all levels' of the Syrian Armed Forces, including at the tactical level. Dvornikov said the advisors trained Syrian troops, 'helped Kurdish and other patriotic formations', and took 'the very most active part [*samoe aktivnoe uchastie*] in the preparation of military actions'.[53] Other reports paint a similar picture of pervasive Russian military advisory presence in allied ground units in Syria.[54] Among other things, advisers have worked to introduce Russian military know-how into pro-regime forces, including through training, support, mentoring and the directing of military actions. They have also, as previously mentioned, trained Syrian officers and soldiers in the use of Russian military equipment, including artillery,[55] and participated in military actions on the frontlines, where Russian media has credited them with coordinating Syrian tactical actions. Numerous advisors have been killed or wounded in armed engagements.[56]

Russian military officials have argued that the military advisors have been key to the success of the military intervention. In his aforementioned interview, Dvornikov said Russian advisors permitted 'the destruction of the infrastructure of the supply channels of the terrorists, to seize the initiative and to go on the offensive',[57] whilst chief of the General Staff Valeriy Gerasimov in a 2017 address to the Academy of Military Sciences said that

---

[52] Anon, 'Chechenskii batal'on voennoy politsii v Sirii gotovili lyudi iz "Al'fi" i GRU' [The Chechen battalion of military police in Syria were prepared by people from 'Alfa' and the GRU], *RBC*, 16 March 2017, https://www.rbc.ru/politics/16/03/2017/58c6c2ba9a79470d5909cdf6.

[53] Yurii Gavrilov, 'Siriya: russkii grom' [Syria: Russian thunder], *Rossiiskaya Gazeta*, 23 March 2016, https://rg.ru/2016/03/23/aleksandr-dvornikov-dejstviia-rf-v-korne-perelomili-situaciiu-v-sirii.html.

[54] Anon, 'Russian Military Advisors Work with all Syrian Army Units—Russian General Staff', *TASS*, 27 December 2017, https://tass.com/world/983232; Anon, 'V luchshikh traditsiyakh russkogo voinstva' [In the best traditions of the Russian warrior], *Krasnaya zvezda*, 29 June 2018, http://redstar.ru/v-luchshih-traditsiyah-russkogo-voinstva/?print=print.

[55] Grebenyuk, 'Esli imya tebe komandir'.

[56] Anon, 'V Sirii pogibli chetvero rossiiskikh voennykh sovetnikov', *Interfax*, 27 May 2018, https://www.interfax.ru/world/614489.

[57] Gavrilov, 'Siriya: russkii grom'.

'under the leadership of Russian military advisors and with the incessant support of the airplanes of Russia's Aerospace Forces, large armed formations were destroyed in the provinces of Latakia, Aleppo and Damascus. Control was established over Palmyra.'[58] To be sure there is an element of political communication in this, since the lauding of military advisors and the aerospace forces signals that it was Russian military know-how and superior military technology that turned the tide of the civil war rather than the fighting prowess of Syrian ground forces, Iranian auxiliaries, and Russian private military contractors. But even so, the system of military advisors appears to have provided an edge to the pro-regime forces. Dima Adamsky suggests that the advisors gave Russia a means to coordinate the activities of the motley of pro-regime forces at multiple levels, including in frontline tactical units. In doing so, they drew on a new command and control system that had been developed within the Russian Armed Forces and tested in military exercises in previous years. Adamsky deems that advisors were a crucial component of the Russian command and control architecture as embodied by the Command Post of the Grouping of Forces, located at the Kheimim airbase. The Command Post functioned to coordinate the 'activities of the Russian Forces with the Syrian Army, Republican Guard, and local and foreign militias.'[59] Subordinated to the Command Post were Operational Groups of Advisers, which were embedded with tactical units and enabled the coordination of military action at the tactical level.[60]

Private Military Companies

A number of Russian PMCs have operated in Syria during the civil war. One was the Slavonic Corps which deployed to Syria in 2013, long before the Russian intervention in September 2015. Consisting of ex-soldiers, among others, it was tasked with securing oil resources near the eastern city of Deir ez-Zor, but the venture failed and the Corps was soon disbanded, with a number of its members being arrested when they returned to Russia on charges of involvement in mercenary activity.

After the start of the Russian intervention, reports surfaced that another group of Russian military contractors had been deployed to Syria. This was the Wagner Group, which had previously deployed to eastern Ukraine where

[58] Valeriy V. Gerasimov, 'Sovremennye voyny i aktual'nye voprosy oborony strany' [Contemporary wars and current issues regarding the country's defense], *Vestnik Akademii Voyennykh Nauk* 2, 59 (2017): 13.
[59] Adamsky, 'Russian Lessons Learned from the Operation in Syria', 389.
[60] Ibid.

it had supported pro-Russian separatists in their war with Ukrainian government forces. Now it was operating in Syria from what appeared to be highly
exposed positions. In December 2015, nine Russian military contractors were
reportedly killed in a mortar attack on their base.[61]

The Wagner Group is a nominally independent entity, albeit one with
highly murky funding and ownership lines, yet has well documented ties
with the Russian military establishment.[62] As Russian investigative journalists have reported, Wagner fighters have been trained in facilities attached to
the training grounds of Russia's 10th GRU Spetsnaz brigade in Molkino in
southern Russia.[63] Some of its fighters have been flown to Russia onboard
Russian military airplanes after being wounded[64] and they are kitted with
Russian military gear.[65] Their apparent commander, Dmitry Utkin, is a GRU
spetsnaz veteran, and other Wagner fighters whose names and backgrounds
have come to light are also veterans of Russia's elite *spetsnaz* formations or
other units in the Russian military. Timothy Thomas has queried whether
Wagner should be considered a private military company at all, since it, in
contrast to other private military companies, is involved not only in military
support and training, but in direct combat operations. Perhaps a better term,
he suggests, is 'illegal armed formation'.

The Wagner Group in Syria certainly seems to be operating as an extension
of the Russian military in the form of a semi-covert entity that grants Russian
officials a degree of plausible deniability. To be sure, that plausible deniability,
as Kimberly Marten has pointed out, has waned over time as the relationship between Wagner and Russian officialdom has been increasingly difficult
to hide, yet it seems to have played an important role periodically during
the Russian intervention in Syria. Whilst the details of their activities remain
shrouded in some secrecy, it seems that Wagner fighters are tasked with more
dangerous missions in what may be termed a case of risk outsourcing. Judging
by media reports and research publications, Wagner has been used in risky
forward ground operations and has been deployed in areas that have been
more exposed to enemy action. One Russian investigative journalists, Iliya
Rozhdestvenskiy, reported that Wagner operatives claimed that they played

---

[61] Thomas Grove, 'Up to Nine Russian Contractors Die in Syria, Experts Say', *The Wall Street Journal*, 18
December 2015, https://www.wsj.com/articles/up-to-nine-russian-contractors-die-in-syria-experts-say–
1450467757.
[62] Kimberly Marten, 'Russia's Use of Semi-State Security Forces: the Case of the Wagner Group', *Post-
Soviet Affairs* 35, 3 (2018): 181–204.
[63] Anon, 'Raskryto soderzhanie trudovykh dogovorov naemnikov ChVK Vagnera', *Lenta.ru*,
23 November 2018, https://lenta.ru/news/2018/11/23/vagner/; Dennis Korotkov, 'Spisok Vagnera',
Fontanka.ru, 21 August 2017, https://www.fontanka.ru/2017/08/18/075/.
[64] Marten, 'Russia's Use of Semi-State Security Forces'.
[65] Ibid.

a leading role in the second battle for Palmyra in 2016, although Syrian forces were given most of the credit for the success in retaking the city.[66] Whilst it is not possible to verify such claims, casualties have indeed been far higher among Wagner contractors than among regular Russian forces, which lends credence to the suggestion that the PMC has been used as a surrogate for regular units in high-risk missions. Many of Wagner's casualties were incurred when the force, acting together with Syrian pro-regime forces, attacked a Kurdish-led opposition position near Deir ez-Zour in February 2018 where around thirty American troops were embedded.[67] This brought down massive US air attacks on their forces, killing an estimated 200–300 members of the attacking force.[68] Even if this incident is discounted, casualties appear to have been higher in the Wagner Group than in regular units.

The clash outside Deir ez-Zour suggests that the plausible deniability offered by the Wagner Group may have come at the price of reduced control for Russian authorities. There are different versions of the event, but it seems that the attack on the Kurdish position was initiated independently by Wagner, without endorsement from the regular chain of command. Indeed, as American bombs rained down on the contractors, Russian officers washed their hands, refraining from assisting the contractors and denying responsibility for them to their US counterparts. The event appears to have embarrassed the Russian government, and, as Martens suggests, may be one reason why Wagner's presence in Syria dwindled after it.[69]

The Strategic Functions of Russia's Ground-based Contingent

Russia's range of assets employed in Syria and the small scale of its overall presence in the country reflect its strategizing. In Syria, Russia has sought to square bold ambitions with a limited appetite for risk. Its apparent ambitions include expanding its influence in the Middle East and the eastern Mediterranean, neutralizing post-Soviet Jihadis, forcing the West to recognize Russia as a major stakeholder in the resolution of the Syrian Civil War and opening

[66] Iliya Rozhdestvenskiy, 'Prizraki voyny: kak v Sirii poyavilas' rossiiskaya chastnaya armiya', *RBC*, 25 August 2016, https://www.rbc.ru/magazine/2016/09/57bac4309a79476d978e850d.

[67] Thomas Gibson-Neff, 'How a 4-Hour Battle between Russian Mercenaries and U.S. Commandos Unfolded in Syria', *The New York Times*, May 24 2018, https://www.nytimes.com/2018/05/24/world/middleeast/american-commandos-russian-mercenaries-syria.html.

[68] Marten, 'Russia's Use of Semi-State Security Forces'.

[69] Marten, 'Russia's Use of Semi-State Security Forces'.

up a new front in its ongoing conflict with the West. These apparent objectives have been matched with only limited military deployment. In terms of boots on the ground, Russia has reportedly deployed only around 3,000 regular troops and 2,000 military contractors at any given time.

The decision to conduct a military intervention on a shoestring has several probable reasons. One is casualty aversion, as the Russian public—although initially favourable to the intervention—is understood to have little stomach for a large number of regular Russian casualties. Another is a fear of quagmire, with Russia mindful of the problems likely to arise if its ground forces are allowed to displace those of the Syrian regime and its Iranian and other backers. That situation would recall what presented during the Soviet-Afghan War (1979–1989) when Soviet ground forces displaced and further demoralized Afghan Army units and came to draw much of the Mujahidin fire. A third likely reason is escalation management, as Russia is keen to keep its footprint relatively slight in a civil war where numerous regional and world powers have important stakes in order to reduce the risk of escalatory incidents and deliberate escalatory reactions.

Aware of the West's disappetite for expanding its engagement in Syria, Russia has bet on its ability to achieve great impact with limited means, partly because it has at its disposal advanced 4CISR technology and highly trained military personnel. If the Soviet 40th Army that went into Afghanistan consisted largely of blue-collar soldiers, and the Red Army that confronted the Second World War–Nazi Wehrmacht comprised largely peasant soldiers, Russia's small contingent in Syria is highly professionalized and trained. By various accounts, it appears to be agile, capable of implementing what Russian military thinkers have termed 'non-standard' methods, and reacting to and learning from emerging situations.[70]

This small, highly trained, professional, and technologically well-supported contingent seems to serve six key functions in overall Russian force employment in Syria.

Aerial Support

Russian reconnaissance teams, specifically *spetsnaz*, have supplied ground intelligence, guided Russian airpower and assessed the impact of bombardment in support of Russian aerial operations. Limitations in aerial reconnaissance and in the reliability of intelligence supplied by other pro-regime forces

[70] Thomas, 'Russian Lessons Learned in Syria'.

appear to have made the participation of Russian ground troops necessary. Thereby, Russian aerial action in Syria has been predicated on the presence of ground personnel, notwithstanding official Russian attempts initially to downplay the role of ground-based special forces in the conflict.

Base Security

Naval infantry battalion tactical groups, bolstered by paratroopers and reportedly *spetsnaz*-turned-military police, have helped to fortify Russia's military bases in Tartus and Kheimim. The security has not been watertight, as opposition forces have carried out numerous drone attacks against the bases, claiming a number of casualties.[71] But the ground-based naval forces have been essential for ensuring the relatively smooth functioning of the two bases as well access to them by Russian supply lines. The naval infantry, parts of which were deployed to the Crimea in the 2014 Russian takeover of the Ukrainian peninsula, has again shown its importance as a key component of Russia's rapid reaction capabilities in Syria, enabling Russia to gain and secure footholds in the country.

High-value Tasks

Both naval infantry and other elite Russian troops have been used for high-value and high-risk tasks. *Spetsnaz* forces are reported to have carried out targeted killings and other actions behind enemy lines, and at least one naval infantryman participated in an operation to rescue a Russian pilot downed by a Turkish F-16 (in which he was killed). In addition, Wagner contractors appear to have been used as the sharp end of the stick in ground attacks. The presence of Russian elite forces—many of them either *spetsnaz* or former *spetsnaz*—has equipped the Russian contingent with a means to undertake high-impact missions. Both the Wagner Group and Russian Military Police units appear to have comprised large numbers of former *spetsnaz* soldiers, indicating the premium that Russian military planners have placed on this military background and training.

---

[71] Anon, 'Russia Says 13 Drones Used in Attack on Its Air Base, Naval Facility in Syria', *RFERL*, 8 January 2018, https://www.rferl.org/a/syria-russia-says-drones-used-attack-bases/28963399.html; Anon, 'Russia Repels 3rd Drone Attack on Syrian base', *The Moscow Times*, August 12 2019, https://www.themoscowtimes.com/2019/08/12/russia-repels-3rd-drone-attack-on-syrian-base-a66807; Anon, 'Unidentified Drone Downed at Distance from Russian Hmeymim Base in Syria', *TASS*, 3 February 2019. https://tass.com/defense/1115979.

Ally Coordination

Russia's airpower, material strength, and diplomatic clout have given it an obvious leadership role among the pro-Assad forces in Syria, which comprise Syrian army units, Iranian proxies, Hezbollah troops, and local militias. When Russian forces first intervened in Syria, they were reportedly surprised by the weak combat capability of the pro-regime forces, and Russian military commentators have stressed the difficulties that have emerged over working with them. However, over time Russia appears to have reaped some successes in doing so, which is often credited to its system of military advisors and C4ISR technology which have supported operational and tactical coordination across multiple units. Arguably, too, the deployment of military police battalions has given Russia a means to impose authority over pro-regime forces in post-violence contexts such as Aleppo, promoting discipline among patrolling troops.

Capacity building

Russian military advisors have sought to build the capacity of allied forces, both through formal training and mentoring and through assistance during operations. Some of this has been tactical training—as in the case of artillery troops trained in mobility and deception. Other activity has focused on weapons and equipment handling, with Russian advisors instructing Syrians in the use of Russian-supplied materiel. In addition, Russian officers have helped to reform the Syrian armed forces and supported some of its most combat-capable units such as the so-called Tiger Forces, the 25th Special Missions Forces, commanded by brigadier-general Suheil Salman al-Hassan. It should be borne in mind that the state of the Syrian armed forces upon the Russian entry into the war appears to have been very poor, having been decimated by mass desertions and plummeting morale. Whilst official sources have certainly overstated the effectiveness of Russian capacity building, significant gains seem to have been made, as illustrated by the success of the retaking of Aleppo in autumn 2016.

Deniability, Deterrence, and Escalation Management

Russian ground assets in Syria may be understood as operating on a spectrum from covert to overt and deniable to official action. If the Wagner Group and *spetsnaz* can be located at one end of this spectrum, the military police forces find themselves at the other end as the most public face of the current incarnation of the Russian contingent in Syria. Among other things, the military police have been used in public relations efforts, including the distribution of humanitarian aid, escorting foreign visitors, and conducting patrols in highly mediatized events.

The use of forces with varying degrees of deniability has allowed Russia to deny official involvement in some politically sensitive military actions whilst demonstrating official commitment to other elements of its operation. This has meant that it has been able to engage and interact with adversaries and other stakeholders from a range of political positions. Austin Carson views covertness as an important tool of escalation management in military interventions, as it has the potential to blunt foreign adversaries' reactions by maintaining a fiction of official non-involvement into which the others can choose to buy.[72] Certainly the February 2018 event involving the Wagner Group contractors demonstrates the ability to help keep a case of mass bloodshed from escalating into a major diplomatic incident through official denial, even though the Wagner fighters until that point had been evidently supported by the Russian military establishment.

Overt military action, meanwhile, can be used to demonstrate resolve,[73] upping the stakes of armed intervention. This may have a restraining effect as it appears to have had with the deployment of military police in areas contested by Turkey and Turkish-backed militias, credited with reducing local hostilities. Russia, having undertaken high-visibility actions during the intervention—including a flight around the British Isles, aerial bombardments, and the launching of Kalibr cruise missiles from the Caspian Sea[74]—has been mindful of the manner in which the optics of its intervention demonstrates resolve to foreign adversaries who have themselves vacillated in their intent to use force in Syria.

In other words, Russia uses overt and (semi-)covert action in part arguably to control escalatory dynamics vis-à-vis local adversaries and foreign stakeholders in Syria. Its constellation of ground-based forces, which have acted with different degrees of covertness and overtness, have offered a tool for navigating the local and international political dynamics of a civil war in which a mesh of local armed actors and foreign backers are engaged.

[72] Austin Carson, *Secret Wars. Covert Conflict in International Politics* (Princeton, NJ and Oxford: Princeton University Press, 2018).
[73] Carson, *Secret Wars*, 14.
[74] Anon, 'Russian Missiles 'Hit IS in Syria from Caspian Sea', *BBC*, 7 October 2015, https://www.bbc.com/news/world-middle-east-34465425.

Conclusion

Russia's force employment in Syria represents an historical anomaly in Russian warfare. The Ground Forces, the historic centrepiece of Russia's Armed Forces, have played a secondary role to the aerospace forces in the Syrian campaign. Instead, Russian military action in the Middle Eastern country has been based largely on long-range and close-air support along with other indirect military assistance to Bashar al-Assad's Baathist regime. This has allowed Russia to minimize immediate contact between its regular forces and the Syrian armed opposition, keeping regular losses low whilst exploiting opportunities to test novel C4ISR technology and non-contact weaponry. Indeed, to a significant extent, Russia's Syrian intervention has involved a decoupling of Russian troops from Syrian battlefields as well as a division of labour between Russian forces and Syrian regime and other auxiliaries. If the former have provided support, the latter have done the bulk of the close-quarter fighting.

With that said, the Russian ground-based contingent has been crucial to Russian force employment in Syria. Beyond aerial support and base security, Russia has relied on ground-based troops to conduct high-value missions, coordinate a panoply of allied forces, build the capacity of these forces, and promote post-violence de-escalation and consolidation. Russian private military contractors have reportedly offered an important asset in ground operations, taking the bulk of Russian casualties off official registers.

Russia's ground-based deployment in Syria has been limited in scope but high in impact. It has facilitated—and multiplied the impact of—air strikes and artillery and supported Russia's management of relations with friend and foe. Much as in eastern Ukraine and to some extent in Georgia, Chechnya, and Afghanistan, Moscow has collaborated with local forces. The ground-based contingent has been a key enabler in this, allowing Russia to force project through local assets brought under Russian—however limited and imperfect—leadership. In Syria, Russia has deployed mainly elite units—including the naval infantry and special operations forces—as well as former elite troops dispatched as part of the military police and the Wagner Group. Highly trained and tactically agile, many of these forces have seemed well suited to the Syrian context.

The Russian ground-based deployment has not been free of problems. One problem has been the unexpectedly low quality of Syrian and other local forces, and the need to rely on a host of auxiliary forces in addition to the Syrian armed forces. Another has been mission creep, which has led Russia

to concede a larger ground presence in the form of military police battalions, deployed apparently to secure and stabilize areas retaken by pro-regime
troops. High casualties among private military contractors, further, reveals
that low official death tolls has hinged partly on the possibility to outsource
risk to private entrepreneurs. But even so, the Russian contingent in Syria
has demonstrated its ability to achieve considerable impact with a limited
force projection, relying on limited ground-forces, effective collaboration
with local and regional assets and a Western disappetite for conflict expansion
to turn the tide of the civil war in favour of the Syrian regime.

15

The Role of Israel's Ground Forces in Israel's Wars

Eado Hecht and Eitan Shamir

Introduction

Israel has been at war continuously for a hundred years—constant low-intensity warfare punctuated by brief periods of medium to high-intensity warfare. Politically and strategically these variations in intensity of fighting are connected—the results of one often leading directly to the other.

Israeli ground forces have borne the brunt of this fighting. This chapter will attempt to provide a synopsis of the evolution of those forces, focusing on the period following Israel's war for independence from 1949 until today. The chapter includes five sections: the first three sections will provide a background on the origins of the Israeli ground forces, Israel's perception of the threats, and the role of the ground forces in its strategy for dealing with those threats. The core of the chapter will trace the changing doctrines and organization as they were adapted to the acquisition of new capabilities, lessons derived from combat, and the changing political and strategic situation within which they were fought.

Background

Israel has been at war since 1920, 28 years before is official establishment. From 1920 until 1948 the Jewish population of Palestine fought to create the conditions to establish a state. Their military forces evolved gradually from a small private security organization, to a national underground defence force that focused on defending Palestinian Jewish communities against attacks by their Palestinian Arab neighbours, and guerrilla operations against British rule, to an overt semi-regular army fighting to define the borders of the new state and connect the separate concentrations of Jewish communities

Eado Hecht and Eitan Shamir, *The Role of Israel's Ground Forces in Israel's Wars*. In: *Advanced Land Warfare*. Edited by Mikael Weissmann and Niklas Nilsson, Oxford University Press. © Eado Hecht and Eitan Shamir (2023).
DOI: 10.1093/oso/9780192857422.003.0015

distributed across the country against Palestinian Arab mostly irregular forces and then against the invading regular armies of a number of Arab states. Officially Israel's war for independence ended in spring 1949 in a series of Armistice Agreements with neighbouring Arab states, but low-intensity fighting continued unabated. To the Israeli leadership it was clear that it was just a matter of time before there was a high-intensity Arab offensive to correct the catastrophe of 1948 and therefore Israel must invest in a military powerful enough to defeat such an offensive, even one conducted by a coalition of Arab states.

Although initially based entirely on infantry, the Jewish defence organizations quickly learned the utility of using armoured vehicles—albeit initially improvised ones—aircraft, and ships, so that by the summer of 1948, even though it was still mostly infantry, it was altogether an all-domain force conducting independent aerial and naval actions as well as providing aerial and naval support to ground battles. However, during the war for independence and after it, it was clear the main threats to Israel and the main response to those threats were on land and, therefore, the ground forces were the centre of Israel's military. By the summer of 1948 the ground forces had diversified to include tanks, armoured personnel carriers, artillery, and combat-engineer vehicles—although the majority were still worn-down second-hand or improvised. The understanding that Israel's main line of defence were the ground forces and that the aerial and naval forces' main mission was to support those ground forces remained into the 1990s.

Continuous low-intensity fighting between Israel and its enemies was occasionally punctuated by brief rounds of high-intensity fighting. As they fought each other, the rival armies grew in size and competed in acquiring the latest technologies. Older technologies were gradually phased out and up-to-date equipment was procured—most imported but some designed and manufactured indigenously. Throughout, all the rivals remained—and remain—dependent on foreign support to continue fighting.

Although the rival air and naval forces participated, the majority of the fighting was between the rival ground forces. However, gradually, as new technologies and the ability to procure them developed, a shift occurred between the relative roles of Israel's ground forces and air forces, reducing use of the former and increasing use of the latter. This shift, and the vociferous debate it initiated, continues today within the Israeli security community.

Perception of the Threat

From 1948 onward, Israeli ground forces had to contend with two connected but separate threats. In Israeli national security doctrine these were termed the Routine Threat and the Fundamental Threat.[1]

The Routine Threat consisted of the constant low-intensity harassment by Palestinians and by the surrounding Arab states. This was intended to achieve small changes in the borders determined by the 1949 Armistice Agreements, induce Jews to leave Israel, and prepare the strategic situation for the next round.[2]

The Fundamental Threat consisted of a major high-intensity offensive to capture terrain and perhaps 'throw the Jews into the sea'.

Israel has a much smaller population than most single Arab states, certainly less than a combination of Arabs states. It has no geographic depth to absorb a major enemy ground offensive.[3] Initially, virtually the entire population lived within ordinary artillery range of at least one border—some within small-arms range. The successes of 1967 pushed enemy artillery away from Israel's centre, but the Arabs soon compensated with the acquisition of long-range rockets and missiles that covered, even more so today with improved rocket technology, the entire depth of Israel.[4]

Conversely, the Arab states are enormous and populous—much too big to be conquered by an army of a size Israel can create. Furthermore, they can maintain large standing armies, many of them equivalent or superior in size to that of Israel's military potential at full mobilization. Thus, those Arab states that border Israel can transit from ceasefire deployment to offensive in a relatively short time, leaving Israel with only a minimum of warning time.

Neither Israel nor its enemies can manufacture for themselves all the hardware and supplies necessary for war. Therefore, both sides are dependent on

[1] Motti Golani, *There Will Be War Next Summer ...*, (Tel Aviv: Maarachot, 1997), 18 (Hebrew); Moshe Dayan, 'Retribution Operations as a Means to Maintain the Peace', Speech to IDF Commanders Conference, July 1955, published in the IDF journal *Monthly Summary*, August 1955 (Hebrew).

[2] Thus, from the 1949 Armistice until October 1956, approximately 265 Israeli civilians and nearly 200 Israeli soldiers were killed in Arab attacks; while an additional 70 were killed in IDF retaliation operations, Yehuda Wallach (chief editor), *Carta's Atlas of Israel: The First Years 1948–1960* (Jerusalem: Carta, 1978), 113 (Hebrew).

[3] Israel within the 1949 Armistice borders is some 420 km in length, about 105 km across at the widest point and less than 16 km at its narrowest point. The country is bordered by Lebanon to the north, Syria to the northeast, Jordan to the east, Egypt to the southwest and the Mediterranean Sea to the west.

[4] 'Missiles and Rockets of Hezbollah', CSIS Missile Defense Project, 10 August, 2021, https://missilethreat.csis.org/country/hezbollahs-rocket-arsenal/; on the Hamas missile arsenal, see report by Jonathan Marcus 'Israel-Gaza Violence: The Strength And Limitations of Hamas Arsenal', BBC News, 12 May 2021, https://www.bbc.com/news/world-middle-east-57092245.

the willingness of foreign suppliers to provide or sell them enough to fight. The stocks accumulated before each escalation might be quickly depleted, allowing foreign powers to pressure or support their local dependents.

Israel's National Security Strategy

Unable to maintain a standing army big enough to defeat any full-scale enemy invasion, Israel had no choice but to create a large reserve force. In fact, up to 80 per cent of Israel's fighting power have usually been reserves.

Surprise by the enemy means that the small standing army must fight to hold a numerically superior enemy at all costs until enough reserves are mobilized and reach the battlefield. It is naturally preferable to mobilize before a threat consummates but mobilizing mistakenly carries severe economic penalties. Furthermore, as mobilization of the reserves freezes the economy, it is incumbent on Israel to win its wars quickly so as to return reservists to their homes and productive employment.

The shallowness of Israeli territory means any enemy success immediately impacts Israeli civilians and national infrastructure.

Also pushing Israel to achieve a quick decision is the political pressure that grows on it to desist the longer a war lasts—pressure that might prevent achieving the required military goals. However, given the total asymmetry in size and political power, Israel can never hope to achieve a total military victory following which its enemies could not recover and attack again. Each war might be brief, but the overall conflict is protracted and intractable.[5]

All of the above factors dictate the need for brief aggressive campaigns to achieve the required military result: push the battle away from Israel's civilians into enemy territory; maximal destruction of enemy military capabilities to enable Israeli reservists to go home; and to delay the next escalation because the defeated enemy is deterred and requires time to rebuild his army; and, in some cases, the taking of ground to be used as a bargaining chip for the post-war ceasefire negotiations.[6]

---

[5] David Ben-Gurion, *Uniqueness and Purpose* (Tel Aviv: Maarachot, 1971), 219 (Hebrew).

[6] Yitschak Ben-Israel, *Israel Defense Doctrine* (Tel Aviv: Modan & Misrad Habitachon, 2013), 59–67 (Hebrew). In 1967, the Israeli government decided it would return virtually all the territory taken in the war (except ancient Jerusalem) in return for a comprehensive peace treaty. However, after the Arab decision to refuse making peace and even to refuse negotiating (the Khartoum Declaration, 1 September 1967) some Israelis gradually begin demanding that the territory taken, or at least a significant part of it, remain in Israeli hands.

The required military result has dictated that the ground forces be the heart of the Israel Defense Forces (IDF). Air power was essential and its unique attributes made it easier to mobilize to full strength, but it could not, on its own, achieve the military results required either against the Routine Threats or against the Fundamental Threats. The IDF was therefore initially designed as a ground-forces army, with air and naval components in support. They might have independent missions (prevent enemy air forces and navies from bypassing the Israeli ground forces to strike Israeli civilians and infrastructure), but their main effort was to assist the ground forces achieve their missions.

The Routine Threat required constant action, but the Fundamental Threat was deemed more dangerous. It was therefore decided that the focus of the ground forces equipment, organization, and training would be against a major invasion (termed Fundamental Security), but that the constant border-fighting (termed Routine Security) would not be ignored so that all units of all arms, but infantry especially, and both active and reserve units would also have a secondary capability for these operations and serve rotations in border-security missions.

Evolution of the Ground Forces

The driving forces behind the evolution of Israel's ground forces in doctrines, composition, and organization included the discovery or conception of new ideas (adopted only after vigorous debates), the acquisition of new capabilities (new technologies or improved versions of older equipment), identifying problems exposed in combat, and adapting to changes in the perception of enemy capabilities and doctrines and the political background of the wars. This evolution can be roughly divided into five main periods, although the actual transition was never on an exact single day or year.

1948–1956: From Infantry to Armour

From December 1947 until mid-May 1948, the IDF ground forces rapidly evolved from an underground militia into a conventional infantry force of twelve brigades. From mid-May, the newly established state of Israel managed to procure a small number of tanks and artillery to support the infantry force. By October 1948 it had acquired enough armoured trucks and buses and armoured personnel vehicles as well as unarmoured cross-terrain vehicles to simultaneously carry up to three of its infantry brigades in long, rapid advances, exploiting gaps in the deployment of the Arab forces in southern and northern Israel, to quickly capture ground and surround portions of the Arab forces, which were then attacked in converging manoeuvres that caused them to flee or surrender.

After the war, mechanized infantry supported by artillery and a small number of tanks were the centrepiece of ground-force operational doctrine. Partially this was a continuation of the successes of late 1948, but partially also because of the limitations in the quantity and quality of the equipment procured—mostly discarded worn-out Western Second World War equipment.[7] Thus, in the early 1950s, exercises with Sherman tanks usually ended with most halted because of mechanical failures.[8] One senior Israeli general even suggested that the only way to use the tanks was to have them carried behind the infantry on truck-towed flat-bed semi-trailers, unload them just short of the objectives in order for them to provide direct-fire support for the infantry attack, and then load them again on the trucks to be carried to the next battlefield.[9]

In May 1956 the source-barrier was finally broken with France agreeing to sell Israel first-hand or at least fairly new second-hand weapons. A concentrated effort, several months long, refurbishing the older tanks and other heavy equipment provided a new capability which was proven for the first time in Israeli tank units crossing the Sinai with a minimum of breakdowns. The 1956 Suez War operational plan was to break through Egyptian defences with infantry night attacks supported by artillery and then send the mechanized infantry through the corridors to exploit into the depth of Sinai supported by tanks used as direct-fire mobile artillery. This was a magnified repeat of the last operations conducted in 1948. In fact, the newly refurbished tank units quickly took the lead, defeated Egyptian tank units and assaulted through Egyptian fortifications with the mechanized infantry following in their wake to mop-up behind them.[10] This contrasted with failures by some of the infantry and mechanized infantry units in their missions.

---

[7] David Eshel, *Chariots of the Desert: The Story of the Israeli Armoured Corps* (London: Brassey, 1989), 5.

[8] Diary of the Bureau of the Chief of Staff, quoted on the Yad Lashiryon website: https://yadlashiryon.com/armour_wars/sinai-war/.

[9] Amiad Brezner, *Noble Stallions: The Development and Evolution of the IDF Armour from the End of the Independence War to the Sinai Campaign* (Tel Aviv: Ministry of Defense, Ma'arachot, 1999), 259 (Hebrew). Years later Moshe Dayan, then Israel's Chief of Staff said in interview: 'For me, the infantry was the "queen of battle" and the function of everyone else was to assist her ...' Cited in Yaakov Erez and Ilan Kfir, *Conversations with Moshe Dayan* (Tel Aviv: Masada 1981), 42 (Hebrew).

[10] Brezner, *Noble Stallions*, 322–332, 409–411.

1957–1973: The IDF Blitzkrieg

One of the major lessons of the Suez War was the usefulness of tanks in achieving the operational objective of relatively fast long-range manoeuvres through difficult terrain complemented by a powerful assault capability, whilst exposing a minimum of personnel to enemy fire.[11]

The 1956 Suez War added another major source for the procurement of arms—Britain. Also, in the early 1960s West Germany agreed to sell second-hand but not worn-out tanks to Israel. The lessons of the Suez War and the availability of better, especially more mechanically reliable, tanks focused the Israeli ground force build-up over the next decade on increasing the tank forces from 5 battalions in 1956 to 20 battalions in 1967. The total number of infantry battalions, approximately 50,[12] was not reduced, but more of them, eight instead of four, were provided with armoured personnel carriers (APCs), to enable them to follow the tanks. Extra APCs were also purchased to enable temporary, mission-specific, conversions of infantry and paratroop battalions into mechanized infantry.

Whereas in the 1950s only infantry were employed in the constant skirmishes along the borders, in the 1960s tanks too were employed. The tanks conducted precise long-range fire on enemy targets without crossing the border, in lieu of using artillery or aircraft because using these was considered too escalatory. In addition to stealthy infantry retaliation raids, a number of tank and mechanized infantry raids were also conducted.[13]

The increased use of the tanks in actual combat exposed various deficiencies in crew training. Again too many tanks broke down, but a review showed the problem was not the tanks themselves but training crews to maintain them. The most glaring and embarrassing failures were in tactical use and gunnery accuracy. Following a particularly embarrassing failure, in which 89 rounds were fired at two Syrian tanks at a range of 1,200 meters and not one hit, the armoured corps underwent a major revision of tactics and tactical and mechanical training.[14]

---

[11] Eli Michelson, 'The IDF Process of Lesson Learning from the Sinai Campaign, November 1957—May 1957', PhD dissertation, Hebrew University in Jerusalem, January 2019, 208. (Hebrew); Brezner, *Noble Stallions*, 408, 422.

[12] This includes all types and qualities of infantry: regulars and reserves, paratroops, 'leg'-infantry, mechanized infantry, static border defence infantry, etc.

[13] Benny Morris, *Righteous Victims: A History of the Zionist-Arab Conflict, 1881–1998* (New York: Knopf Doubleday Publishing Group, 2011), 30; Tal interview with Mordechai Bar-On and Pinchas Ginosar at: http://in.bgu.ac.il/bgi/iyunim/10/2.pdf (Hebrew).

[14] Amiad Brezner, *And Fire Shall Precede Him: The Development and Evolution of the IDF Armour from the End of the Sinai War to the Six Day War* (Tel Aviv: Ministry of Defense, Ma'arachot and Modan Publishers, 2017), 146–149, 164–168 (Hebrew); Shabtai Teveth, *Exposed in the Turret* (Tel-Aviv: Schoken Publishing, 1968), 84–86, 108–117, 126–128 (Hebrew).

The revision led to the spectacular successes of the tank units in the 1967 war. Tank forces lead the way in almost every attack in almost every terrain, including hilly and built-up terrain. Wherever possible the tank units attempted to bypass enemy obstacles and fortifications through terrain deemed by the enemy to be unpassable for traffic and therefore less strongly defended. The tanks broke through enemy defence lines, followed by the infantry who cleared enemy positions whilst the tank units continued to advance.[15] Only where there were not enough tanks, or the terrain proved exceedingly difficult to cross, did the infantry lead.

The typical attack began with the tanks moving into positions at the extreme range from the objectives of their guns and sniping at all observed targets (tanks, anti-tank guns, other heavy weapons, and bunkers) on the objective or around it. After all these had been destroyed, the artillery fired concentrated barrages to suppress the enemy whilst engineers cleared paths through obstacles. Then the tanks rushed through these paths and to the depth of the enemy deployment to engage enemy reserves (usually tanks) and exploit the operational rear of the theatre. Meanwhile, the infantry following the tanks cleared the enemy positions left behind along the route of advance to enable the safe passage of supply convoys and reserve units. Israeli artillery was mostly towed and therefore participated only in the initial breakthrough battles—not being able to keep up with the tanks or APC-mounted infantry during the exploitation that followed.

The lessons of 1967 enhanced the lessons of 1956 so that the Israeli ground forces from 1968 to 1973 further focused on increasing their tank and mechanized infantry units. Again the infantry was not reduced, but the tank arm was increased to almost 50 battalions. Furthermore, doctrinal changes placed even more focus on tanks being the main arm in almost every conceivable situation and organizational changes were made to further this concept.[16]

From 1949 until 1968, Israeli ground forces were organized in mixed-arm brigades: infantry/paratrooper brigades were all-arm minus tanks; mechanized infantry had two APC-mounted infantry battalions and one tank battalion, but otherwise were identical to infantry brigades except for additional necessary logistic and ordnance functions; armoured brigades had two tank battalions and one mechanized infantry battalion and all the other arms as well. Divisions were task-forces—a permanent headquarters receiving

---

[15] *Operation Order Nakhshonim*, IDF General Staff—Operations Branch, 4 June 1967 (Hebrew).
[16] Gunther E. Rothenberg: *The Anatomy of the Israeli Army* (London: B.T. Batsford, 1979), 159–160.

units as required by the current mission. In each arm there were also a number of independent battalions to be used to reinforce brigades of whatever type if required by their mission.

In 1968 the armoured forces were reorganized to suit the new doctrine. All divisions became permanent armoured divisions: two tank brigades, one armoured infantry brigade, an artillery group, and a permanent complement of engineers, plus reconnaissance and logistics units. The tank brigades had the same number of tanks as previously divided between three (instead of two) tank battalions and the former infantry battalion was disbanded, its companies being permanently attached, one each, to the tank battalions. The armoured infantry brigade maintained the original structure of mechanized infantry brigades. Non-mechanized infantry brigades, including the paratroopers, were maintained as independent brigades to be allocated according to requirements. Some of the independent regular infantry received APCs to replace trucks, but otherwise remained the same.

The new doctrine was challenged in the 1973 war and various deficiencies were found. The surprise offensive, which caught Israel unprepared with only small forces at the front facing enormous odds, affected a proper appreciation of many of the lessons of that war. The Syrian offensive was focused on an armoured attack supported by an enormous artillery forces, whereas the Egyptian offensive focused on infantry saturated with anti-tank weapons and supported by an even bigger artillery force and tanks.

The initial Israeli lessons were: first, a need to increase the total size of the ground forces and especially of the regular component so that a surprise attack would not have such a numerical advantage in the future; second, that the reliance of the ground forces on the air force for fire support was mistaken and required an increase in the quantity and mobility of the artillery arm; and third, that the emphasis on tanks as the leading ground arm had been correct in general but exaggerated in practice so that better combined-arms training was required; fourth, that the mechanized brigades had in fact failed as an organization—they had usually employed their tank battalions only with the infantry providing just ancillary support, henceforth all brigades would be single-arm, tanks or infantry, and combined-arms task-forces would be created on a mission by mission basis by cross-attaching battalions between brigades.[17]

[17] Richard Gabriel, *Operation Peace for Galilee* (New York: Hill and Wang, 1984), 16–29.

1974–Mid 1990s: 'The Manoeuvre Crisis'

By 1979, a crash implementation of the lessons of the 1973 war increased the overall size of the Israeli ground forces from one regular division plus a number of independent brigades and five reserve divisions (some of the independent regular brigades belonged to reserve divisions when these were activated) to three regular armoured divisions plus a few independent brigades, nine reserve armoured divisions, a reserve paratroop division, and a reserve infantry division. The mechanized brigades were abolished and all the tanks were concentrated in tank brigades (three per armoured division), but all infantry units, including the paratroops, were equipped with APCs. By 1982 Israeli ground forces numbered approximately 90 tank battalions, 80 infantry battalions of all types, and 80 artillery battalions (each artillery battalion was increased from 12 guns in 1973 to 18)—a 1.8 increase in tank battalions, a 1.6 increase in infantry, and a 2.4 increase in artillery.[18] The entire artillery force was re-equipped with new self-propelled 155 mm guns instead of the previous assortment of various calibres of towed guns and improvised self-propelled artillery—much of it mortars.[19]

In 1982 this mostly armoured force attacked into Lebanon, fought Palestinian regular and guerrilla forces and the Syrian army in hilly and mountainous terrain dotted with numerous small to medium sized villages and towns and a number of cities. Despite the terrain, on most axes of attack the Israeli forces led the advance with tanks supported by mobile artillery. Rather than employ infantry to lead the offensive through the difficult terrain, it was employed only where tanks failed, and then also usually to support the tanks by outflanking enemy positions through terrain impassable to tanks to attack them from the rear and open the route for the tanks, clearing houses along routes, etc. Only on one axis did infantry lead the advance. The preference to lead with tanks in hilly to mountainous terrain was contrary to accepted knowledge, however the Israelis preferred it because, although the tank units would have more difficulty and take more vehicle casualties, the total number of human casualties would be lower.

---

[18] Actually, the increase in artillery was greater because an undisclosed number of the 55 artillery battalions that theoretically existed in 1973 had in fact been slated for closure and were therefore not combat-capable during that war. See Ze'ev Schiff and Ehud Yaari, *Israel's Lebanon War* (New York: Touchstone, 1985); Gabriel, *Operation Peace for Galilee*; Brigadier General (retired) Arieh Mizrahi (commander of the Israeli artillery arm in 1982), *Lecture on the Israeli Artillery in Operation 'Peace for Galilee'*, The Israel Military History Society, November 2012.

[19] By June 1977, Israel had replaced its materiel and increased its order of battle by: artillery 100%, tanks 50%, APCs 800%. Yitzhak Rabin, Service Notebook (Tel Aviv: Ma'ariv 1979), 505 (Hebrew).

From the late 1970s a new problem and a new technology entered the doctrinal discussions of the Israeli ground forces.

The new problem was that the size of the rival armies was continuously growing whereas the size of the theatres of operation were not. Given the Israeli preference to outmanoeuvre enemy forces rather than attack them head-on, this was a major issue—enemy forces were creating gapless defence lines echeloned in depth. Thus, as a typical example, the Syrian army in 1948 had employed three brigades, in 1967 9 brigades, in 1973 25 brigades and by the mid-1980s had 60 brigades available to it. The Golan Heights theatre had not grown by one centimetre. The new situation was termed 'the manoeuvre crisis' and 'the saturated battlefield'. It was clear that no matter how much Israel improved its ground force equipment and training, fighting through such dense and deep arrays would entail casualties too heavy and a duration that would be too long for Israel to achieve its typical strategic goals—annihilating enemy forces and capturing terrain to serve as bargaining chips for the expected post-war diplomacy.[20]

The new technology of accurate long-range munitions seemed to offer a solution. In terms of defence, they would enable the Israeli forces to start to destroy attacking enemy units long before they reached positions from which they could engage the Israeli ground forces awaiting them on the border. Also, because of their range, a few such units, centrally located, could engage enemy units on different sectors of the front merely by shifting their aim and could thus rapidly reinforce outnumbered Israeli defenders long before other reserve ground units could be sent to them. In terms of offence, the new munitions would enable Israeli forces to drastically diminish dense enemy units prior to approaching them, engaging them with tanks and infantry, and attempting to attack through them. It would be a radical enhancement of the previous Israeli attack tactic of first using precise tank fire at the extreme range as described above and, if successful, could be used to create corridors for manoeuvre. That manoeuvre would be conducted according to the same ideas used in 1967, 1973, and 1982—thrust to the depth, surround, and destroy enemy forces.[21]

---

[20] For examples of this discussion see: Dov Tamari, 'Reflections on Tactics', Ma'arachot 273–274 (May–June 1980) (Hebrew); Azar Gat, 'On the Manoeuvre Crisis', Ma'arachot 275 (August 1980) (Hebrew); Benny Mem, 'On the Margins of Reflections on Tactics', Ma'arachot 275 (August 1980) (Hebrew).

[21] Meir Finkel, *The Israeli General Staff: Learning Methods, Planning Processes, Organziational Rational* (Moshav Ben Shemen: Ministry of Defense, Ma'arachot—Modan Publishing House, 2020), 255–260 (Hebrew).

Mid 1990s–2006: The Impact of the IDF Revolution in Military Affairs on the Ground Forces

No war broke out in the mid 1990s up to the mid 2000s, enabling an assessment of the capabilities and limitations of the new ideas and technology. However, meanwhile, in the 1980s, Israel suffered a financial crisis that compelled a drastic reduction in defence spending, followed by a gradual reduction in the overall size of the IDF, including the ground forces.

Gradually, as the technology improved, and inspired by the conduct and results of the Kuwait War (1991) and Kosovo War (1999), the new concept evolved into a belief that the thinning process itself would be sufficient to defeat the enemy—ground manoeuvre would be superfluous or only minimal to prove the point achieved by standoff fire alone. For the casualty-sensitive Israeli leadership, it seemed like a perfect solution.[22]

Another cause for change in the size and composition of the ground forces was the shift in perception of the threat. Whereas until 1973 the threat was perceived as a balance between Fundamental Security and Routine Security, with Fundamental Security deemed more dangerous and so the focus of attention in weapons procurement and training, Routine Security was still being catered to—especially in strategic planning and thinking. After the traumatic 1973 war, Routine Security, although still required, was virtually ignored in procurement, training, and strategic planning and thinking. However, gradually, as the durability of the 1979 peace treaty with Egypt was accepted and following the destruction of Iraq's military might in the 1991 Kuwait War and the demise of the Soviet Union as a superpower backer of Israel's enemies, and the euphoria of the peace process with the Palestinians set-in, the perception of the threat changed: the Fundamental Threat had receded and only the Routine Threat remained. The future would entail almost only counter-guerrilla and counter-terrorist operations. Furthermore, the constantly improving technologies of accurate long-range munitions would ensure that if a Fundamental Threat did emerge, major ground manoeuvres were almost unnecessary. Enemy offensives would be defeated, the enemy army destroyed, and the enemy state would be cowed into submission by fire only.

The exact numbers have not been published, but it is clear that the Israeli army today is much smaller than it was 30 years ago—entire divisions and brigades have been cancelled. Furthermore, tank and artillery units have

---

[22] Pnina Shuker, 'The Perception of Societal Sensitivity to Casualties and Its Impact on War Management: Israel in the Second Lebanon War (2006)', (Ramat Gan: Bar Ilan University, MA Thesis, 2013 (Hebrew).

been cut the most, with some of the cuts being used to create new infantry units—especially in the regular army: in 1973 the regular army had two infantry brigades, in 1985 it had four and today it has five plus a commando brigade as well as a plethora of independent infantry battalions and various special operations units equivalent to at least two more brigades. This shift in composition expresses the shift of the focus from mechanized manoeuvre warfare against state armies to counter-guerrilla and counter-terrorist warfare.[23]

Initially it seemed that this direction was justified—fighting in the 1990s and early 2000s was indeed only against non-state forces conducting terrorist or guerrilla campaigns against Israel. From 2000 to 2006 the Palestinian Terror Offensive was essentially defeated by an infantry-focused counter-terror and counter-guerrilla campaign. Tank and artillery units were hurriedly retrained to improve their infantry skills and used as second-rate infantry on secondary missions. Regular warfare training for both the regular army and the reserves ceased almost completely. Even when it became clear that major operations were necessary to clear areas controlled by Palestinian terrorist groups, they were mostly in urban terrain. No artillery was used and only small numbers of tanks were used to support infantry through a minimum of direct short-range fire. The use of APCs was not an operational manoeuvre, but used to enable the infantry to move short distances through Palestinian small-arms fire until they dismounted to fight from house-to-house.

However, the 2006 war against Hezbollah in southern Lebanon shocked the IDF. Following the Israeli withdrawal from its southern Lebanon buffer-zone in 2000 and whilst the Israeli ground forces were fighting and converting to an almost purely counter-guerrilla and counter-terrorist force from 2000 to 2006, Hezbollah was evolving in the opposite direction—equipping, organizing, and training to conduct regular warfare. By 2006 they had begun to acquire the capability to conduct company-sized regular defensive battles. Initially Israel responded to another, but more disastrous, Hezbollah attack on an Israeli border patrol with air strikes only. It attempted to employ the new concept of victory through accurate standoff fire to defeat Hezbollah. Hezbollah responded with an indiscriminate rocket-artillery bombardment

---

[23] 'All You Wanted to Know About the Multi-Year Plan—Gideon', IDF Website Board, 26 July 2015; http://www.idf.il/1133-22449-he/Dover.aspx (Hebrew). As Amir Rapaport reports: 'The truth is that the IDF has been engaged in an effort to adapt to the wars of the present and the future for some time: since 1985, the number of tanks was reduced by 75%, the number of aircraft was reduced by 50% and the number of UAVs—Unmanned Airborne Vehicles—increased by 400%. The number of reservists was cut down by hundreds of thousands'. See Amir Rapaport, 'The New Multi-Year Plan of the IDF and the Agreement with Iran', Israel Defense, 9 September 2015. According to IISS 'Military Balance' reports, from 1989 to 2020, the number of standing armoured brigades was reduced from 6 to 3 and the reserve armoured brigades were reduced from 18 to 9.

of Israeli towns and villages. After Israel's fire-only strategy failed to compel Hezbollah to desist from firing its rockets into Israel, Israel was forced to employ its ground forces to capture the territory from which the majority of those rockets were being launched. The result of the ground incursion can best be summed-up in the words of an Israeli paratrooper battalion commander after the war: 'I entered Lebanon to arrest terrorists and collided with a regular army ...'[24] Having not trained for this type of combat for several years, the Israeli ground forces suffered one tactical setback after another (in one example a whole Israeli division needed more than 36 hours and suffered dozens of casualties to cross a few kilometres of ground whilst faced with the equivalent of one infantry company with a few anti-tank missile launchers).[25] Gradually through sheer persistence, superior numbers, and superior firepower, the Israelis wore down the Hezbollah ground units facing them and managed to achieve some territorial gains but much less than had been planned. However, the ground operations did force Hezbollah to expose more of its personnel, who had spent the fires-only phase of the Israeli offensive hidden in underground bunkers, civilian houses, and the dense woods and scrub of southern Lebanon. Finally, accumulation of attrition and the exhaustion of Hezbollah forces, led to an inability to reinforce or resupply their forward forces, and brought them to agree to a ceasefire.

2006–Present: Continuous Debate on the Role of Ground Forces

The major initial lesson of the 2006 war was that the Israeli ground forces had lost their ability to conduct regular warfare, whereas the irregular enemy was simultaneously acquiring exactly that ability.[26] The central message for the IDF ground forces over the next few years was: 'back to the basics'.[27] The ground forces entered a period of intensive training to reacquire the lost

---

[24] Conversation with one of the current authors, October 2006.

[25] See Avi Kober, 'The Israel Defense Forces in the Second Lebanon War: Why the Poor Performance?' *Journal of Strategic Studies* 31, 1 (2008): 3–40.

[26] Itay Brun, 'Where has the Manoeuvre Disappeared?', Ma'arachot 420–421 (September 2008): 4–14 (Hebrew); Victor, 'This Is How the IDF Subordinated Itself to Fire and Abandoned Manoeuvre', Ma'arachot 415 (November 2007): 4–9 (Hebrew); Ron Tira, 'Has the IDF Given Up on Manoeuvre?', Ma'arachot 453 (February 2014): 14–17 (Hebrew). On the shift by Hezbollah toward regular warfare see, for example, in summer 2012, it was reported that Hezbollah had conducted an Iranian-mentored exercise in which 10,000 fighters practised defensive battles and attacks to capture portions of northern Israel. N. Yahav, 'Hezbollah Conducted 10000 Man Exercise', *Walla News*, http://news.walla.co.il/item/2560837 (Hebrew). Hezbollah's operations during the Syrian Civil War against Syrian rebels have provided it much experience in conducting such operations.

[27] See Scott C. Farquhar, ed., *Back to Basics: A Study of the Second Lebanon War and Operation Cast Lead*, (Fort Leavenworth: Combat Studies Institute Press, US Army Combined Armed Center, 2009), 5–20.

skills. The next major ground operation was against Hamas in Gaza in early 2009. Hamas had been trying to follow the path of Hezbollah, but had not yet achieved the same capability when its escalating rocket fire against Israeli civilians triggered an Israeli offensive that, as in Lebanon 2006, began with precision fire strikes and was followed by a ground offensive that employed three infantry brigades reinforced with tank units and a tank brigade to surround Gaza City, whilst capturing a number of smaller suburbs and villages around it. In November 2012 another Israeli offensive, responding to an escalation of Hamas rocket fire and cross-border raids, involved only a week of heavy standoff fire that achieved the coveted strategic result—a drastic reduction in Palestinian attacks over the following fifteen months. This success lulled the Israeli security leadership to believe that the new methods had finally become effective. However, the next escalation, in summer 2014, conducted at first only with fire, once again required the employment of large ground forces (three divisions) conducting combined-arms regular warfare. It also showed, at least to some, that even if fire alone could perhaps achieve the desired final outcome, the enemy had learned to absorb this fire sufficiently to prolong the war considerably—the war of the summer of 2014 against a vastly inferior military force lasted 50 days as compared to the 6 days of 1967 and the 19 days of 1973 in which the IDF faced considerably stronger enemies. Conversely, although much longer, both the war against Hezbollah in 2006 and against Gaza in 2014, cost Israel only a very small fraction of the casualties suffered in those shorter more intense wars.[28]

However, although some in the IDF read the lessons of the 2006, 2009, and 2014 wars as requiring the rebuilding and maintenance of a massed manoeuvre capability similar in principle if not in details to the capability of the past, many in the IDF continued to believe that the new technologies had changed the reality of the battlefield to an extent that made ground manoeuvre, or at least massed mechanized ground manoeuvre, archaic. At present there are three schools in the IDF debating what Israel's doctrine vis-à-vis its current and near-future threats should be:[29]

No manoeuvre: Boosting defence systems against missiles and rockets and conducting offensive action with only accurate standoff fire based on target information collected by increasingly sophisticated intelligence

---

[28] Eitan Shamir and Eado Hecht, 'Gaza 2014: Israel's Attrition vs Hamas' Exhaustion', *Parameters* 44, 4 (2014): 10.

[29] Brigadier General Eran Ortal describes three schools of thought in the IDF that are similar to the schools we describe here, however whilst Ortal's presentation is aimed to show why one is superior over the other two, we have tried to maintain an objective position. Eran Ortal, 'Turn on the Light, Put out the Fire', Between the Poles—Land Forces Part B 31–32, 53–69 (Hebrew).

technologies. Since Israel knows in advance where and against whom it will fight—only the when is unknown—Israeli intelligence continuously collects targets to create long lists that are as up-to-date as possible on the day an escalation flares-up.[30] These are to be struck immediately on the first few days of each escalation—the exact targets chosen depending on the strategy and operational goals. One problem exposed in past operations has been that within a few days all targets on the lists have been hit and often the damage caused is not enough to induce the enemy to desist from further fighting. Furthermore, not all enemy assets are stationary, so that shortly after the war begins targets move and the targeting information collected previously might become irrelevant. However, proponents of this concept, especially the air force and intelligence, have invested heavily in creating the ability to rapidly detect and strike previously unknown targets. In May 2014 the Air Force Commander claimed that the Israeli Air Force had multiplied its ability to strike targets at a rate four times greater than in the Second Lebanon War.[31] In theory, the ability to continuously detect, identify, and precisely strike hundreds of worthwhile targets per day, should suffice to bring any enemy to its knees. Ground forces are to be used only to defend the border against enemy incursions and conduct armed-police style counter-guerrilla and counter-terrorist operations with a minimum of heavy weapons.

Limited 'sophisticated' manoeuvre: Essentially similar to the first school, this school believes that modern intelligence technology cannot completely replace a human presence in the heart of the enemy's deployment areas, and therefore adds the employment of a multitude of small infantry teams, assisted by 'swarms' of small remotely-piloted aircraft, to help uncover the enemy's positions, communicating these by network technology to aircraft and ground-launchers that would then destroy these targets within seconds or mere minutes by standoff fire from afar. A test unit was set up a few years ago and has been conducting experiments and exercises to validate the concept. During the May 2021 escalation of fighting with Gaza, the unit was deployed on the border of Gaza and conducted actual operations to detect and destroy

---

[30] See for example: Nir Dvori, 'Observation Posts, Listening and Satellite Photographs: Thus was Created the IDF's Target-List in Gaza', *N12*, 19 May 2021, https://www.mako.co.il/news-military/2021_q2/Article-a0deea7d7d38971027.htm (Hebrew).

[31] Ron Ben Yishai, 'The Air Force is More Lethal than ever', *YNET News*, 5 May 2014, https://www.ynet.co.il/articles/0,7340,L-4515955,00.html (Hebrew); Amir Bokhbot, 'Every Syrian Aircraft that Crosses the Border Will Be Shot Down', *Walla News*, 22 May 2014, https://news.walla.co.il/item/2748999 (Hebrew).

Palestinian rocket-launchers and guided anti-tank missile launchers firing into Israel.[32]

Improved Traditional Manoeuvre: This school argues that the above capabilities and concepts might indeed suffice in some scenarios, that is, when the enemy is not determined to fight and wishes only to achieve a very limited goal. However, when fighting a strong enemy who has set for himself a more ambitious goal and is therefore determined and ready to fight, they would, as shown in the past, fail or take a very long time to succeed—time that would be exploited by the enemy to fire thousands, even tens of thousands, of rockets on to Israeli civilians.[33] In these scenarios, only the employment of massive forces, conducting traditional fire and movement combined-arms operations with large formations would be able to disrupt the enemy's strategy and annihilate enough of his forces within a reasonable time-frame so as to convince him to give up his political and strategic goals. In these scenarios, the above concepts would only be a preliminary preparation or a supporting action to the coup-de-grâce of the mass manoeuvre conducted with tanks and infantry units supported by the new sophisticated fire capabilities. The new fire capabilities would be employed by the addition of an 'expose/attack' company to each infantry and tank combined-arms battalion team. The 'expose/attack' company would employ small drones to scout the terrain ahead and around the battalion in order to discover enemy forces or equipment and direct precise fires onto them. Depending on the characteristics of the target, the fire would be a short-range portable guided missile carried by the unit itself or a longer-range guided artillery missile or an air strike. After the initial fires had destroyed or suppressed the enemy, the battalion would assault the position with infantry and tanks.[34]

Whilst the IDF has stated in its latest strategy document that it is committed to land manoeuvre,[35] IDF Chief of Staff, General Aviv Kochavi (2019 – 2023) is a strong advocate of option B. This is clearly expressed in Kochavi's

[32] Amir Bokhbot, 'Flocks of IDF Quadcopters Participated in the Fighting in Gaza and the Rules of the Game are Expected to Change', *Walla News*, 5 June 2021, https://news.walla.co.il/item/3439695 (Hebrew).

[33] Hezbollah alone is assessed as having at least 130,000 rockets in its arsenal. 'Missiles and Rockets of Hezbollah', *Missile Threat: CSIS Missile Defense Project*, https://missilethreat.csis.org/country/hezbollahs-rocket-arsenal/.

[34] Tsach Moshe, 'The IDF must manoeuvre wisely', Between the Poles—Land Forces Part B 31–32, 159–174 (Hebrew). Bokhbot, 'Flocks of IDF quadcopters'.

[35] Michael Herzog, 'The IDF Strategy Goes Public' *Policy Watch 2479*, The Washington Institute, 28 August 2015, http://www.washingtoninstitute.org/policy-analysis/view/new-idf-strategy-goes-public. This commitment was repeated in the 2018 document 'IDF Strategy', https://www.idf.il/media/34416/strategy.pdf (Hebrew).

five-year force development plan 'Momentum' (*Tnufa*) which aspires to create a networked military with an emphasis on rapid multiple target detection and allocation of sensors to munitions. Kochavi's most notable manifestation of his concept is the experimental unit dubbed '*the Ghost Unit*', which combines special operations forces and elite infantry with enhanced cyber and drone capabilities in addition to soldiers from the air force and field intelligence. These forces are trained to infiltrate enemy territory to locate targets, communicate their precise location to a long-range weapon-launcher firing precise munitions to destroy them.[36] It remains to be seen if the next IDF Chief of Staff will continue Kochavi's direction or will opt for one of the other two options presented above.

The shift of focus from manoeuvre to precision standoff fire has been accompanied by a shift in the role of the reserves from the primary force to a secondary one. Furthermore, despite the lessons of the Second Lebanon War, this shift has been followed with a reduction in training for the reserves.[37]

In any case, it should be noted that currently the IDF still fields a notable armour fleet of about 1,450 tanks (Merkava Mk 3 and Mk 4) and all its infantry brigades are equipped with APCs, whether of the heavy models (Namer, Achzarit) or the older M113 which IDF plans to replace with an advanced and better protected local design, the Eytan. The vast majority of the tanks are in reserve units, whereas the heavy APCs are more evenly distributed between regular infantry and reserve infantry units.

Conclusion

The geostrategic situation and the nature of Israel's enemies dictated that the IDF was developed and centred primarily on land forces. A reflection of this reality was the fact that its Chief of Staff—actually the commander-in-chief of all Israel's armed forces, land, sea, and air—has always, with one exception only,[38] been selected from the land forces and was generally considered to also be the specific direct operational commander of the land forces.[39]

---

[36] Yoav Limor, 'Ghost People', *Israel Today*, 20 August 2020, at: https://www.israelhayom.co.il/article/793027 (Hebrew); Tal Ram Lev, 'A Drone to Each Platoon Commander and Quickly Employing Fire Support: This is How the Future of the Ground Forces is Going to Look', *Ma'ariv Online*, 18 March 2021, https://www.maariv.co.il/news/military/Article-828550 (Hebrew).

[37] Dotan Druck, 'The Reserves Will Hold: Changes in the Israeli Defense Forces' Operational Concept' *The RUSI Journal* 166, 4 (2021), 40–50.

[38] Chief of Staff Dan Halutz from the Air Force was IDF Chief of Staff from 2005 to 2007. Some of the failures of the Second Lebanon War, rightly or wrongly, were attributed to his lack of knowledge in land warfare.

[39] In the 1980s a Ground Forces Headquarters was established to conduct force build-up, but operational control of the land forces is still directly in the hands of the IDF Chief of Staff.

The IDF evolved from an underground guerilla fighting force before the establishment of the state, to infantry brigades upon its establishment and then gradually transitioned to armour divisions conducting high tempo mobile combined arms operations.

During the 1990s, new technologies enabled more precise standoff fire. A new concept gradually evolved preferring reliance on fire from afar over land manoeuvre. This development coupled with a reduction of the Fundamental Threat (a major ground invasion) led to a decline in the IDF's ability to conduct large mobile combined-arms operations. In the 2006 Second Lebanon War, the deliberate degradation of ground manoeuvre capabilities was exposed by Israeli tactical failures against Hezbollah. Following the poor demonstration of land forces in this war there was a renewed focus on reacquiring ground manoeuvre capabilities. However, the role of the land forces in the IDF is still subject to debate to this day.[40] On the one side, the proponents of improving standoff fire technologies as a substitute for manoeuvre argued that manoeuvre should be limited and sensor-saturated in order to rapidly discover enemy locations and pass them on to the fire-forces. On the other side, a 'back to the basics' approach that argued that, although the new technologies did improve fire capabilities and should be added to the traditional manoeuvre units to increase their lethality, they did not enable fire to fully replace aggressive large-force manoeuvres that found and defeated the enemy whilst conquering ground. The IDF latest multiyear force build up plan *Tenufa* (Momentum) seems to be an attempt to find a middle ground between these two approaches.

[40] Reflections of this debate: Gabi Siboni and Yuval Bazak, 'The IDF "Victory Doctrine": The Need for an Updated Doctrine', The Jerusalem Institute for Strategy and Security (JISS), 14 June 2021, https://jiss.org.il/en/siboni-idf-victory-doctrine-the-need-for-an-updated-doctrine/; Yair Golan and Gal Perl Finkel, 'On Manoeuvre and Training for it', Beyn HaMaarachot—IDF Digital Journal, 22 October 2020, https://www.idf.il/media/77423/%D7%91%D7%99%D7%9F-%D7%94%D7%9E%D7%A2%D7%A8%D7%9B%D7%95%D7%AA-%D7%92%D7%95%D7%9C%D7%9F-%D7%95%D7%A4%D7%A8%D7%9C-%D7%A4%D7%99%D7%A0%D7%A7%D7%9C.pdf (Hebrew); Ofer Shelakh, 'He who flees from manoeuvre will not win', Institute for National Security Studies (INSS), 12 August 2021, https://www.inss.org.il/he/wp-content/uploads/sites/2/2021/08/%D7%A4%D7%A8%D7%A1%D7%95%D7%9D-%D7%9E%D7%99%D7%95%D7%97%D7%93-1108212.pdf. (Hebrew).

16

Tactics and Trade-Offs

The Evolution of Manoeuvre in the British Army

Alex Neads and David J. Galbreath

Introduction

The future trajectory of land warfare in the United Kingdom stands at a cross-roads. For decades, the British Army has striven to become a reliable and enthusiastic proponent of US-led digital transformation, quietly adapting expensive US concepts to suit British budgets and organizational preferences through its own 'manoeuvrist approach' to operations. In so doing, the UK has widely been seen as a bridge between the Pentagon and European armies, especially during the defining conflicts in Iraq and Afghanistan. Indeed, the desire to maintain operational currency and tactical interoperability with the US military lies at the heart of British defence policy, even as the UK has increasingly struggled to afford the full spectrum of capabilities these doctrines necessitate. Now, with the character of warfare evolving once again, this old paradox presents new challenges as the British Army attempts to rejuvenate its warfighting capabilities in a fashion fit for the future.

On the one hand, the UK Ministry of Defence's new *Integrated Operating Concept* mirrors the essential contours of the USA's *Multi-Domain Operations*, aiming to buttress the utility of British military power through a shift in emphasis toward information and meaning, underpinned by broader and deeper cross-governmental operational integration. Such concepts are reflected in the British Army's *Land Operations* doctrine, which posits 'integrated action' as fundamental to land manoeuvre. On the other hand, the British Army's ageing fleet of conventional platforms—from main battle tanks and infantry fighting vehicles to artillery systems and communication suites—are now in urgent need of re-capitalization, raising profound questions about where the technological crux of future tactical capability should lie. This chapter reveals the complex trade-offs and path dependencies

Alex Neads and David J. Galbreath, *Tactics and Trade-Offs*. In: *Advanced Land Warfare*. Edited by Mikael Weissmann and Niklas Nilsson, Oxford University Press. © Alex Neads and David J. Galbreath (2023). DOI: 10.1093/oso/9780192857422.003.0016

inherent in implementing the British Army's emergent approach to land warfare. It examines recent debates within the British profession of arms on doctrine and acquisitions to explore the ongoing development of British military tactical practice. At heart, these discussions illuminate an uncomfortable interaction between martial concepts and material realities, strategic ambition and financial constraint in the construction of British land power—with attendant implications for future tactical and operational realities.

The chapter proceeds in the following way. The first section examines the importation of manoeuvre warfare doctrines into the British Army from the USA, via NATO. The second section explores the subsequent development of these ideas in the early post-Cold War period and into the British Army's operations in Iraq and Afghanistan, focusing specifically on the development of the material capabilities associated with 'force transformation'—and the financial and organizational challenges these presented. The third section then turns to the British Army's efforts to regenerate manoeuvre doctrine and rationalize force structures for the post-Afghanistan era through the Army 2020 reforms, shaping present options for the future in so doing. Finally, the chapter turns to examine the opportunities and constraints for future manoeuvre presented by the path-dependent evolution of British military doctrine and capability. The chapter concludes that whilst the British Army's understanding of manoeuvre has been heavily shaped by its most significant ally, the USA, the implementation of these concepts has been defined by the unique politics of British defence—and above all, a particular blend of organizational preferences, cultural attitudes, and financial constraints. Moreover, these peculiarities of British defence now appear to be shaping the reality of British military manoeuvre more than ever, potentially leading to either a divergence between British and US constructions of manoeuvre—or else a gap between British doctrine on paper and British capabilities in practice.

Importing Manoeuvre into British Military Thought

Since the 1980s, the British Army has successively imported US concepts of land manoeuvre, progressively adapting American military ideas and practices to suit British budgets and cultural preferences. This emulation made good sense in the context of Britain's Cold War defence policy, which sought to tie the US into European defence whilst maintaining enough independent capability to safeguard British interests, hedging between the USA and the continent. Consequently, NATO operations in northern Europe became a central concern for British defence policy, especially during the later Cold

War after Britain had largely divested herself of Empire.[1] Accordingly, the British Army's current doctrinal thinking on manoeuvre has its roots in the organizational change undertaken by the US Army following the Vietnam War. In 1976, the US Army Training and Doctrine Command (TRADOC) published a new operational doctrine called *Active Defense*, intended to refocus the US military on countering the Soviet threat in Europe. TRADOC itself had been established to help revitalize the US Army in the aftermath of the Vietnam War and the introduction of the All-Volunteer Force model, and its first commander General William DePuy viewed *Active Defense* first and foremost as a means to improve the US Army's collective training and professional education standards. However, the doctrine also advocated a firepower-heavy positional style of fighting which aroused significant controversy and professional debate, ultimately culminating in the development and adoption of an alternative concept of manoeuvre warfare in the publication of the US *AirLand Battle* doctrine in 1982.[2]

These debates chimed with reform-minded British officers on the other side of the Atlantic, themselves preoccupied with the British Army's own lack of preparedness to meet the forces of the Warsaw Pact. In particular, the British commander of NATO's Northern Army Group (NORTHAG), General Sir Nigel Bagnall, had become simultaneously concerned with the relative decline in the training and equipment of the British and allied divisions in northern Germany when compared with the much larger and increasingly modernized Soviet forces. Bagnall believed that a lightning Warsaw Pact campaign conducted in the style of the Soviet's Second World War Operational Manoeuvre Groups might rapidly overrun NORTHAG, creating a strategic *fait accompli* before NATO's civilian leadership could agree on an effective political response. Bagnall thus sought to bog down any prospective Soviet thrust, abandoning NORTHAG's previous positional 'forward defence' posture in favour of a new twin-track approach. This saw NORTHAG's air component tasked with targeting Soviet second echelon forces in depth to prevent them from reaching the battle area, whilst the British troops under Bagnall's command were restructured and re-equipped to undertake mobile counter-attacks against the Soviet first echelon, which would now have to fight alone.[3]

[1] Andrew Dorman, 'Reconciling Britain to Europe in the Next Millennium: The Evolution of British Defense Policy in the Post-Cold War Era', *Defense Analysis*, 17, 2 (2001): 188–91.
[2] Richard Lock-Pullan, 'How to Rethink War: Conceptual Innovation and AirLand Battle Doctrine', *Journal of Strategic Studies*, 28, 4 (2005): 679–702.
[3] Andrew Dorman, 'A Peculiarly British Revolution: Missing the Point or Just Avoiding Change?' in *Reassessing the Revolution in Military Affairs: Transformation, Evolution and Lessons Learnt*, edited by Jeffrey Collins and Andrew Futter (Basingstoke: Palgrave Macmillan, 2015), 33–50.

NORTHAG's focus on deep strike and operational manoeuvre developed into the NATO doctrine of *Follow-On Forces Attack*, which, despite some differences in detail, shared a common intellectual pedigree with US *AirLand Battle*. Moreover, Bagnall's elevation to Chief of the General Staff confirmed the British Army on the same developmental trajectory as its US interlocutor, leaving a lasting impression on British military practice. Indeed, manoeuvre warfare enthusiasts in the British Army leaned heavily on US doctrine in their own thinking throughout the late Cold War period, notwithstanding some of Bagnall's own reservations.[4] Brigadier Richard Simpkin's influential book *Race to the Swift*, for example, propounded the importance of concepts like tempo, momentum, and simultaneity for the British Army of the Rhine, underpinned by a recognition of the importance of information processing and rapid decision–action cycles to creating operational advantage.[5] As in the USA, the British adoption of manoeuvre warfare also provided the British Army with a template for organizational change and technological modernization, evident in the publication the UK's first higher military doctrine, initially called *British Military Doctrine* and subsequently *British Defence Doctrine* after its capstone concepts were adopted by the Royal Navy and Royal Air Force. Importantly, the 1997 edition confirmed 'the Manoeuvrist Approach' as the cornerstone of this new British way in war, alongside 'Mission Command' as the British view on *Auftragstaktik*-style command and control (C2).[6]

The Manoeuvrist Approach broadly mirrored US ideas about manoeuvre warfare, but also reflected some uniquely British accommodations. Shortly after the publication of this new doctrine, Assistant Chief of the Defence Staff Major General John Kiszely observed how the prevailing NATO definition of manoeuvre as 'the employment of forces on the battlefield through movement in combination with fire, or fire potential, to achieve a position of advantage in respect to the enemy in order to accomplish the mission' sat uncomfortably between two divergent schools of thought: one which viewed manoeuvre as little more than the conduct of fire and movement, and its alternate, more abstract, understanding as the adroit creation and exploitation of leverage to produce a disproportionate effect on the enemy.[7] At least at first, the Manoeuvrist Approach sought to span both these poles whilst simultaneously leaning towards the latter. Defined as 'an attitude of mind' focused

---

[4] Ibid.: 38–44.

[5] Richard Simpkin, *Race to the Swift: Thoughts on Twenty-First Century Warfare* (London: Brassey's Defence, 1985).

[6] See John Kiszely, 'The Meaning of Manoeuvre', *RUSI Journal* 143, 6 (1998): 37.

[7] Ibid.: 36.

on the use of guile to shatter the enemy's will and cohesion (as opposed to simply eroding his fighting forces in the field), the Manoeuvrist Approach was presented as a manner of fighting favouring the quantitively weaker but qualitatively more capable belligerent, reliant on precision, flexibility, and joint and combined arms integration rather than mass. Although the British Army would now 'fight to move' rather than 'move to fight', Kiszely was nontheless quick to recognize the enduring importance of tactical attrition to British military operations—both for enabling operational manoeuvre and as a fall-back when manoeuvre was seen as too risky.[8]

If the British military's instinct, therefore, was to view the Manoeuvrist Approach as primarily an operational level concept with limited direct bearing on the messy tactical realities of combat, the end of the Cold War began to challenge this perspective. In principle, British defence policy continued to be guided by Cold War assumptions during the early 1990s, even despite the fall of the Berlin Wall. The acquisition of Challenger 2 main battle tanks for the British Army, for example, reflected a direct continuation of previous capability requirements. The British Army likewise traded command of NORTHAG on its disbandment for stewardship of the new Allied Rapid Reaction Corps (ARRC—initially ACE/RRC), which was seen as its spiritual successor in NATO. However, this conceptual stasis belied a significant upheaval in British defence policy. All three services were subjected to sweeping financial cuts and downsizings, described as 'traumatic' by one commentator, as the government of the day attempted to wring an economic dividend from the absence of a clear military threat. Moreover, neither the USA's growing distance from Europe, nor initial efforts to build an EU defence infrastructure in their absence, suited the UK's traditional preferences.[9]

Toward the end of the 1990s, British foreign policy gained a new sense of direction under Prime Minister Tony Blair, leading to a new emphasis on expeditionary capabilities. By the time of New Labour's new Strategic Defence Review (SDR) in 1998, the British Army's complement of main battle tanks had already been reduced by 45 per cent on Cold War levels. Although the SDR retained the ability to deploy heavy armour at divisional strength, increasing prominence was given to the development of rapidly deployable light and medium-weight brigades suitable for limited interventions and Peace Support Operations (PSOs).[10] This placed a renewed premium on the further adoption of joint operations, underpinned by greater digitization

---

[8] Ibid.: 37–39. See also, John Kiszely, 'The British Army and Approaches to Warfare since 1945', *Journal of Strategic Studies* 19, 4 (1996): 179–206

[9] Dorman, 'Reconciling Britain to Europe': 191–194.

[10] British Government, *Strategic Defence Review: Modern Forces for the Modern World* (London: HMSO, 1998).

and a general trend toward technological modernization. Here, New Labour actively sought to reconcile the US and European facets of British defence policy, arguing that modernized European forces would enable the EU to share the burden of regional security whilst remaining available to NATO (and thus subject to US veto) in the event of a major war—confirming Britain's own view of itself as a bridge between the USA and Europe in the process. Moreover, British experience in Bosnia and especially Kosovo— where European forces were forced to rely on US air power owing to a lack of modern capability—lent further credence to the need for more agile and technologically advanced forces.[11] Further adoption of the US-inspired Revolution in Military Affairs thus became a core feature of British efforts to balance NATO and the EU, whilst also enabling its own interventionist 'ethical foreign policy'.

Even so, manoeuvrism itself remained a somewhat contentious topic among some British officers, even at the cusp of the new millennium. On the one hand, the 1991 Gulf War confirmed to British officers the importance of maintaining interoperability with developing US warfighting concepts, especially given the relative prominence the Americans afforded the British Army compared with other less modernized allied contingents. Indeed, the *Follow-On Forces Attack* plan developed from NORTHAG's reforms had provided the building block for coalition planning in the Gulf.[12] On the other hand, some dogmatic officers were beginning to view the idea of manoeuvre as the antithesis of attrition, leading laggards to deride the Manoeuvrist Approach as a dangerous myth and lampoon reformers for ostensibly believing that a mastery of manoeuvre might prevent the need for bloodshed altogether.[13] Writing in 1998, Royal Marine Brigadier Robert Fry observed that the British military had traditionally been too small to bother much with grand operational concepts like manoeuvre, quipping that 'whilst we are all manoeuvrists now, we seem to have reached this position independently from our history in modern warfare'.[14] Fry himself concluded that whilst technological modernization made manoeuvre doctrine viable, and the need to remain compatible with US practices made it desirable, manoeuvre and attrition should not be seen as polar opposites, given that 'an element of attrition is a necessary precondition to successful manoeuvre'.[15] Moreover, state-on-state warfare continued to be seen as the Army's *raison d'être* by senior officers,

---

[11] Dorman, 'Reconciling Britain to Europe': 194–198.
[12] Dorman, 'A Peculiarly British Revolution': 44–45.
[13] Ibid.: 46; Kiszely, 'The Meaning of Manoeuvre': 38.
[14] Robert Fry, 'The Meaning of Manoeuvre', *RUSI Journal* 143, 6 (1998): 41.
[15] Ibid.: 42.

notwithstanding the new-found emphasis on PSOs. Although the 2001 edition of *British Defence Doctrine* confirmed the centrality of the Manoeuvrist Approach to *all* operations, the head of the UK's Joint Doctrine and Concepts Centre also reiterated that 'all those who wear the uniform of the UK's Armed Forces must be prepared to deliver lethal force and, if necessary, die for whatever legitimate cause the UK is fighting'—just in case there was any doubt as to what the Manoeuvrist Approach actually involved.[16]

Between Warfighting Capability and Counter-insurgency

The September of 2001 set in train two parallel and ultimately divergent processes, which would confirm the supremacy of manoeuvrist thinking in the British Army but also simultaneously undermine the technological and organizational foundations of the British military's conventional modernization efforts. That month marked both the 9/11 terrorist attacks on the Twin Towers in New York, and also the publication of the US Department of Defense's *Quadrennial Defense Review* (QDR).[17] The latter document signalled a further evolution in US military thinking on manoeuvre, which, when combined with the geostrategic implications of 9/11, confirmed and accelerated the British Army's trajectory of reform. At the same time, however, the so-called Global War on Terror which followed the 9/11 attacks saw the British Army embroiled in a series of protracted counterinsurgency operations in Iraq and Afghanistan, placing both the Army as an institution and wider British defence policy under significant pressure, ultimately calling into question the financial and organizational viability of US-inspired military modernization.

Admittedly, the publication of the US QDR in 2001 was far from the first move toward greater trans-Atlantic interoperability in military capabilities. The Americans, for their part, have been encouraging European military modernization since at least the 1999 NATO summit in Washington, and the UK's 1997 SDR had likewise begun to reorient the British Army's capabilities in line with US digitization agendas. The SDR had seen the creation of the UK's Joint Rapid Reaction Force, used to much effect in Sierra Leone in 2000. The British Army was similarly moving toward the acquisition of more transportable and expeditionary capabilities prior to the QDR, with

---

[16] Major General Anthony A. Milton, 'British Defence Doctrine and the British Approach to Military Operations', *RUSI Journal* 146, 6 (2001): 42.
[17] *Quadrennial Defense Review Report, September 30, 2001*, US Department of Defense (Washington, DC: DoD, 2001).

the initiation of the TRACER and MRAV programmes to acquire new reconnaissance and multi-role armoured vehicles via US and European consortia respectively. Nonetheless, profound acceptance of digitization in the British officer corps prior to the QDR has been described as lacklustre, and even some US officers continued to view the RMA as something of a fad.[18]

Importantly, the QDR—following shortly after both a change of US government and 9/11—introduced a new language of defence 'transformation' into military doctrine. Although still fundamentally manoeuvrist in character, the transformation agenda focused attention on the emerging material capabilities through which manoeuvrist principles could be applied to their fullest extent. Here, US transformation can be seen as the product of three intersecting elements: Network-Centric Warfare (NCW), Effects-Based Operations (EBO), and expeditionary force structures. The adoption of each of these in the British Army would require further doctrinal changes, but critically also material acquisitions.[19]

In the UK, this shift began immediately after 9/11, with the publication of the 'New Chapter' to the SDR focusing primarily on responding to international terrorism. Then, shortly after the 2003 Iraq War, the MoD published a further Defence White Paper entitled *Delivering Security in a Changing World*, which sought to significantly reshape the force structure of the British Army in line with the US model of digital transformation. This embedded a new brigade structure built around two heavy brigades, three medium-weight brigades, and a light brigade; enough to undertake two minor contingency tasks or one short major conflict.[20] This represented a conscious shift away from heavy armoured units toward medium-weight forces, to be equipped with a new vehicle system procured under the Future Rapid Effects System (or FRES). The FRES programme was intended to provide a family of medium-weight, air mobile armoured vehicles, equipped with modern sensors and digital connectivity. Not only would their medium weight make them much easier to deploy and support on expeditionary operations, but the use of a common chassis to produce various different specialist vehicles (where previously multiple entirely different platforms had been acquired) would simplify fleet management and generate cost savings. As such, the FRES programme replaced both the TRACER programme, which the US had lost interest in

---

[18] David Galbreath, 'Western European Armed Forces and the Modernisation Agenda: Following or Falling Behind?', *Defence Studies* 14, 4 (2014): 398–402.

[19] Theo Farrell, 'The Dynamics of British Military Transformation', *International Affairs* 84, 4 (2008): 777–779.

[20] Ibid.: 798–800.

by 2001, and MRAV, from which the UK unilaterally withdrew in 2003 over concerns about its weight.[21]

This shift toward medium forces represented a calculated risk, highlighting the extent to which the British Army had internalized the idea of expeditionary manoeuvre based around digitally modernized forces. The shift to medium armour was accompanied by the acquisition of both Apache attack helicopters and man-portable Javelin anti-tank missiles, which were considered to somewhat offset the reduction in heavy armour in a conventional war in both the deep and close battle.[22] Even so, the British Army recognized that by converting heavy brigades with Warrior IFV and Challenger 2 into medium-weight formations equipped with FRES it was, as Theo Farrell has argued, 'consciously sacrificing combat power for increased mobility'. Nonetheless, Farrell concluded that this 'move to medium weight was not forced on the army by civilian policy-makers', but instead reflected a considered judgement about the likely character of future conflict, notwithstanding the desire to maintain a minimal divisional capability to retain a degree of 'full-spectrum' credibility in the eyes of the US Army.[23]

That said, if the British Army embraced transformation as the latest evolution of operational manoeuvre, it also sought to adapt some of its core principles just as it had done with the translation of manoeuvre warfare into the Manoeuvrist Approach. This can be seen in the British response both to NCW and EBO. Although British officers recognized the importance of digital communications, ISTAR capabilities, and long-range fires, the acquisition of profound levels of digital communications equipment necessary for a network-*centric* doctrine was considered unaffordable—even if procured in an incremental fashion. Moreover, whilst the British experience of the 2003 invasion of Iraq confirmed the importance of force transformation, British officers remained somewhat sceptical about the cost-effectiveness of aspects of digitization. This was especially true of systems like the US blue-force tracker, intended to provide a real-time 'common operating picture' of friendly and enemy locations to assist with the planning and execution of integrated operations, but which had actually provided less than seamless situational awareness about friendly forces movements let alone enemy dispositions.[24] Moreover, the increasingly centralized and hierarchical command structure produced by high-levels of digitization sat ill at ease with

---

[21] Ibid.: 800–801; *Obsolescent and Outgunned: The British Army's Armoured Vehicle Capability, Fifth Report of Session 2019–21*, Defence Select Committee (London: House of Commons, 2021), 47–49.
[22] *Delivering Security in a Changing World: Future Capabilities* (London: Ministry of Defence, 2004), 8.
[23] Farrell, 'The Dynamics of British Military Transformation': 800–804.
[24] Ibid.: 784–787.

the doctrine of delegated Mission Command the British Army had already internalized as part of its Manoeuvrist Approach, ultimately leading to concern about the possibility of digitally-enabled micromanagement of tactical commanders.[25] Hence, although the UK invested in both tactical and operational/strategic digital communications, most notably in the Skynet 5 satellite communications system and the Bowman family of digital radios and tactical information systems, the result was nonetheless a watering-down of NCW into a more affordable and palatable hybrid of existing processes and digital change, described as Network Enabled Capability (NEC).[26]

If anything, the British adoption of EBO was even more limited, at least initially. The British military recognized the utility of planning in terms of capabilities and effects (and the more fluid planning and force structures this implied), but the British Army was nonetheless uncomfortable with the increasingly scientific, technocratic, and metricized approach to operational planning EBO had produced in the US Army. Instead, the British military chose to view its own 'effects-based approach to operations' (EBAO) as an opportunity to develop greater interdepartmental involvement in campaign design—especially in the context of the British Army's growing counterinsurgency commitments in Iraq and Afghanistan. EBAO thus morphed into the idea of a 'Comprehensive Approach' to operations, encompassing both kinetic and non-kinetic effects—although the concept initially struggled to gain traction beyond the MoD.[27] Although the British Army has undoubtedly continued to internalize elements of EBO, particularly around the routine assessment of the effect caused by its kinetic activities and in the idea of a synergistic relationship between violent and non-violent military activities in achieving desired end-states, the direct lineage of EBO is less visible in British military practice than with manoeuvre warfare or digitization. Even so, the combination of EBAO, NEC, and light- and medium-weight expeditionary force structures optimized for joint operations at brigade level represented a significant evolution of the British Army's understanding of warfare from the initial adoption of the Manoeuvrist Approach as an operational level concept dependent at least in part on localized tactical attrition.

Unfortunately, the generation of major capabilities for even this adapted version of transformation was significantly undermined by the British Army's parallel commitment to expeditionary operations in Iraq and Afghanistan,

---

[25] Ibid.: 788–789; see also Paul Cornish, 'Cry "Havoc!" and Let Slip the Managers of War': The Strategic, Military and Moral Hazards of Micro-Managed Warfare (London: Strategic and Combat Studies Institute, 2006).

[26] Farrell, 'The Dynamics of British Military Transformation': 784–789.

[27] Ibid.: 790–798.

which placed all aspects of UK defence policy under considerable strain. In principle, the British commitment to counterinsurgency (COIN) in Iraq and Afghanistan was not inimical to the vision of warfare advanced by transformation. Indeed, much of the equipment procured specifically for expeditionary use in Afghanistan, in particular, relied heavily on digital networking and precision technology to identify and target insurgents, reflecting the core approach to warfighting at the heart of RMA and force transformation.[28] However, the rift between Europe and the USA generated by the 2003 Iraq War—and the British decision to follow the US trajectory—significantly undermined the balance between NATO and the EU in military modernization envisaged by New Labour at St Malo, with long-term implications for procurement.[29] Moreover, the British Army's lack of modern protected mobility vehicles saw light troops in Land Rovers suffer sustained casualties from IEDs in Iraq, leading to a domestic public reaction against MoD procurement policies and a further erosion of political support for the conflict.[30] The ensuing need to procure a spate of urgent operational requirements (UORs) to prosecute the campaigns in Iraq and Afghanistan—and allay political fallout from casualties—added further strain to the MoD equipment budget, at the expense of other long-term modernization programmes.

In normal circumstances, the cost of UORs were met from the treasury reserve rather than by the MoD, protecting core procurement programmes and in-year budgets. However, the extent of the UOR bill needed to equip the armed forces for COIN, combined with the scale of individual acquisition programmes such as protected mobility vehicles, created suspicions in the treasury that UORs were being used for routine procurement by stealth and the decision that UORs above a certain threshold must be met in part by the MoD's own funds.[31] Meanwhile, the cost of procuring the expected medium-weight FRES vehicle fleet had spiralled. Changes to the design (in part in response to greater force protection requirements arising from recent operational experience) also delayed the project, and resulted in a significant increase in the vehicle's weight—from 17 tonnes to somewhere in the region of 25–32 tonnes—leading to concern that the ensuing platform would be too heavy to be transported in the C-130 Hercules; the workhorse of the RAF air

---

[28] Jon R. Lindsay, 'Reinventing the Revolution: Technological Visions, Counterinsurgent Criticism, and the Rise of Special Operations', *Journal of Strategic Studies* 36, 3 (2013): 422–453.

[29] Jolyon Howarth, 'France, Britain and the Euro-Atlantic Crisis', *Survival* 45, 4 (2003): 173–192.

[30] On the so-called 'Wootton Bassett Phenomenon', see K. Neil Jenkings, Nick Megoran, Rachel Woodward, and Daniel Bos, 'Wootton Bassett and the Political Spaces of Remembrance and Mourning', *Area* 44, 3 (2012): 356–363; Michael Freeden, 'The Politics of Ceremony: The Wootton Bassett Phenomenon', *Journal of Political Ideologies* 16, 1 (2011): 1–10.

[31] Paul Cornish and Andrew Dorman, 'Blair's Wars and Brown's Budgets: From Strategic Defence Review to Strategic Decay in Less than a Decade', *International Affairs* 85, 2 (2009): 259–260.

mobility fleet at the time. By 2008, the MoD's equipment budget deficit sat at around £2 billion, leading to fears of further personnel cuts as the 2010 Strategic Defence and Security Review (SDSR) loomed. Against this backdrop, FRES became the stone cast aside to stem the tide. The bulk of the programme was effectively cancelled in 2008, leaving the Army without an obvious medium-weight capability despite its centrality to emerging force structure and doctrine.[32]

By then, the British Army had moved from its threat-focused structure and doctrine of the late Cold War, through a period of 'capability'-focused transformation, to become almost by default an army overwhelmingly preoccupied with campaigning in Afghanistan. A schism in the officer corps was also becoming apparent between those advocating for the adoption of COIN-type interventions as the armed forces' primary mission-set, and those who wanted to retain a semblance of so-called 'full-spectrum' warfighting capabilities. This debate had both intra- and inter-service dynamics, encompassing genuine professional disagreement over the future trajectory of warfare alongside organizational politics over resource allocations in the face of national austerity. Advocates of the 'New Wars', influenced by senior officers such as General Sir Rupert Smith, viewed the sort of interventions witnessed since the end of the Cold War and culminating in the protracted insurgencies in Iraq and Afghanistan as *the* template for future 'wars amongst the people'. Accordingly, they argued for a permanent realignment of force structures, training, doctrine, and equipment toward COIN, PSOs, and 'small war' at the expense of heavy armoured forces. This vision seemed to better reflect the reality of recent campaign experience, but would also safeguard the Army at the expense of the Royal Navy and RAF, which, as little more than adjuncts for the Army's force projection, would no longer require expensive high-end warfighting platforms.[33]

Within the Army and beyond, however, a rival school of thought continued to view COIN as an aberration rather than the rule, and maintained that all three services must retain the ability to conduct high-intensity manoeuvre against the armed forces of a rival peer state. Importantly, this was seen as essential not just to protecting the UK's national interests in the future, but also to maintaining credibility and relevance with key allies and alliances—most notably, the USA.[34] Certainly, whilst the experience of campaigning in

---

[32] *Obsolescent and Outgunned*, 48–9; Farrell, 'The Dynamics of British Military Transformation': 800–801; Galbreath, 'Following or Falling Behind?': 407; Cornish and Dorman, 'Blair's Wars and Brown's Budgets': 258.

[33] Cornish & Dorman, 'Blair's Wars and Brown's Budgets': 255–258; see also Rupert Smith, *The Utility of Force: The Art of War in the Modern World* (London: Penguin, 2006).

[34] David Blagden, 'Strategic Thinking for the Age of Austerity', *RUSI Journal* 154, 6 (2009): 60–66.

Afghanistan continues to loom large in the British Army's collective memory, the ideal of 'conventional' warfare against another state military remains the 'gold-standard' of professional military practice within the British Army's organizational culture. Despite the significant tactical adjustments made in response to COIN, for instance, the values against which promotions, appointments, training, and doctrine operated continued to be rooted in the Manoeuvrist Approach and the pre-requisites of manoeuvre warfare.[35]

In the event, the resultant outcome in the 2010 SDSR was something of a fudge. On the one hand, state failure and the increasingly 'hybrid' merger of state and non-state threats were identified as the likely character of future conflict. On the other, core capabilities highlighted by the transformation agenda as necessary for modern manoeuvre warfare, such as ISTAR, were advanced as essential for meeting these hybrid threats— over and above population-centric manpower. The size of the expeditionary forces the UK would expect to deploy were also scaled down, although defence planning assumptions maintained the ability to deploy a small division of three brigades in extremis. The primary building block of the Army would become the multi-role brigade, the centrepiece of which would be two medium-weight armoured vehicles rescued from the ruins of the FRES programme—the Scout Specialist Vehicle and the FRES Utility Vehicle. Concomitantly, however, heavy armour, armoured infantry, and self-propelled artillery would be dramatically reduced.[36]

Indeed, it is clear that the difficulties of transformation were themselves the product of path dependent processes rooted in the impact of short-term contingencies on the British Army's long-term decision-making. Farrell, for example, has argued that the British Army's efforts at emulating US transformation were conditioned by a mixture of operational exigency, pre-existing organizational culture, domestic politics, and limited means. Of these factors, however, budget appears to have been by far the most constraining, exacerbated by the cost of campaigning. Writing in 2008, for example, Farrell argued that 'budget problems are unlikely significantly to affect the direction of, let alone derail, British military transformation'.[37] Inasmuch as the further reorganization of the British Army precipitated by the 2010 Strategic Defence and Security Review maintained the focus on the medium-weight forces first

---

[35] Sergio Catignani, '"Getting COIN" at the Tactical Level in Afghanistan: Reassessing Counter-Insurgency Adaptation in the British Army', *Journal of Strategic Studies* 35, 4 (2012): 513–539; Sergio Catignani, 'Coping with Knowledge: Organizational Learning in the British Army?', *Journal of Strategic Studies* 37, 1 (2014): 30–64.

[36] *Securing Britain in an Age of Uncertainty: The Strategic Defence and Security Review*, British Government (London: HMSO, 2010).

[37] Farrell, 'The Dynamics of British Military Transformation': 807.

selected in the late 1990s, and continued within the vision of digitization, he was right. Even so, as Cornish and Dorman have argued, austerity has meant that the treasury has become the ultimate arbiter of defence policy, shaping both the force structure of the British Army, and with it, the doctrinal understandings of manoeuvre that can reasonably be achieved.[38]

Reinventing Manoeuvre after Afghanistan: Army 2020 and Organizational Change

The programme of reforms initiated by the 2010 SDSR was known as Future Force 2020, with the direction of change for land forces set by the Army 2020 programme released in 2012 and further updated in 2013.[39] Army 2020 was primarily intended to reorient the British Army away from counterinsurgency in anticipation of the eventual drawdown of British forces in Afghanistan, against a backdrop of significant fiscal austerity and the need to make further cost savings across defence. Army 2020 was therefore a concerted attempt to reshape the force in the light of immediate challenges, but the trajectory of this programme has fundamentally shaped the options for, and understanding of, future manoeuvre presently being grappled with in the British Army today.

Organizationally, Army 2020 envisaged a new model for force generation, in part driven by the significant strain that had been placed on the Army during the recent surge in operational tempo. This saw the Army divided into a Reactive Force, intended to provide a high readiness capability for deterrence and contingency tasks where the Army's main conventional warfighting assets would be held, and a reactive force, which was to provide troops for follow-on roulements in a major intervention as well as other enduring overseas tasks and standing domestic commitments. The reactive force was to be comprised of three armoured infantry brigades, together with 16 Air Assault Brigade, which would rotate through a three-phase readiness cycle to provide one armoured infantry brigade and one air assault battlegroup at high readiness at any given time, whilst still retaining the ability to deploy a divisional sized force in extremis. In so doing, the Army 2020 plan maintained the Army's previous heavy–light split, but sought to provide greater flexibility by including medium-weight forces alongside existing formations.[40]

---

[38] Cornish and Dorman, 'Blair's Wars and Brown's Budgets': 248–249.
[39] *Securing Britain in an Age of Uncertainty*; *Transforming the British Army: Modernising to Face an Unpredictable Future* (Andover: British Army, 2012); *Transforming the British Army: An Update* (Andover: British Army, 2013).
[40] *Transforming the British Army*: 4–6; *Transforming the British Army: An Update*: 6–13.

In the Reactive Force, each armoured infantry brigade was restructured to include a 'heavy protected mobility' battalion equipped initially with Mastiff, a mine resistant protected patrol vehicle acquired as a UOR for Afghanistan, until replaced with the medium-weight Utility Vehicle to be procured from the ruins of the FRES programme. Armoured reconnaissance regiments would likewise be equipped with the Scout Specialist Vehicle, procurement of which remained funded after the closure of FRES. Light infantry brigades would retain various protected mobility and patrol vehicles initially procured for counterinsurgency operations. Army 2020 also continued to emphasize the importance of digitization, ISTAR, precision fires, and joint interoperability, which had featured prominently in US concepts of force transformation. Indeed, the Army 2020 plan was initially released in 2012 under the title of 'Transforming the British Army'.[41] Moreover, the re-organization of the Army's structure into reactive and adaptable forces optimized for action at brigade level was in continuity with earlier British thinking on contingency operations, and Army 2020 explicitly emphasized the importance of using the adaptable force to conduct capacity building and post-conflict reconstruction activities 'upstream' and 'downstream' of any major expeditionary intervention.[42]

It is clear, however, that much of this planning was a compromise driven by the need for austerity savings. The SDSR itself described the previous equipment plan as 'unaffordable', carrying an estimated unfunded liability of £38 billion across defence out to 2020, and Army 2020 documents themselves noted 'the financial imperatives facing the Army to play its part in bringing the Ministry of Defence's budget back into balance'.[43] The Army 2020 plan was accompanied by a further downsizing of the Army's regular establishment by 12,000 troops, to be offset by a major investment in reserve forces through the accompanying Army Reserve 2020 plan. This latter element was intended to make reserve forces more deployable and usable by improving their training, equipment, and size, integrating them more closely into the regular force. In many respects, this reflected the recent experience of counterinsurgency operations, where extensive use had been made of reservists to augment regular units as a tactical reserve, as opposed to their traditional Cold War role as a strategic reserve. Nonetheless, the combination of downsizing and investment in reserves was not universally welcomed by senior

---

[41] *Transforming the British Army*: 4–5.
[42] *Transforming the British Army*, 2; *Transforming the British Army: An Update*, 21. See also, Robert Johnson, 'Upstream Engagement and Downstream Entanglements: The Assumptions, Opportunities, and Threats of Partnering', *Small Wars & Insurgencies* 25, 3 (2014): 647–668.
[43] *Securing Britain in an Age of Uncertainty*: 15; *Transforming the British Army: An Update*: 2.

officers, with the incumbent Chief of the General Staff describing it as a risky 'finger in the wind thing' imposed by politicians bent on cuts.[44]

Consequently, this structure was revised in important ways following the 2015 SDSR, under what became known as Army 2020 Refine. By 2015, the UK's strategic environment was perceived quite differently from that faced by the 2010 SDSR. Russia's unexpected seizure of Crimea and the ensuing conflict in Eastern Ukraine, in particular, focused attention on Russia as a renewed concern for European security and the pacing threat for UK defence assumptions to plan against. Moreover, the US pivot to Asia driven by China's growing bullishness in the Indo-Pacific, although somewhat abated by Russian revanchism, also served to focus British attention on the rejuvenation of its military capabilities for high-intensity inter-state warfare.[45] Indeed, the USA had itself begun to embark on a further programme of military 'off-set', aimed at developing the next-generation of technological capabilities in the face of an increasingly modernized Russia and China. Here, a particular emphasis was placed on unmanned systems and the military applications of artificial intelligence and machine learning software, together with the doctrinal concepts required to effectively deploy such technologies.[46] This clearer conventional threat not only provided a planning focus for land capabilities and doctrine, but was enough to stabilize the defence budget and create expectations in the British Army of a future funding uplift to support greater digital modernization and major equipment acquisitions.[47]

Importantly, Army 2020 Refine aimed at regenerating the Army's capabilities for divisional warfighting. The deployment of a division to a US-led coalition during a major warfighting campaign was now seen by the British Army as the minimum amount required to retain command of the ARRC, the smallest capability considered to be of credible independent value to the USA, and simultaneously also the maximum size of force the British Army could hope to deploy and maintain in the field for any sustained length of time. Importantly, this division was primarily expected to be drawn from two armoured infantry brigades (down from three) and two new medium-weight 'Strike Brigades', which were to be equipped with a mixture of Ajax—the turreted, tracked, medium-weight 'light tank' equipped with a 40 mm cannon

---

[44] Patrick Bury and Sergio Catignani, 'Future Reserves 2020, the British Army and the Politics of Military Innovation During the Cameron Era', *International Affairs* 95, 3 (2019): 696 and passim.

[45] *National Security Strategy and Strategic Defence and Security Review 2015: A Secure and Prosperous United Kingdom*, British Government (London: HMSO, 2015).

[46] See Gian Gentile, Michael Shurkin, Alexandra T. Evans, Michelle Grisé, Mark Hvizda, and Rebecca Jensen, *A History of the Third Offset, 2014–2018* (Santa Monica: RAND, 2021).

[47] Ewen MacAskill, 'Does the UK Really Need to Increase its Defence Spending?', *The Guardian*, 22 January 2018, https://www.theguardian.com/politics/2018/jan/22/does-the-uk-really-need-to-increase-its-defence-spending-russia/.

developed under Scout SV programme—and a new Mechanised Infantry Vehicle (MIV)—the conceptual inheritor of the Utility Vehicle requirement. The platform selected to fulfil this role was Boxer, a medium-weight wheeled armoured personnel carrier, which had been developed from the MRAV programme the UK had originally withdrawn from in 2003.[48] Here, this vision for Army modernization can be seen as a logical extension of many of the core ideas about the evolution of manoeuvre initially adopted in the UK versions of EBAO, network enablement, and force transformation, continuing the same emphasis on expeditionary deployment, flexible force structures, and digital connectivity.

In particular, the renewed shift toward a medium-weight armoured force can be seen as a direct response to the expanding depth of the battlespace resulting from the profusion of modern sensor and precision fires capabilities, especially when augmented by UAVs. The diffusion of these capabilities to Russia, together with Russian bastions in Kaliningrad and on the eastern borders of Poland, potentially allowed Russian firepower to reach across northwest Europe.[49] In an Article 5 scenario, this reach could render the pre-positioning of heavy armour in forward-mounted locations in Europe immediately vulnerable, simultaneously placing at risk the limited rail and road routes able to transport heavy tracked armour east, thereby degrading or slowing the UK's theatre-entry capabilities with heavy equipment. Wheeled armour, in contrast, might be able to self-drive significant distances across western Europe using a plethora of more dispersed minor routes, in order to rapidly congregate in the theatre of operations with greater security and survivability. In this vein, the French intervention in Mali in 2013 aroused significant interest and admiration in the British Army, in part because of the apparent deployability of medium-weight wheeled armoured vehicles such as the French VBCI. Indeed, some French force elements had been re-deployed to Mali from operations in the Ivory Coast, driving some 1,300 km in convoy from Abidjan to Bamako to enter their new theatre of operations, ostensibly confirming the utility of medium-weight capabilities.[50] The creation of a dedicated medium-weight force thus represented a culmination of the British Army's longstanding ambition to develop a more expeditionary armoured capability and still maintain the Army's focus on force transformation.

[48] *National Security Strategy and Strategic Defence and Security Review 2015* (London: HM Government, 2015), 31; *Obsolescent and Outgunned*: 44–52.

[49] Stephan Frühling and Guillaume Lasconjarias, 'NATO, A2/AD and the Kaliningrad Challenge', *Survival* 58, 2 (2016): 95–116.

[50] Jack Watling and Justin Bronk, 'Strike: From Concept to Force', RUSI Occasional Paper, June 2019, https://static.rusi.org/201906_op_strike_web.pdf.

Although wheeled armour tends to have less tactical mobility in difficult going than tracked armour, the operational mobility of wheeled medium armour may also provide greater tactical flexibility on an expanded and more dispersed future battlespace. The profusion of UAVs, electro-magnetic, and space-based sensors is expected to render the future battlefield far more transparent, whilst the extended range of such systems will concomitantly expand the physical scale of tactical space. Simultaneously, the diffusion of long-range precision strike technologies (including loitering munitions capable of selecting their own targets fired by conventional tube artillery), combined with the rapidity of identification-to-firing cycles enabled by modern digital communications, could make the prospect of being targeted by enemy forces much more fatal—even at far greater ranges. Much of future manoeuvre is therefore expected to take place 'in the deep', as each side attempts to use its long-range target acquisition and fires capabilities to 'shape' the enemy's ability to mass forces to their own advantage in the close battle. This idea has been likened by one senior German officer to age-of-sail naval battlefleets trading shots and jockeying for position before coming to quarters.[51]

The combined impact of these developments is to place a far higher premium on deception (physical and electronic) and on unmanned systems in order to reduce the risk to friendly forces and conceal intent, but also on tactical dispersion. By dispersing forces into smaller packets over much greater distances, Western armies hope to keep most force elements below the size threshold at which targeting by enemy artillery is worthwhile, given that doing so involves unmasking valuable guns or missile launchers and thereby exposing them to counter-battery fire. Dispersion might also allow commanders to hide high-value or 'signature' equipment amid the 'noise' of widely distributed small force elements, thereby concealing their true intentions. Even so, the ultimate defeat of determined enemy units and the seizure and holding of ground is still expected to require close combat at some point—which in turn will likely require these dispersed forces to concentrate mass against the enemy at a critical point. Moreover, this must be done very swiftly in order to prevent any remaining enemy indirect fire assets from destroying vulnerable densely packed formations, or the enemy similarly massing his dispersed forces to respond, and will likely require rapid dispersal after tactical engagements in order to protect friendly forces from enemy defensive fires.[52]

[51] Frank Leidenberger, 'How Allies Will Manoeuvre Beyond 2025', RUSI Land Warfare Conference presentation, 21 June 2018, https://www.youtube.com/watch?v=zMCdP8UYL_g/.

[52] See for example, *Joint Concept Note 1/17: Future Force Concept*, Ministry of Defence (Shrivenham: Development, Concepts and Doctrine Centre, 2017).

In this future dispersed battlespace, therefore, small units must be capable both of hiding statically for long periods and sustaining themselves, before undertaking very rapid movement over extended distances in order to survive and fight. Such a situation can be likened to atoms of a fixed amount of gas spreading out to fill a container. As the volume of the container expands—analogous to the expansion of tactical space—each atom, or dispersed force element, must move more rapidly to fill the available space. Consequently, wheeled medium armour might offer significant benefits over heavy tracked armour in this future environment. Not only are wheeled vehicles capable of travelling longer distances at higher speeds than tracked vehicles (particularly but not exclusively on roads or tracks), they tend to have a much lower breakdown rate, therefore requiring a smaller logistic footprint than tracked vehicles for a given milage—further improving their discreteness and survivability in a future high-intensity battlefield.[53] Moreover, the British Army's Strike Brigade concept was also explicitly advanced as a vehicle for further digitization, facilitating the profound levels of rapid and secure information exchange at the distance required to enable a more dispersed operating concept. Boxer, for example, has been described as a digitally-enabled 'node' hosting the Army's new digital 'backbone'—to be acquired via the Land Environment Tactical Communication and Information Systems (LE TacCIS) programme, known as Project Morpheus, that will replace Bowman—with sufficient power and space to enable incremental future upgrades, including the incorporation of future unmanned vehicles.[54]

More broadly, the British Army's capstone operational doctrine also developed in line with this information-centric vision of warfare, as has its supporting force structures. In the latest version of Army Doctrine Publication *Land Operations*, the Manoeuvrist Approach remains the central construct guiding land operations, but the management of information and the utility of non-kinetic effects (that is, non-violent actions, or actions that threaten but fall short of actual violence) are now portrayed as central to manoeuvrism. The Manoeuvrist Approach, for example, is still seen as an attitude of mind, but its execution requires a detailed understanding of the enemy's vulnerabilities, which in turn enables the commander to manipulate the enemy's understanding, perception, and behaviour in favourable ways.

---

[53] For a discussion, see Watling and Bronk, 'Strike: From Concept to Force': 15–20; John Matsumura, John Gordon IV, Randall Steeb, Scott Boston, Caitlin Lee, Phillip Padilla, and John Parmentola, *Assessing Tracked and Wheeled Vehicles for Australian Mounted Combat Operations* (Santa Monica: RAND, 2017), 27.

[54] 'Written Evidence Submitted by the Ministry of Defence', House of Commons Defence Committee Inquiry: Progress in Delivering the British Army's Armoured Vehicle Capability, AVF0016, 28 September 2020, https://committees.parliament.uk/writtenevidence/12523/pdf/; 'Guidance: LE TacCIS Programme', Ministry of Defence, 1 October 2020, https://www.gov.uk/guidance/le-taccis-programme/.

Although warfare is still seen as inherently violent, the concept of behavioural change and the use of information and narrative to shape 'audience' perception is increasingly central to the British Army's doctrinal understanding of manoeuvre, reflecting both the experience of 'winning hearts and minds' in COIN and a growing concern about the use of online media to manipulate domestic public opinion. This concept is epitomized by the idea of Integrated Action, which now sits alongside the Manoeuvrist Approach and Mission Command as a fundamental tenet of land doctrine.[55]

Integrated Action is defined as 'the application of the full range of lethal and non-lethal capabilities to change and maintain the understanding and behaviour of audiences to achieve a successful outcome'.[56] This concept is therefore a response to the perceived importance of information to manoeuvre on the digitally enabled battlefield, as the Chief of the General Staff's introduction to Integrated Action makes clear:

> in this complex and dynamic environment manoeuvre has to take account of a much broader audience than simply the 'enemy'. A new idea is therefore required—this is called Integrated Action. It is a unifying doctrine that requires commanders first to identify their outcome; second to study all of the audiences that are relevant to the attainment of the outcome; third to analyse the effects that need to be imparted on the relevant audience; before determining the best mix of capabilities, from soft through to hard power, required to impart effect onto those audiences to achieve the outcome.[57]

The concept thus reflects both the longstanding drive toward both joint (i.e. inter-service) integration first advocated in *AirLand Battle*, and the holistic cross-government involvement in the use of military forces advocated by the Comprehensive Approach—subsequently renamed the Integrated Approach and reflected in the scope and titling of the 2021 Integrated Review and in the accompanying Integrated Operating Concept—but also the effects-based approach to military campaign planning originating in US EBO.[58]

It has also been mirrored in the development of the British Army's force structure. From the beginning, Army 2020 sought to draw together and expand the Army's capabilities for information and intelligence gathering and

[55] *Army Doctrine Publication AC 71940: Land Operations*, British Army (Warminster: Land Warfare Development Centre, 2017), 4–1–5–4.

[56] *Army Doctrine Publication AC 71940*: 2.

[57] *Army Doctrine Publication AC 71940*: i.

[58] *Global Britain in a Competitive Age: The Integrated Review of Security, Defence, Development and Foreign Policy*, British Government (London: HMSO, 2021); *Integrated Operating Concept*, Ministry of Defence (Shrivenham: Development, Concepts and Doctrine Centre, 2021).

exploitation, creating a dedicated Intelligence, Surveillance and Reconnaissance Brigade and a Security Assistance Group containing media and psychological operations specialists.[59] This latter formation subsequently grew to become 77th Brigade, with expanded 'cyber' capabilities for 'behavioural change'. More recently, both brigades have been subordinated to the newly formed 6th Division alongside 1 Signal Brigade, bringing together the bulk of the Army's capabilities for 'Information Manoeuvre and Unconventional Warfare'.[60] However, these changes in force structure and doctrine have not been universally welcomed among the British officer corps, and the procurement of sufficient capabilities to make them work *in budget* has presented a significant challenge to this future vision of manoeuvre—as the recent Integrated Review has highlighted.

From Army 2020 to Integrated Manoeuvre

The idea of information manoeuvre and the incorporation of Integrated Action represents a radical departure from the original concept of the Manoeuvrist Approach as developed in the 1990s, in which manoeuvre remained an essentially enemy-focused activity reliant at least in part on bloody attrition. This shift away from a conventional platform and battle-centric understanding of warfare has elicited significant resistance both within and beyond the British officer corps—not least because of the MoD's inability to actually generate the medium armour and advanced technological capabilities required to enact it. In the run up to the (delayed) 2021 Integrated Review (and in subsequent reactions to its conclusions), much of this debate has centred around the ongoing utility of heavy armour in high-intensity warfare, based largely in observations of the fighting in the Donbass (prior to 2022) and Nagorno-Karabakh.

The dramatic and widely reported use of Turkish-supplied UAVs by Azerbaijan to destroy Armenian armoured forces during that conflict has frequently been described as a harbinger of the end of heavy armour on the battlefield. Equally, the employment of UAVs, electronic sensors, and modern ISTAR equipment to direct long-range fires was also a defining feature of the conflict in the Donbass in the years prior to the 2022 invaison, where Russian-backed separatists made extensive use of UAVs for

[59] *Transforming the British Army: Modernising to Face an Unpredictable Future*: 4; *Transforming the British Army: An Update*: 10–13.

[60] See Simon Goldstein, 'A British Perspective on Information Manoeuvre', *DefStrat Magazine*, 28 July 2020, https://www.defstrat.com/magazine_articles/a-british-perspective-on-information-manoeuvre/.

reconnaissance, artillery spotting, and electronic warfare. In one notorious incident, Russian-backed forces were able to rapidly defeat elements of two Ukrainian Army mechanized brigades massing in an assembly area near Zelenopillya, using UAVs to jam Ukrainian tactical communications systems before cuing a strike by multiple-launch rocket systems that purportedly destroyed two battalions' worth of combat vehicles in the space of a few minutes.[61] In many respects, however, Azerbaijani success owed as much to the modern sensor suites Azerbaijan was able to employ, together with their ability to effectively link them to various types of 'shooter' in a timely fashion, as to the decisive use of UAVs themselves.[62] In a similar fashion, the wider experience of fighting in Eastern Ukraine during the years before Russia's 2022 invasion appeared far more equivocal with regard to the utility of conventional armour than might first appear. Both sides made extensive use of upgraded and obsolescent heavy and medium armour to conduct offensive manoeuvre in open country, and to support more attritional fighting in urban centres. Indeed, in the latter environment, armoured vehicles appeared to retain significant utility—providing they were not subject to conventional aerial attack from attack helicopters or ground attack aircraft.[63] Moreover, this attritional type of street fighting seems likely to become the dominant form of urban warfare in the future, notwithstanding wishful thinking to the contrary.[64]

Consequently, the shift to a medium-weight capability at the expense of traditional heavy armour elicited significant criticism—especially after the 2021 Integrated Review confirmed the effective demise of the British Army's conventional armoured capability without an immediately serviceable medium-weight alternative. The UK's existing armoured brigades are built around the combination of the Challenger 2 main battle tank and the

[61] Amos Fox, 'The Russian-Ukrainian War: Understanding the Dust Clouds on the Battlefield', *Modern War Institute*, 17 January 2017, https://mwi.usma.edu/russian-ukrainian-war-understanding-dust-clouds-battlefield/.

[62] Jack Watling, 'The Key to Armenia's Tank Losses: The Sensors, Not the Shooters', *RUSI Defence Systems*, 6 October 2020, https://rusieurope.eu/publication/rusi-defence-systems/key-armenia-tank-losses-sensors-not-shooters/.

[63] See Amos Fox, '"Cyborgs at Little Stalingrad": A Brief History of the Battles of the Donetsk Airport, 26 May 2014 to 21 January 2015', Land warfare paper 125 (Arlington: Association of the United States Army, Institute of Land Warfare, May 2019), https://www.ausa.org/sites/default/files/publications/LWP-125-Cyborgs-at-Little-Stalingrad-A-Brief-History-of-the-Battle-of-the-Donetsk-Airport.pdf; Oksana Kovalenko and Galina Titish, 'Brigade Commander Yevgeny Moysyuk about Airport, Raid and Features of Enemy Action', *Ukrainian Pravda*, 12 February 2016, https://www.pravda.com.ua/articles/2016/02/12/7098744/; Amos C. Fox and Andrew J. Rossow, *Making Sense of Russian Hybrid Warfare: A Brief Assessment of the Russo–Ukrainian War*, Land Warfare Paper 112 (Arlington: Association of the United States Army, Institute of Land Warfare, March 2017), https://www.ausa.org/sites/default/files/publications/LWP-112-Making-Sense-of-Russian-Hybrid-Warfare-A-Brief-Assessment-of-the-Russo-Ukrainian-War.pdf; John M. Cantin, H. David Pendleton, and Jon Moilanen, 'Threat Tactics Report: Russia', TRADOC G-2 ACE Threats Integration, version 1.1, October 2015, https://info.publicintelligence.net/USArmy-RussiaTactics.pdf.

[64] Anthony King, *Urban Warfare in the Twenty-First Century* (Cambridge: Polity, 2021).

Warrior infantry fighting vehicle (IFV), both of which have barely been upgraded since they entered service in the late 1980s/early 1990s and are bordering on obsolescence. The Challenger Life Extension Programme (LEP) was intended to modernize the tank's turret to create 'Challenger 3', including an upgraded 120 mm smoothbore gun capable of firing more modern and penetrating types of ammunition, alongside new digital communication and battle management systems. Critically, new sight systems will provide separate day and night sights for commander and gunner, improving the rapidity of target acquisition. However, under the Defence Command Paper which followed the Integrated Review, only 148 Challenger 3 upgrades will be funded, resulting the retirement of approximately 35 per cent of the current Main Battle Tank fleet.[65]

Simultaneously, the Integrated Review also cut funding for the Warrior Capability Sustainment Programme (CSP), resulting in the planned withdrawal of this tracked IFV by the middle of the coming decade. Like the LEP, the CSP had been expected to upgrade the turret, optics, and main armament of the Warrior, including a new 40 mm cannon.[66] In theory, Ajax will by then be in service. However, Ajax is not intended as a direct replacement for Warrior in armoured infantry brigades, but was instead destined to equip armoured cavalry regiments of the new Strike Brigades as a 'light tank', alongside a turretless reconnaissance variant called Ares. Instead, mechanized infantry in these Strike brigades were to be equipped with the wheeled and turretless Boxer armoured personnel carrier procured under the MIV programme.[67] Commenting on the withdrawal of Warrior, Brigadier John Clark, the British Army's head of strategy, said that the Army was 'under no illusions' as to the difference between Boxer and Warrior, 'but there are other ways in which you can deliver the overall effects of the suite that we have at the moment'. Nonetheless, Clark also admitted that the Army was simultaneously 'working out what more we might be able to do in order to make [Boxer] more IFV-like', suggesting that the draw-down in heavy armour is already expected to produce a capability gap.[68]

[65] *Obsolescent and Outgunned*; 'Challenger 3 vs Challenger 2: How Does the Upgraded Tank Compare to its Predecessor?', British Forces Broadcasting Service, Forces Net, 13 May 2021, https://www.forces.net/news/challenger-3-vs-challenger-2-how-does-upgraded-tank-compare-its/; CP 411: Defence in a Competitive Age, *Ministry of Defence* (London: HMSO, 2021), 54.

[66] Ibid; Harry Lye, 'Lockheed Martin UK Cuts 158 Jobs as Warrior Decision Bites', *Army Technology*, 12 April 2021, https://www.army-technology.com/news/lockheed-martin-warrior-jobs/.

[67] 'Written Evidence Submitted by the Ministry of Defence': 5; 'First Ares Armoured Vehicles Delivered to the Army', British Army, Army press release, 27 July 2020, https://www.army.mod.uk/news-and-events/news/2020/07/first-ares-armoured-vehicles-delivered-to-the-army/.

[68] Harry Lye, 'British Army Outlines How Boxer Will Fill Warrior Capability Gap', *Army Technology*, 7 May 2021, https://www.army-technology.com/features/british-army-outlines-how-boxer-will-fill-warrior-capability-gap/.

Importantly, advocates of the retention of conventional heavy armour have argued that whilst these heavy tracked vehicles are undoubtedly less operationally mobile than a (wheeled) medium-weight capability, they also have significant advantages in the ability to defeat modernized Soviet-era tanks, which both Russia and many other states retain in service in significant numbers. Moreover, the extensive use of both applique explosive reactive armour and active protection systems retro-fitted to ageing Soviet-era tanks has improved the survivability of Russian armour. These defensive aids were felt to be particularly effective against anti-tank missiles and other man-portable anti-armour projectiles, but far less so against modern high-calibre tank ammunition.[69]

To add insult to injury, it appears that the acquisition of both Ajax and Boxer are not without issue. Boxer is a mature platform already in service with a number of other nations, but the projected delivery schedule is very slow, and does not currently match the rate of withdrawal of existing armoured vehicles.[70] Meanwhile, it has recently emerged that Ajax—although already in production—suffers from a major design flaw which produces excessive noise and vibration in the turreted version. This issue is sufficiently bad that it had begun to cause deafness in soldiers assigned to trial the vehicle, resulting first in significant time and speed limits being placed on vehicle operations to protect the crews, and latterly the MoD's refusal to accept the vehicle into service. To date, only fourteen of the Ares variant have been accepted into service, and ministers have been forced to deny rumours that the project will be cancelled.[71]

Critics of the shift to medium armour—including the chair of the Defence Select Committee—have complained that even if serviceable, Ajax is effectively too heavy to be considered a medium-weight vehicle. At 43 tonnes, Ajax is both heavier than Warrior and too heavy to be easily transported by air at scale.[72] This raised significant questions about the strategic mobility of the planned Strike Brigade concept, as the tracked Ajax and Ares vehicles intended to equip the forward recce and close support elements of the force would have significantly less operational mobility than the mechanized

---

[69] 'Written Evidence Submitted by the Ministry of Defence': 10; Ben Barry, 'British Army Heavy Division Comes Up Light', *IISS Military Balance Blog*, 8 January 2021, https://www.iiss.org/blogs/military-balance/2021/01/british-army-heavy-division/.

[70] Andrew Chuter, 'British Army Wants More Punch in its Boxer Vehicle Fleet', *DefenceNews*, 6 April 2021, https://www.defensenews.com/global/europe/2021/04/06/british-army-wants-more-punch-in-its-boxer-vehicle-fleet/.

[71] Helen Warrell, 'Defects with UK Army's New Tank go back to 2019, Minister Admits', *Financial Times*, 16 June 2021, https://www.ft.com/content/8be0a6e5-f75c-4ef8-9b44-2c2950c1a6f9/.

[72] Ibid.; Mark Hookham and John Collingridge, 'Tanks Too Heavy to Fly in One Piece', *The Times*, 5 February 2017, https://www.thetimes.co.uk/article/tanks-too-heavy-to-fly-in-one-piece-3nbc5m2jw/.

infantry they were expected to operate with, and would still need to be moved into theatre by modified Light Equipment Transporters, Heavy Equipment Transporters, rail, or (with difficulty) by air, just as with heavy armour. To complicate matters further, Boxer is not presently equipped with any under-armour anti-tank capability (although the infantry dismounts it carries are, and in principle such a capability might be added on later), raising fears that the Army will be significantly under-equipped to meet a Russian tank division.[73]

To address these issues, the Integrated Review heralded a number of further structural changes, outlined in the subsequent Defence Command Paper.[74] First, the Army will group together its attack aviation to create a combat aviation brigade comprised of Apache and Wildcat Lynx helicopters. Like the evolution of the strike concept, this can also be seen as a response to the threat of heavy armour, and in many regards is the conceptual inheritor of the late Cold War Follow-On Forces Attack concept, in which aviation was used to break up enemy armoured formations deep in their rear areas. This may in principle go some way to improve the Army's anti-tank capabilities in the deep, but it will likely come at the expense of the close fight, as attack helicopters will no longer be available to support manoeuvre brigades directly. It also reflects the growing air threat to military helicopters posed by layered air defence systems and A2/AD capabilities, necessitating their massing in order to plan operations in greater detail and ensure survivability through the use of other assets. Traditionally, attack helicopters have avoided enemy air defences by low 'nap-of-earth' flying where 'ground clutter' and terrain masking conceals them from hostile radar tracking. However, by flying low helicopters become highly vulnerable to ground fire from small-arms, anti-aircraft artillery, and man-portable air defence weapons, which have all proliferated in recent years.[75] Moreover, the British Army has very limited amounts of its own air defence systems, and so would likely rely heavily on allies and the RAF to protect both its ground troops and attack helicopters from enemy air attack in the event of a major war.

Even with the new combat aviation brigade combat team, however, the future force will only be able to call on two armoured ground manoeuvre brigades; a single armoured infantry brigade (redesignated as a 'heavy brigade combat team') and an 'interim manoeuvre support brigade' which

[73] Barry, 'Heavy Division Comes Up Light'; Watling and Bronk, 'Strike: From Concept to Force': 16–19.

[74] *Defence in a Competitive Age*: 51–54.

[75] Jack Watling and Justin Bronk, 'Maximising the Utility of the British Army's Combat Aviation', RUSI Occasional Paper, April 2021, https://rusieurope.eu/sites/default/files/236_op_uk_aviation_capabilities_final_web_version.pdf, 11–32.

will eventually develop into a 'deep recce strike brigade combat team' combining Ajax with artillery and multiple launch missile systems. In the 'heavy brigade', Challenger 3 will be accompanied by Warrior until replaced by some combination of either Ajax or Boxer, once in service.[76] The interim manoeuvre support brigade initially appeared to reflect the much-criticized Strike Brigade concept, equipped with Ajax, Boxer, and Foxhound (a protected mobility patrol vehicle); its title likely an indication of the difficulty the Army has faced in developing the Strike concept absent the vehicles with which to staff it. Although its predecessor concept—the Joint Medium Weight Capability concept—was touted as 'platform agnostic', delays to the creation of Strike Brigades prior to the Integrated Review appear to have been caused by a lack of actual vehicles, and the doctrine remains unconfirmed.[77] Instead, the successor deep recce strike brigade combat team appears orientated toward the conduct of long-range precision fires rather than close combat, and may lack any organic mechanized infantry. It will also rely on the modernization or replacement of the Army's existing GMLRS and self-propelled 155 mm artillery (the tracked AS-90), both of which are now ageing, against a backdrop of wider concern at the Army's significant lack of conventional artillery.[78]

This reduction will potentially call into question the validity of the UK's divisional capability in the eyes of key allies. As a single division, moreover, it is unlikely that the British Army would be able to sustain its armoured divisional capability in the field for a protracted period of time, as follow-on roulements would not be trained or equipped with the same vehicle platforms or doctrinal concepts. The extent to which a British Government would be willing to risk the country's solitary warfighting capability in a conflagration short of existential threat to the UK also remains an open question. Indeed, even the ability to field this limited capability depends on the continuation of a series of troubled acquisition programmes and promised future procurement, without which, in the words of one commentator, the 'UK will have to write a sick note to Nato explaining the problem'.[79]

[76] 'Written Evidence Submitted by the Ministry of Defence': 5–6; *Defence in a Competitive Age*: 51–54.

[77] Farrell, 'Dynamics of British Military Transformation': 801; Written answer to Question UIN 27961 by James Heappey MP, Parliamentary Under-Secretary (Ministry of Defence), 16 March 2020, https://questions-statements.parliament.uk/written-questions/detail/2020-03-11/27961/.

[78] Jack Watling, 'The Future of Fires: Maximising the UK's Tactical and Operational Firepower', RUSI Occasional Paper, November 2019, https://static.rusi.org/op_201911_future_of_fires_watling_web_0.pdf.

[79] Barry, 'Heavy Division Comes Up Light'; Warrell, 'Defects with UK Army's New Tank'. See also, 'How the British Army will Fight in the Future', British Army, British Army media video, 3 June 2011, https://www.youtube.com/watch?v=kedBlURaRaE.

Whilst these changes do continue the Army's focus on deeper, ISTAR-led manoeuvre enabled by digitization and long-range strike capabilities, the extent to which its medium-weight expeditionary vision remains intact is unclear. This latest restructuring might yet herald a return to the British Army's previous heavy–light split and the conceptual abandonment of medium manoeuvre, notwithstanding current procurement programmes, or instead lead to a gradual blending of both into something entirely new. Historically, the background of key decision-makers has provided an important indicator in the future trajectory of British Army reforms, with cap-badge and career experience going some way to accounting for senior commanders' organizational inclinations.[80] The medium-weight concept was heavily associated with the incumbent Chief of the Defence Staff between 2018–2021, General Sir Nicholas Carter, whose command experience has been shaped by operations in the Balkans and Afghanistan. Following a period in command of NATO forces in southern Afghanistan in 2009–2010, Cater went on to serve as Director General Land Warfare as Army 2020 was developed in 2011. His subsequent elevation to Chief of the General Staff and then Chief of Defence Staff confirmed the Army's direction of travel during the last decade.[81] In contrast, his immediate successor as head of the British Army, General Sir Mark Carleton-Smith, rose to prominence serving with British Special Forces.

Carter and Carleton-Smith both outwardly described the need to modernize the Army in broadly similar terms, talking of the need to create an 'agile manoeuvre division' and withdraw 'sunset' capabilities to make space for new 'sunrise' capabilities. Carleton-Smith similarly argued that the Army is on the cusp of a 'Midway moment', witnessing the greatest shift in warfare since the move from 'hay nets to fuel cans'. Both officers also suggested that whilst heavy armour is not yet obsolete, its days are numbered.[82] At the same time, however, the Integrated Review precipitated a major re-organization of the Army's capabilities toward 'sub-threshold' hybrid threats. The review precipitated the creation a new Ranger Regiment to train and accompany partners and proxies overseas, alongside the creation of a new Security Force

---

[80] Keith Macdonald, 'Black Mafia, Loggies and Going for the Stars: The Military Elite Revisited', *Sociological Review* 52, 1 (2004): 106–135.

[81] 'Chief of the Defence Staff: General Sir Nick Carter GCB CBE DSO ADC Gen', official biography, n.d., Ministry of Defence, https://www.gov.uk/government/people/nicholas-patrick-carter.

[82] Con Coughlin, 'Tanks Risk Becoming "Difficult and Dangerous" on Battlefield, Warns Head of British Army', *The Telegraph*, 1 June 2021, https://www.telegraph.co.uk/news/2021/06/01/exclusive-tanks-risk-becoming-difficult-dangerous-battlefield/; Nick Carter, 'Chief of Defence Staff Speech: RUSI Annual Lecture', Ministry of Defence press release, 17 December 2020, https://www.gov.uk/government/speeches/chief-of-defence-staff-at-rusi-annual-lecture; 'Transforming the British Army: A conversation with General Sir Mark Carleton-Smith', Public interview, Atlantic Council, 14 May 2021, https://www.atlanticcouncil.org/event/transforming-the-british-army-a-conversation-with-general-sir-mark-carleton-smith/.

Assistance Brigade.[83] This development is in part a logical extension of the Army's focus on overseas capacity building and 'upstream' conflict prevention in evidence since the 2010 SDSR and the release of the first Building Stability Overseas Strategy. Equally, though, the explicit modelling of the new Ranger force on the US Green Berets extends a wider process of so-called 'special forcification' underway in a number of Western armies, and likely reflected Carleton-Smith's own personal background as well as continuation of the doctrinal shift toward Integrated Action and information manoeuvre.[84]

Here, the accompanying Integrated Operating Concept has also adopted a new typology in which British forces will no longer be 'deployed on operations' or at home (and therefore in a state of relative peace), but instead constantly 'operating' short of war, creating a new binary between 'operating' and 'warfighting'. Such a change reflects the more neo-realist language of constant 'strategic competition' presented in the Integrated Review, and in principle creates an escalatory spectrum between a hypothetical non-operational peacetime, competitive 'operating' (at home or abroad), and full-scale 'warfighting'.[85] To a certain extent, the British Army has little choice but to find ways of making its light infantry more useful, given the prohibitive costs of modern (armoured) mechanization, and 'operating' reflects this reality. Indeed, it is now possible to envision a future in which the British Army's combat units all become increasingly functionally specialized in some way, even its ubiquitous light-role infantry. On the other hand, this latest evolution may transpire to be little more than a fillip to conceal the real degradation of British military power at the harder end of the spectrum of conflict. Either way, the expanded focus on 'operators' and 'constant operating' will likely place additional pressure on the Army's reducing number of personnel. In any eventuality, the ongoing modernization of the British Army will experience an uncomfortable hiatus in the coming decade, as the force attempts to compensate for the promise of 'jam tomorrow' contained in the Integrated Review, even as it consumes the last of yesterday's ration—with significant attendant consequences for the vision of manoeuvre that the British Army can practically employ.

Russia's renewed invasion of Ukriane in early 2022 has only served to underline the tensions in British capabilities and doctrine, potentially serving as a further inflection point. The initial fighting appears to have confirmed the

---

[83] *Defence in a Competitive Age*: 52–53; 'Transforming the British Army'.

[84] *Building Stability Overseas Strategy*, Ministry of Defence (London: MOD/DFID/FCO, 2011); *Securing Britain in an Age of Uncertainty*: 44–45; Anthony C. King, 'Close Quarters Battle: Urban Combat and "Special Forcification"', *Armed Forces & Society* 42, 2 (2016): 276–300.

[85] *Integrated Operating Concept*: passim; 'Transforming the British Army'; *Defence in a Competitive Age*: passim.

importance of long-range fires, digital information gathering and targeting, but also the enduring possibilities for tactical manoeuvre, giving succour to advocates of both traditional heavy armour and lighter modernized forces. Importantly, the war has also been presented both as a clarion call for renewing the UK's conventional land manouevre capabilities, and simultaneously also as a justification for allowing such capabilities to atrophy further. On the one hand, the conflict has focused attention in defence ministries across Europe on the threat posed by Russian revanchism, underscoring the argument for British military recapitalization. On the other hand, the conflict may also serve to erode Russia's warfighting capacity, perhaps for decades to come, potentially enabling a shift in UK strategic focus east of Suez and away from continental defence.

Conclusion

Since its inception in the 1980s, the development of manoeuvre doctrine in the British Army has been heavily influenced by US military thought. Simultaneously, though, its evolution has also been shaped by a series of unique British peculiarities, rooted in the British Army's distinct organizational culture, strategic environment, and financial means. This has resulted in significant adaptation and alteration of US concepts to suit the British context and preferences; a process of translation that can be seen from the very inception of manoeuvre warfare principles in British military doctrine with the conversion of *AirLand Battle* ideas into the *Manoeuvreist Approach* via NATO. It can also be seen in the subsequent internalization of the salient features of US 'transformation', with some non-trivial watering down, in the British Army's language and practices of network-enabled operations, the effects-based approach to operations, and a focus on (at first, brigade-level) expeditionary force structures.

Importantly, whilst the parallel development of manoeuvreist thinking in the UK demonstrates the continued importance placed on the trans-Atlantic alliance by successive generations of senior British officers, it also highlights some of the fundamental differences between British and American military practice—and the underlying constraints that explain them. Undoubtedly, where British adoption of US ideas about manoeuvre has been limited, this can partly be attributed to differences in British and US professional military culture and wider institutional processes—as with British hesitancy over the centralization and scientification of C2 and planning promulgated by US doctrine in the early 2000s. Increasingly, though, these differences can be attributed to the more limited financial means available to the British Army compared with its US cousin, which has placed some aspects of capability beyond British reach (certainly at scale), thereby necessitating either adaptation or partial adoption of US military practice. Arguably, moreover, this divergence has grown more acute in recent years—perhaps hastened, or at least laid bare by, recent campaigning—as the British military has struggled to adapt both to the demands of counterinsurgency *and* longer-term force modernization focused on inter-state conflicts.

Since the end of the Cold War, British conceptions of manoeuvre have gradually shifted emphasis away from the centrality of platform-dependent, heavily armoured tactical attrition to place a greater premium on the role of information, organizational interconnectedness and speed of action. Such a change mirrors wider shifts in the understanding of conflict itself, now typically perceived as something holistic and by nature complex, and therefore requiring equally holistic responses. This is reflected in the importance placed on narrative, audience perceptions, and behavioural change in the vision of manoeuvre put forward by the recent *Integrated Operating Concept* and the idea of 'Integrated Action' embedded in UK Land Operations doctrine. Here, emerging British practices match the emphasis on interconnectivity and informational networking seen in US *Multi-Domain Operations*, but appear to place less weight on the hard coercive aspects of the military instrument than in some allies' understanding of future manoeuvre. This shift can also be seen in the British Army's longstanding efforts to develop more flexible and expeditionary medium armoured forces in lieu of its Cold War heavy armour, which might have simultaneously allowed the British Army to realize cost efficiencies from modernization—the holy grail of having-your-cake-and-eating it.

However, this latest evolution of manoeuvre thinking in British concepts and doctrine has revealed significant tensions in the British officer corps, which have been especially apparent in debates over the procurement of the material capabilities required to practically enact them. In many respects, recent efforts to acquire new medium-weight armoured vehicles represent the culmination of a long process of digitization and structural decentralization, with roots at least as far back as the turn of the new millennium. However, the challenges first of FRES and then of Ajax and Boxer procurement have called aspects of this vision of manoeuvre into question, leading first to the demise of Strike Brigades as a doctrinal concept, and more recently for calls to re-invest in heavy armour. Indeed, the agonies of capability procurement have only served to exacerbate discomfort in elements of the British establishment at the relegation of 'traditional' combined-arms manoeuvreist ideas, leading to a series of power struggles between the first generation manoeuvre advocates of the 1990s, and the proponents of a further digital shift. The roots of this schism can arguably be seen as early as the 2010 defence review in debates over the importance of conventional 'full-spectrum' capabilities versus more limited 'wars amongst the people' and the associated rasion d'être of the British Army.

The recent emphasis placed on 'hybrid' conflict and sub-threshold 'operating' is a product of this evolutionary dialogue, if one that seems likely to bring these contradictions to a head. Undoubtedly, the vision of the British Army stemming from the Integrated Review stands in continuity with the force's longstanding trajectory of professionalization, digitization, and specialization. Equally, the emphasis placed on operations in the 'grey zone' between peace and declared war provides a rationale for British military employment that is achievable with modestly sized and lighter-weight forces, allowing the Army to bridge the conceptual gap as new platforms and capabilities are brought into service. Yet the implicit refocus toward 'traditional' inter-state adversaries this latest policy embraces sits ill at ease with the trials and tribulations of recent British efforts to rejuvenate 'conventional' warfighting capabilities at the divisional level. With the simultaneous demise of Strike Brigades as originally envisaged, and reduced funding for existing heavy armour life-extensions, the growing centrality of digital information to espoused doctrines of manoeuvre rather than large numbers of armoured platforms seems likely to add fuel to sceptics' fire at a perceived decline in British military hard power. Even so, these changes may yet provide the foundation for the next evolution in British military manoeuvre, creating Donald Rumsfeld's proverbial leaner, faster, meaner (and cheaper) force, fit to meet the next evolution in the character of warfare. Whether the British Army's physical manoeuvre capabilities will be perceived in that light by her allies and adversaries, however, remains to be seen.

17

Caught between a Rock and a Hard Place

The French Army, Expeditionary Warfare, and the Return of Strategic Competition

Olivier Schmitt and Elie Tenenbaum

Introduction

The Army has been at the centre of the transformation of the French armed forces since the end of the Cold War. The French Army was at the centre of the debates on professionalization through the 1990s, because land forces represented, more than any other service, the 'nation in arms' that had become both a legacy of the 1789 revolution and a bulwark against the insurrectionism that characterized parts of the professional army in the early 1960s. Concerned that France could not pull its weight in the Atlantic Alliance and not back claims of enhanced 'European' influence, President Chirac in 1996–1997 opted for full professionalization.

In March 1996, the French Ministry of Defence released a document announcing a new format for the armed forces to be completed by 2015, and the draft of the new Military Programming Law for the 1997–2002 period was sent to the Parliament in June 1996. The 'model 2015' announced a radical transformation of the format of the French armed forces, which had to go from 525,000 to 396,000 troops. In order to partially compensate for such a drastic reduction, it was planned to reorganize reserves, integrate them further into the forces, but also cut the numbers of reservists from 500,000 to 100,000.

Of all the services, the Army was supposed to absorb the most important cuts, going from 239,000 to 136,000 soldiers, and closing 44 regiments (from 129 to 85 regiments). This also had consequences for the overall force structure: the 'Rapid Action Force' and the 'Armored Army Corps' were dissolved as organizational structures and reorganized into a 'Land Action Force' whose headquarters were to command a total of 11 brigades. Logistics was also regrouped in a 'Land Logistic Force'.

Olivier Schmitt and Elie Tenenbaum, *Caught between a Rock and a Hard Place*. In: *Advanced Land Warfare*. Edited by Mikael Weissmann and Niklas Nilsson, Oxford University Press. © Olivier Schmitt and Elie Tenenbaum (2023). DOI: 10.1093/oso/9780192857422.003.0017

Transformations would not stop there. As soon as Nicolas Sarkozy was elected in 2007, he initiated three major converging initiatives related to French defence policy: a new defence white paper (published in 2008) to replace the obsolete 1994 paper, a new Military Programming Law for 2009–2014, and the application to the French MoD of the 'General Review of Public Policies' (*Révision Générale des Politiques Publiques—RGPP*), a reform inspired by the New Public Management ideology and designed to reduce public spending by improving the efficiency of the State. Coupled with the 2009–2014 Military Programming Law, the 2008 white paper abandoned 'Model 2015', which had been the official goal since the decision to professionalize the armed forces a decade earlier. The Army's 'operational contract' (meaning the number of troops the Army is supposed to be able to deploy all the time) was reduced from 50,000 to 35,000 troops.

The key organizational challenge for the French Army in the late 1990s/early 2000s was to integrate and digest the consequences of professionalization, including the changes in the relationship to authority that it triggered within the forces. Some years later, when the USA drove the new wave of 'transformation', the French Army was again put to the test: it had to integrate into a 'joint' information technology architecture and simultaneously define its own distinct service footprint in the shape of an expeditionary warfighting capacity[1]. After three decades of expeditionary warfare, the French Army is once again evolving, and pivoting toward high-intensity warfare.

However, those changes were shaped by some core cultural features of the army. During the Cold War, the French army was divided between the metropolitan or 'metro' army, based in north-eastern France and Germany that was preparing for war against the Soviet Union, on the one hand, and the so-called 'colo' troops (short for 'colonial'[2]), used for light interventions, mostly in Africa, on the other. The metropolitan army, 200,000 strong, included the cavalry, artillery, and infantry units required for high-intensity warfare in a NATO context. Culturally and symbolically, although they comprised the vast majority of the Army, these troops were considered less prestigious than the colonial troops: condemned to a secondary role in nuclear deterrence (compared with the Navy or the Air Force), seemingly old-fashioned and revering outmoded traditions. In contrast, the so-called 'colo' troops were used in expeditionary operations (in particular in African

---

[1] Theo Farrell, Sten Rynning, and Terry Terriff, *Transforming Military Power since the Cold War* (Cambridge: Cambridge University Press, 2013).

[2] These units are the heir of the *Armée d'Afrique*, which designates the French army troops stationed in French colonies in Africa since the colonization in the nineteenth century.

countries such as Chad, Central African Republic, Zaire/DRC, but also in Lebanon), and were deployed abroad as part of the pre-positioned forces France sustained in several African countries. It was composed of allegedly more prestigious units, coming from the Foreign Legion or the *Troupes de Marines* (Marine Units)[3].

The end of the Cold War drastically transformed this model of two parallel armies, by triggering a transformation process toward expeditionary warfare. This of course favoured the 'colo', which could claim previous expertise in the so-called 'OPEXs' (*operations extérieures*). This domination also manifests itself in the command positions. Of the six chiefs of general staff coming from the army since 1991, four had a career in the foreign legion or the *Troupes de Marine* (Generals Kelche, Bentegeat, Lecointre, and Burkhard). To some degree, the post-Cold War trajectory of the French Army can be read as a convergence of the metropolitan army toward the expeditionary warfare model favoured by the 'colo'.

This chapter explores the transformations of the French Army, and its impact on army tactics, broadly understood. The first section discusses the importance of foreign interventions for the Army, and details some lessons learned of three decades of expeditionary warfare. The second section details the institutional, doctrinal, and capability changes in the French Army, whilst the final section looks at some future challenges.

French Military Interventions: An Important Part of the Army Culture

Unlike the Navy and the Air Force, the Army is no longer directly contributing to the French nuclear deterrence capability. Ever since the demise of the Hadès Force in 1997—the 'pre-strategic' nuclear short range missile force operated by the Army—French land forces have had to regularly demonstrate their added value for the national defence strategy. This demonstration has largely taken the form of a decisive contribution to French military interventions abroad. Since 1995, France has participated in over 100 military operations, of different shapes and natures: contributions to peacekeeping operations, training and mentoring missions, election monitoring, and, of course, combat operations, in particular in the Balkans, Afghanistan, the Sahel, and Iraq.[4]

[3] Unlike in the USA, the French Marines are not a separate service but are part of the Army.
[4] Philippe Chapleau and Jean-Marc Marill (eds), *Dictionnaire des Opérations Extérieures de l'Armée Française* (Paris: Nouveau Monde Editions, 2018).

The Army, on average, furnishes 80 per cent of the capabilities required for such interventions. Participation in these military operations has shaped the culture within the Army and fuelled 'a broader military imaginary around the idea of the use of military force as an effective tool of foreign policy'[5]. Indeed, 'military operations are a cornerstone of the French armed forces' identity, quite unlike most of their European counterparts. The military considers itself combat-proven and has integrated a professional warrior ethos of delivering tactical results, regardless of the actual resources at its disposal.'[6] Of course, this cultural predisposition is perfectly aligned with the convergence toward the 'colo' that was described in the introduction.

This culture of expeditionary warfare shapes the career trajectories and broader narratives and memories within the Army, with subsequent organizational consequences, for example favouring operational over analytical profiles when promoting officers, or overvaluing 'field experience' (despite its many documented flaws as a source of knowledge[7]) at the expense of critical thinking in the symbolic hierarchy of valued qualities. This leads to an assessment by a Brigadier-General that France's Army is 'designed for the J3, at the expense of everything else.'[8] In a NATO context, the J3 is the 'operations' directorate of a joint staff, and the comment is meant to emphasize the operational focus of the French Army.

This focus on operations meets a long-standing cultural trait of the French army, shaped by the foundational experiences of infantry and cavalry combat: a fascination for the 'great captains' and the cult of the mission.[9] As such, the French strategic preferences toward military interventions shaped, reinforced, and overlapped with the Army culture.

A Wide-ranging Operational Experience

Over the last 20 years, the French Army had one of the world's richest operational histories, both abroad and at home. Coming out of the 1990s with mixed feelings about its participation in UN Peacekeeping operations, it slowly reduced its engagements with the exception of the French contingent in the United Nations Interim Force in Lebanon (UNIFIL)

---

[5] Alice Pannier and Olivier Schmitt, 'To Fight Another Day: France Between the Fight against Terrorism and Future Warfare', *International Affairs* 95, 4 (2019), 902.

[6] Alice Pannier and Olivier Schmitt, *French Defence Policy since the End of the Cold War* (Abingdon: Routledge, 2020), 127.

[7] James G. March, *The Ambiguities of Experience* (Ithaca, NY: Cornell University Press, 2017).

[8] Interview with a French Brigadier-General, Paris, November 2020.

[9] André Thiéblemont, 'La Culture de l'Armée de Terre à l'Épreuve de la Modernité: l'Imaginaire du Chef et la Sublimation de la Mission', *Cahiers de la Pensée Mili-Terre*, 18 July 2018.

which was reinforced after the 2006 war between Israel and Hezbollah. It remained throughout the period the only UN-led operation in which French land forces committed more than a symbolic number of troops. Stability-oriented multi-national operations remained paramount to French land forces throughout the first decade of the twenty-first century. By 2007 for instance, the NATO-led Kosovo Force (KFOR) was still utilizing 2,500 French soldiers, a contingent second only to Germany's and nearly 20 per cent of the total force. Key to the mission was to gather intelligence to monitor the situation, serve as a deterrent force toward any Serbian aggressive intent, as well as a buffer force between antagonizing communities, especially whenever civil unrest occurred. Hence it remained a singularly low-intensity mission, often tending to merge with riot control.

The second big operation for the French land forces in the 2000s was *Licorne*. The operation was initiated in September 2002 at the height of the crisis in Ivory Coast when the combatants of rebel forces took over the entire northern half of the country. At the request of then-President Laurent Gbagbo, French forces, which were already there under the status of a permanent overseas posting, were set in motion: the mission was to evacuate foreign nationals present in the warzone but mostly to interpose themselves between the two parties in order for dialogue to be renewed. After the signing of the Marcoussis accord in January 2003, the Licorne force—that was to grow to 4,000 by the end of the year—was dedicated to assisting its enforcement. As the political process stalled and Gbagbo tried to reconquer its lost territories by force, the French troops became a hindrance, and eventually a target, provoking a military showdown in November 2004 as well as anti-French riots, repeated in 2005. Despite sporadic fighting episodes, the overall mission remained, as in Kosovo, that of a low intensity, peacekeeping force in a permissive environment until the official end of the operation in 2015.[10]

It was Afghanistan that proved the most challenging experience in the first decade of the twenty-first century. French participation to the US-led operation started small, with an 800-man infantry battalion in charge of securing the Kabul area within NATO's International Security Assistance Force (ISAF). This ISAF contingent was supplemented by 2003 with a small Special Operations Forces (SOF) detachment (Task Force *Arès*), and by 2006 with 500 operational mentoring and liaison teams, embedded with Afghan National Army battalions (*kandak*). French involvement, however, remained relatively benign and low profile until President Nicolas Sarkozy decided to

---

[10] Jacques Aben, 'Licorne ou la guerre si nécessaire, pour maintenir ou imposer la paix', *Stratégique*, 117 (2017): 255–283.

step up its commitment and send a full combat brigade to the Kapisa valley, a key province of the Regional Command East, strategically located on the way to the Pakistani border's Taliban sanctuary. By August 2008, the French Army found itself 'at war' when a party of paratroopers, mostly from the 8th Regiment of the Parachute Marine Infantry, fell in an ambush in the Uzbeen valley and lost ten men. Politically, the battle was a shock for France, and deeply impacted the French Army operational culture[11].

The deteriorating security situation due to the Taliban's increasingly aggressive and sophisticated tactics as well as the influence of the US military approach that was more oriented toward firepower prompted a move away from the traditional 'French touch' (lenient patrolling, increased cultural awareness, and friendly contact with the local population) which was the result of nearly two decades of peacekeeping. The French Army revived the use of combined-arms tactics, with a greater reliance on air and artillery fire support, as testified by a series of offensive operations in 2009–2010 such as 'Dinner Out'—aimed at retaking the Alasay valley in Kapisa. The Afghanistan campaign also prompted a revival of France's counterinsurgency heritage and doctrine, with a new perception of the population as no longer a passive player in the conflict but now understood as the strategic 'centre of gravity' the control of whom was key to winning the war.

Although the period of 'intense fighting' for the French army in Afghanistan was limited to four years (2008–2012) — then President François Hollande having decided to withdraw its troops on an electoral promise— it proved an influential experience for what was to come. Only three weeks after French troops were withdrawn from Afghanistan, they received a call to fight in a new front of the global 'war on terrorism'—a catchphrase that was indeed belatedly adopted by French officials at the time. The Republic of Mali had descended into chaos during 2012 with a Tuareg rebellion taking over the northern half of the country before falling into the hands of Al-Qaeda-related Jihadi groups, whilst the capital in the south was shaken by a military. After six months of severe exactions on the northern population and against the backdrop of Malian historical heritage, the Jihadis went back on the offensive early in January 2013, prompting the French to react by sending in their forces stationed in Ivory Coast and Chad. In three weeks, the French expeditionary corps had repelled, destroyed, or dispersed most of the Jihadi forces and liberated the three cities of the North—Gao, Timbuktu, and Kidal.[12]

---

[11] Jean-Christophe Notin, *La guerre de l'ombre des Français en Afghanistan (1979–2011)* (Paris: Fayard, 2011), 802–844; Christophe Lafaye, *L'Armée française en Afghanistan: le génie au combat (2001–2012)* (Paris: CNRS éditions, 2016), 131–140.

[12] Jean-Christophe Notin, *La guerre de la France au Mali* (Paris: Tallandier, 2014).

Unlike the dull, infantry-based counterinsurgency mission in Afghanistan, operation Serval was a swift cavalry-dominated Reconquista-like manoeuvre, answering to a purely national chain of command. To many officers, it proved a way to emancipate the Army of a growing 'Afghanistan syndrome'. But ghosts of conflict past kept on coming back as, by 2014, Serval was discontinued and replaced by operation *Barkhane*, a long-haul counter terrorism and security assistance mission, spread over five countries—a theatre of 3.5 million sq. km—that was soon to be reminiscent, in a much more overstretched fashion, of Afghanistan in the previous decade. Whilst *Barkhane* started as an agile operation, relying on the combination of a mobile force (light vehicles, attack helicopters, commando teams, and air strikes) designed to track down the remnants of Al-Qaeda in the Islamic Maghreb, it eventually got bogged down in infantry patrolling along the Niger river 'loop', especially around Gao, Ansongo, and Menaka. As the local Jihadi forces patiently increased their support base, especially among the Fulani communities, and refined their guerrilla tactics (with an ever-increasing use of IEDs), French troops found themselves again confronted with the dilemmas of counterinsurgency.

Despite genuine tactical gains, and a gradual increase in troop numbers—from the low 3,000s in 2015 to more than 5,000 by 2020—the French Groupements Tactiques Interarmes (GTIA, typically a battalion-sized, ad hoc combined-arms field unit) found the mission increasingly frustrating. As in Afghanistan, they could 'clear' an area but found it much harder to 'hold' it, not to mention to permanently 'secure' it due to the lack of reliable local forces. As casualties kept increasing and the political situation deteriorated with successive military coups in Bamako from August 2020 onward, French President Emmanuel Macron decided to announce a gradual withdrawal of the troops, the official end of *Barkhane* and a shift of emphasis toward lower-profile cooperation with the local forces.

The Sahel was not the only active front for the Army in the French 'war on terrorism' that started in the 2010s. As the Islamic State grew in strength in Iraq and Syria, France joined in the US-led coalition in the two countries. Although the operation was mostly air-based at the outset, the Army contributed from the beginning by sending in special forces to work with Kurdish militias (Iraqi Peshmerga and Syrian YPG) as well as military advisory teams to the 6th Iraqi Division (TF Monsabert) and the Iraqi Counter Terrorism Service (TF Narvik). From September 2016 onward, French artillery (TF Wagram) also joined in the fight to help reconquer the Iraqi territories held by ISIS, with four CEASAR canons that provided fire support to the Iraqi Army and the Kurdish-dominated Syrian Democratic Forces.

But the most important theatre in terms of deployed troops was France itself. In the wake of the January 2015 attacks, no less than 10,000 troops, almost entirely form the land forces, were mobilized to perform an anti-terrorist internal security mission. At the time of its launch, *Sentinelle* mobilized 15 per cent of the French Army's total operational force, exerting a tremendous impact on capabilities. To make up for the numbers, Army Staff had to reduce training time by up to 30 per cent in 2015 and 2016. This proved to be a stretching but hopefully temporary period as the government adapted quickly and had the Parliament vote on new credits for additional resources to fill the gap. Beyond the numbers game, there have been recurring debates around the genuine purpose of the *Sentinelle* mission. Due to judicial constraints, soldiers are neither allowed to conduct intelligence missions, make arrests, nor to engage in kinetic counter terror operations on metropolitan territory. This is why the major part of the operation initially consisted in rather dull static guard duties in front of high-risk potential targets (train stations, airports, religious buildings, and especially synagogues) that were often criticized for their lack of security added-value. Over time, the Army was granted by civilian leadership to have *Sentinelle* evolve toward a more dynamic posture, with fewer boots on the ground and more 'on-alert', increasing readiness with renewed training for the homeland security contingency crisis.

Fighting terrorism has not been the only mission of the French Army in recent years. Since the Ukraine crisis in 2014, Europe again became a theatre of operation—a decade after the French contribution to KFOR started to reduce. To reassure eastern NATO allies in the face of possible Russian aggression, Paris has decided to demonstrate its solidarity by contributing to an Enhanced forward presence in the Baltic states and Poland by permanently committing a rotational company-size armour unit, including a dozen Leclerc main battle tanks. Even though this commitment remains an order of magnitude inferior to that of German or British land forces, it announces a quite significant turn away from the French Army's mostly expeditionary culture to an increased interest in continental collective defence. As it turns out, France had just taken the lead of the NATO's Very High Readiness Joint Task Force (VJTF) – a rotational responsibility among major NATO allies – when Russia attacked Ukraine on February 24th 2022. Reaction was swift and consistent with commitments as a spearhead battalion of more than 500 men was set up in 72 hours and sent to Constanta, Roumania to reinforce the Alliance's Eastern Flank. It was later decided to turn this into a more permanent mission, baptized *Eagle*, revolving around a French-led armor-dominated multinational battlegroup integrating Belgian troops as well[13].

This reinforcement to the East coincided with the withdrawal of French troops from Mali announced on February 17th, 2022 and the official end to Operation Barkhane eventually proclaimed on November 9th of the same year. Such crossing curves of continental defense and overseas operations will undoubtedly have a lasting impact on the shape and scope of the French Army. As the future of land troops in the Sahel remains in the limbo at the time of writing, it appears clearly that the days of large autonomous operations by the French army in Africa and the Middle East are over. These regions will most probably not be entierely abandoned neither. A middle-ground is more likely to be found in a more-or-less renovated policy of military cooperation— 'operational military partnership' translating in training and security force assistance— essentially provided by the 5,500 'Prepositioned Forces' in Ivory Coast, Senegal, Gabon, Djibouti, and the United Arab Emirates. Each of these are regionally aligned forces, supposed to foster military cooperation within their area of permanent responsibility.

Finally, French overseas dependent territories in the Caribbean, Indian ocean, and the Pacific still mobilize 7,000 military personnel, more than half of them from the Army. Typically a marine infantry mission, these troops are to embody France's sovereignty over disputed lands (such as Foreign Legion postings to the Scattered Islands in the Indian Ocean claimed by Madagascar) and protect these distant territories from illegal activities (such as operation *Harpie*, active since 2008 against illegal gold panners in French Guyana) as well as from "grey zone" encroachments in a Falklands-type scenario. These may find a rejuvenated role in the prospect of a greater French involvement in the Indo-Pacific.

The Force Structure and Main Capabilities

When the first Military Programming Law was voted in 1996, the goal was to reduce a conscript army from 239,000 men to 136,000 by 2015. The deflating trend over the years went beyond this objective as, by 2015, the land force was barely 110,000 soldiers. According to former Army Chief of Staff, Elrick Irastorza, the French Army could now 'fit in the *Stade de France* (national

---

[13] Dossier de Presse 'Mission Aigle', Etat-major des Armées, décembre 2022.

stadium): 80,000 in the stands, the rest on the lawn'.[14] Its territorial footprint had also shrunk, especially after the reform of the 'military map' that led to a reduction by 17 per cent of the armed forces' real estate from 330,000 ha in 2008 to 275,000 in 2017).[15] After 2008, regiments have been geographically regrouped around joint defence bases that can pool and share common support resources.

This downward spiral eventually stopped in 2015, when the sudden need for boots on the ground, especially on national soil after the Paris attacks, led the Parliament to vote for an updated Military Programming Law that stemmed the bleeding. With a stabilized trajectory, General Jean-Pierre Bosser, then Army Chief of Staff in 2016 inaugurated a 'new format' of the land forces, dubbed 'Au Contact' model. This force structure—still valid in 2022—has reintroduced the Division level that had been abandoned in 1999 in the quest for more flexibility in the form of modular brigades. The bulk of the French operational land forces (77,000) now comprise two divisions: the 32,000 strong 1st Division has its headquarters in Besançon, and the HQ of the 26,000 strong 3rd Division is based in Marseille. Together these form a Rapid Reaction Corps, headquartered in Lille, and certified as a NATO Response Force (NRF) land component.

It should be noted that this core capability of more than 55,000 soldiers does not include logistics, IT, reconnaissance. Also not included are special operations forces and army aviation (rotary wing) units, all of which have been set aside in 'specialized commands'. These form capability pools from which resources can be extracted to generate the battalion-sized combined-arms groups (GTIA) that are actually sent on combat operations. The persistence of this organic divide between combat and support units indicates that the 2015 'Divisionary turn' remained an unfinished transformation: French divisions are still mostly resource pools, and not yet fully supported battle-ready units. Force generation is therefore still a complex game, mostly oriented toward expeditionary missions (Figure 17.1).[16]

Each division is composed three brigades, with six to eight regiments per brigade. French brigades are a bit larger than their main European counterparts, such as the German or the British brigades which rarely exceed 5,000 soldiers. Division as well as brigade composition clearly indicates the French preference for light to medium capabilities. This is especially true of

---

[14] 'L'Armée de terre tient dans le Stade de France: 80 000 dans les gradins, le reste sur la pelouse', général d'armée Elrick Irastorza, 18 octobre 2012 à Montpellier.

[15] Rapport d'information n° 661 (2016–2017) de M. Dominique de LEGGE, fait au nom de la commission des finances, déposé le 19 juillet 2017.

[16] François Lamy, « Préparation et Emploi des Forces: Forces Terrestres », avis fait au nom de la commission de la défense nationale et des forces armées sur le projet de loi de finances pour 2016, Tome IV, 8 octobre 2015.

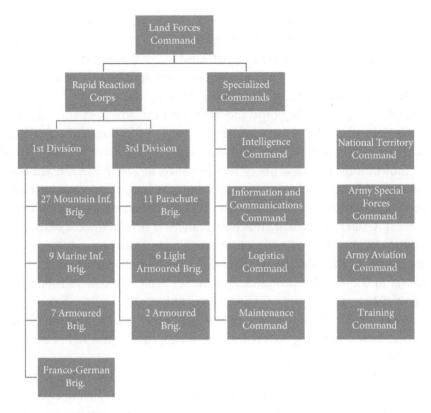

Fig. 17.1 Force structure

the 27th, 9th, 11th, and 6th brigades, which are quite typically dominated by light infantry regiments (alpines, colonials, paratroopers, or Foreign Legion) flanked with one or two cavalry and artillery regiments, usually in wheeled vehicles and lightly armoured, and a couple of combat military engineer units. French infantrymen and military engineers are still mostly mounted on VABs (armoured vanguard vehicle), which has been the workhorse of French operations for more than 40 years. After benefiting from an upgrade in the early 2010, the VAB is finally being replaced by the new Griffon that started to appear in the field from 2020 onward. The light cavalry branch is based on the 15 ton AMX-10 RC and the 4-ton VBL, whilst artillery is mostly ensured by 17 ton CAESAR truck-propelled 155 mm howitzers. These provide the core of the French expeditionary power, perfectly fitted for contingency and expeditionary operations rather than continental and/or high intensity warfare.

The remaining 7th and 2nd armoured brigades are heavier units. Each of them hosts three mechanized infantry regiments, mounted on VBCI, a 30 ton infantry fighting vehicle, on wheels. It is also within these two brigades that one can find four regiments equipped with 50 ton Leclerc Main Battle Tanks. France has only 220 of these, which is 20 less than the number of Leopards in the German army, whilst the French land forces are almost twice the size of Germany's. This scarcity is not anticipated to improve as the total number of MBTs expected to be supported by 2033 is only 200.

Artillery is another key capability that has been stretched since the end of the Cold War, with the aim of being a better fit for expeditionary criteria. From nearly 400 155 mm towed or self-propelled guns at the turn of the millennium, French artillery has dwindled to 77 CAESAR truck-born howitzers. The record is even gloomier as far as long-range rocket artillery is concerned, with only 13 Unitary Rocket Launchers kept at the divisionary level 1st Artillery Regiment. This puts France far behind not only Germany and the UK as well as Italy, Greece, or even Finland and Romania in terms of this capability, which would be much needed in a high intensity scenario, as the war in Ukraine demonstrates. As a matter of fact, French decision in the Fall of 2022 to give-away 18 CAESARs and at least 2 LRU's has deprived the army of a quarter of its artillery capacity. The level of ammunitions stockpiles has also raised worrying concern with what appears to be a rather minimal capacity.

Another well-known gap in the French army lies in their air mobility. Whilst French land forces concentrate most of the rotary wing attack and tactical lift capabilities, they notoriously lack heavy helicopters such as the CH-47 Chinook owned by other European partners such as the UK, but also Italy, Spain, Greece, or the Netherlands and often envied by the French. The deployment of British Chinooks to Mali in support of *Barkhane* since 2018 has only increased the realization of how useful such a capability would be for French forces. Finally, air defence has equally dangerous shortages: the Army is now limited to a couple of hundred short range point defence *Mistral* missiles whilst the rest of the surface-to-air systems operated by the Air Force are also scarce. An upgrade of such capability is now considered more urgent than ever, especially in view of the increasing air threat, particularly at lower altitudes (because of the diffusion of guided rockets, artillery, mortars, and missiles -G-RAMM - and UAVs).[17]

The French Army Doctrine

Tellingly reflecting the operational experience as well as the capability and force structure, the French Army Doctrine of the last 20 years has been mostly oriented toward overseas operations—even though 'Intervention' has only

---

[17] This whole paragraph is based on multiple interviews with French Army staff.

been one of the five strategic functions listed by the successive Defence White Papers and/or Strategic Reviews. The core of the French doctrine is organized around five documents—which are currently being reviewed—summarized in Table 17.1.[18]

In the mid-2000s, the French Army identified a need to organize, hierarchize, and update its core doctrinal documents. The context was a shift in the character of warfare observed in the US-led interventions in Iraq and Afghanistan, and the end of the peacekeeping era that had marked the French army interventions of the 1990s. The core documents were published according to a seemingly logical order. FT-01, published in 2007, describes what the Army perceived as the general operating environment. The document is heavily influenced by its context of production and introduces the concept of 'war amongst the people' borrowed from the British General Rupert Smith, whose book *The Utility of Force* was translated into French in 2007 with a foreword by the then Chief of Staff of the Army, General (OF-9) Bruno Cuche.[19] FT-01 also asserts that new actors now roam the battlefield, including criminal groups and insurgents, and that the 'stabilization' phase of the conflict is now the 'decisive phase' of a military operation. FT-01 thus emphasizes

Table 17.1 Core army doctrinal documents.

Designation	Title	Last update	Content
FT-01	'Winning the battle—leading to peace. Land forces in contemporary and future conflicts'	January 2007	General description of the operating environment and the role of land forces in it.
FT-02	'General Tactics'	August 2009	Core tactical principles and procedures.
FT-03	'The Use of Land Forces in Joint Operations'	July 2015	Place and added value of land forces in joint operations.
FT-04	'The Fundamentals of Combined Arms Operations'	June 2011	Core principles of combined-arms operations.
FT-05	'Commanding in Operations for Tactical Leaders'	January 2011	Command and leadership principles (in particular for junior officers).

[18] The documents are available at this link: https://www.c-dec.terre.defense.gouv.fr/index.php/fr/documents-fondateurs (last accessed 5 May 2022).

[19] See Rupert Smith, *L'Utilité de la Force. L'Art de la Guerre Aujourd'hui*, Paris, Economica, 2007.

that symmetrical conflicts are 'highly unlikely for the next few decades' and that asymmetrical conflicts will be the main operating environment for the Army.

FT-02 identifies three tactical objectives fulfilled by the land forces: coercing the adversary, controlling the domain, and shaping perceptions. To achieve those objectives, the land forces must maintain their legitimacy. Classically for the French army, since it is an inheritance form Marshal Foch and his 'Principles of War', FT-02 lists the three core principles guiding tactical planning: freedom of action, concentration of effort, and economy of means. These three principles are general guidelines for the tactical commander, and not strict planning steps: they must be combined and weighted depending on the situation, and they must serve the element of surprise, which is perceived to be the most effective way to achieve military objectives.

FT-03, FT-04, and FT-05 were published together in 2011, with FT-03 being updated in 2015. They are interesting, because they reveal how the Army sees itself and its added-value in a joint context, and they come close to a French army theory of victory in land operations. FT-03 is described as the 'cornerstone document' for the Army Doctrine, as it articulates the grand strategic guiding papers (such as the 2013 White Book on Defense and Security of the 2017 Strategic Review) of the Army Doctrine. It identifies six key contributions of the Army to a joint operation, due to the peculiar nature of land forces. First, engaging land forces signals resolve and thus increases the credibility of the compellent threats signalled by French political leaders. In rationalist language, the land forces constitute a 'credible commitment'.[20] Second, French land forces are quickly deployable, particularly because of the network of French bases around the world. Third, land forces can modulate the degree of violence they exert, from foot patrols to high-intensity operations, which gives policymakers a number of flexible options. Fourth, land forces can constitute the basis for a multi-national operation. Fifth, they are the ones with the 'boots on the ground', and thus best able to contribute to the resolution of a complex socio-political situation. Sixth, land forces allegedly 'humanize' military power, by being in contact with the population, in line with the strategic assessment that land operations are conducted 'amongst the people'. Together, these six alleged advantages

---

[20] For an academic version of the argument, see Branislav L. Slantchev, *Military Threats. The Costs of Coercion and the Price of Peace* (Cambridge: Cambridge University Press, 2011).

constitute as many claims to expertise and legitimacy for the army in a joint context.

FT-04, in turn, identifies four key functions, which are necessary to successfully conduct military operations: command, master information, operate, and sustain. These four 'key functions' are sustained by eight 'operational functions' (and their associated capabilities): command, command support, intelligence, contact, support, action on perceptions and the operational environment, support to engagement, and logistics. These functions are summarized in Table 17.2.

Overall, the documents constitute an effort to build a cohesive doctrinal foundation for the Army. However, some of them are heavily shaped by the asymmetric wars of the 2000s, and no longer reflect the Army's thinking. In 2016, the French Army published its vision of the future operational environment in which it identified eight 'factors of operational superiority' deemed necessary to succeed on the battlefields of the future.[21] The eight factors are understanding, cooperation, agility, mass, endurance, moral strength, influence, and command performance. This writing reflects a rebalancing toward high-intensity warfare which has been signalled multiple times by army leaders. As such, a new capstone concept of employment was released in autumn 2021 that emphasized new technological environment (heightened digitalization, autonomous systems, etc.), as well as the need a joint multi-domain integration from division to battlegroup levels.

Table 17.2 Key military functions in the army doctrine (adapted from FT-04).

Key functions	Operational functions
Command	Command (HQs)
	Command support (ICTs)
	Intelligence (Intel, geography, meteorology)
Master information	Contact (infantry and mechanized combat)
	Support (air defence, electronic warfare, engineers, etc.)
	Action on perceptions and the operational environment (CIMICs, Influence, propaganda)
Operate	Support to engagement (NRBC, support to mobility)
Sustain	Logistics (Master of fluxes, availability of materials, etc.)

[21] French Army Staff, *Action terrestre future: Demain se gagne aujourd'hui* (Paris: French Army, September 2016).

From Network-centric Warfare to High Intensity

As often in defence policy planning, the main challenges come from adapting a legacy force, perfectly shaped for the missions it has been used to carrying out, to the future needs, which are usually of a different nature. As far the French Army is concerned, the main efforts toward such an adaptation lies in the adoption of information and communications technology to build up an increasingly networked centric combat force ('combat info-valorisé'). This transformation, that began slowly in the 1990s, expanded in scope and pace in the 2010s with the SCORPION programme that is aimed at integrating land combat forces through modernized platforms as well as IT and communications systems. SCORPION's two showcase platforms have been the Griffon, an APC designed to replace the VAB, and the Jaguar, a wheeled reconnaissance tank that will take over from the ageing 10-RCs. SCORPION is not just about new vehicles though, but the interconnectivity that will link them with one another in a digitized vetronic network on the battlefield. The new Contact radio systems, as well as the SCORPION Information System (SIC) will provide the IT support that should help to bring about a radically transformed conception of land forces' tactics and style of manoeuvre. Collaborative combat is the key notion that is supposed to guide these 'SCORPION-ized' forces: fully networked, they should be able to mutually support each other at a distance, flexibly adapt, and reconfigure their battle order in real time.[22]

SCORPION has initially been focused on the 'preferred' medium-capability segment (Griffon and Jaguar), which were the vehicles most needed in operations as well as the ones in most urgent need of modernization (with an average age of more than 30 years in the VAB fleet). Over time, however, the heavier IFVs (VBCI) and MBTs (Leclerc) are to join SCORPION's integrated network to provide a truly full-spectrum modernized army by 2030. Beyond that horizon another layer, the TITAN Project, should bring about new capabilities better suited to high-intensity warfare. Its leading project is certainly the Main Ground Combat System (MGCS) currently being developed with Germany in an industrial joint venture between Nexter, KMW, and Rheinmetall. Due in 2035, the MGCS is mostly considered as a successor to Leclerc and Leopard 2 main battle tanks. By 2040, the

---

[22] Stephanie Pezard, Michael Shurkin, and David Ochmanek, *A Strong Ally Stretched Thin: An Overview of France's Defense Capabilities from a Burdensharing Perspective* (Santa Monica, CA: Rand Corporation, 2021).

Common Indirect Fire System (CIFS) should also provide a successor to the CAESAR 155 mm gun.[23]

Beyond the mere procurement of a new segment of platforms, the French Army's ambition is to brace itself for high-intensity warfare. This notion has been the background of French strategic thinking since the 2017 *Strategic Review* that stressed the return of great power competition and the potential for combat operations in a contested environment due to interference of high end to near-peer competitors. In 2019, the new Army Chief of Staff, General Burkhard, fully embraced this agenda, as it also seemed to fit the winding down of three decades of French and Western interventionism, announcing a soul-searching moment for an institution that had been entirely turned toward stability and contingency operations.

In this perspective, General Burkhard pushed for a 'hardened training', in more realistic conditions, for complex high-intensity multi-domain operations. Project HEMEX-Orion has been the beacon for this new trend, as it expects to hold, by the spring of 2023, a large-scale military exercise at division level with 10,000 soldiers deployed. This is slightly less than is envisioned by the 'Major Engagement' contract (15,000 men) as the French contribution to a typical NATO Major Joint Operation. Orion has the ambition to boost the French Army's credibility in a NATO and collective defence perspective as well as to perform strategic signalling to potential rivals and competitors.[24]

Providing Soldiers for Future Wars

The suspension of national service in 1996, which de facto abolished conscription, was a major change in the interaction between the French armed forces and society. The main strategic justification advanced by then President Jacques Chirac, against the advice of most of the administrative-military elites, was that the core tasks of the French military would be foreign interventions, instead of territorial defence. As such, smaller, professional armed forces would be more militarily effective than a conscription-based force[25]. This professionalization triggered a debate about the alleged 'normalization' of the armed forces and may have contributed to a degree of misunderstanding between the French population and the armed forces, including the

---

[23] *Supériorité Opérationnelle 2030: Plan Stratégique de l'armée de Terre* (Paris: French Army, 2020).

[24] Laurent Lagneau, 'Haute intensité: La France va renouer avec les grandes manœuvres militaires', *Zone militaire*, 26 juin 2021.

[25] Bastien Irondelle, *La Réforme des armées en France. Sociologie de la décision* (Paris: Presses de Sciences Po, 2011).

Army. According to polls, 8 to 9 of 10 French citizens have a good opinion of their armed forces, in stark contrast to the wave of anti-militarism that existed in the 1970s.[26] This positive perception has been consistent across time since the late 1990s. However, media coverage of the armed forces regularly mentions low morale, linked to the fact that the military perceive that their job is not acknowledged in society. There is then a paradox: the French population loves its armed forces but does not know what they are doing. The French soldier is thus a 'misknown soldier'.[27]

In the context of a return to high-intensity warfare, this chasm between the missions and tasks of the Army, and its perception by the broader civilian population might pose a challenge: 'the military specificity being based on defending the country, the meaning of the engagement of those wearing the uniform cannot be understood by their fellow citizens if the fighting dimension is hidden'.[28] There is thus a challenge to communicate about the risks of future warfare (and especially the dramatic lethality that high-intensity warfare implies) whilst still attracting professional soldiers willing to accept these risks. Implicitly, this raises the question of conscription in order the achieve the goals of a proper format for the Army and the need for a greater mass on the battlefield, a solution already adopted by some armed forces in Europe.[29] At the time of writing, there is no political space for such a solution, since discussions of some form of draft are framed as educational, with the aim of facilitating the integration of disfranchised populations. The military purpose is thus not part of the public debate, and the armed forces themselves are comfortable with their all-volunteers model. However, should the international security environment take a drastic turn for the worse, the challenge of providing more mass to the French Army might trigger a renewed discussion about conscription.

Conclusion

The current war in Ukraine illustrates rather than it initiates the advent of a new era of strategic competition and the foreseeable decrease of Western military interventions. This mutation is a key challenge for the identity of the French Army which has been designed, since the end of the Cold War, as

[26] Barbara Jankowski, 'L'Opinion des Francais sur leurs Armées', in Éric Letonturier (ed.), *Guerres, Armées et Communication* (Paris: CNRS Éditions, 2017), 81–98.
[27] Bénédicte Chéron, *Le Soldat Méconnu. Les Français et leurs Armées. État des lieux* (Paris: Armand Colin, 2017).
[28] Ibid., 89.
[29] Elizabeth Braw, 'The Return of the Military Draft', Atlantic Council Issue Brief, February 2017.

a combat-ready expeditionary force best fitted to low- or medium-intensity stability and contingency operations. The new strategic environment is being taken into account and can already be seen in evolving tactics, doctrine, and capability development. This transformation, however, will take time as it challenges both the operational experience and the cultural heritage of a French Army that finds itself, more than ever, at a crossroads for defining its future role in the strategic landscape.[30]

---

[30] Rémy Hémez, 'The French Army at a Crossroads', *Parameters* 47, 1 (2017).

18

Trends in the Land Warfare Capability of Poland and the Visegrád States, 1991–2021

Scott Boston

Introduction

The military ground forces of the members of the Visegrád Group (the Czech Republic, Hungary, Poland, and Slovakia) changed enormously between the end of the Cold War and the present. These militaries had to adjust from functioning as part of the military buffer zone for the Soviet Union to carrying out national mandates in the 1990s and discovering new roles as members of the North Atlantic Treaty Organization (NATO). This chapter considers the ground combat capabilities of the first three former Warsaw Pact countries to join NATO: Poland, the Czech Republic, and Hungary, and the implications of their transition from their Cold War era system to their current systems.

A great deal has been written about the transition experience of the former communist states, including the members of the Visegrád Group. However, much of the focus of past work on transition highlighted the challenges of reforming ministries and developing civil–military relationships compatible with NATO membership. By contrast, this chapter's emphasis is on the implications for combined arms tactics and operations on land, with a bias toward the requirements of state-on-state conflict. Each of the Visegrád states' armies had quite different experiences in pursuit of NATO-compatible forces and in the years that have followed their joining NATO. This short chapter cannot do justice to each country's over 30 years of transition, planning, and real-world experiences. Because this chapter's focus is on the highly demanding case of large-scale land operations, much of the post-NATO membership focus will be on Poland, as, of the countries considered here, the Polish Land Forces have the bulk of the relevant capability. Still, many aspects of the Warsaw Pact and NATO comparisons will hold true across cases.

Scott Boston, *Trends in the Land Warfare Capability of Poland and the Visegrád States, 1991–2021*. In: *Advanced Land Warfare*. Edited by Mikael Weissmann and Niklas Nilsson, Oxford University Press. © Scott Boston (2023). DOI: 10.1093/oso/9780192857422.003.0018

This chapter is presented in three parts. The first part is a brief overview of the transition of these countries from Warsaw Pact member states to NATO membership and their contributions to NATO and other multi-national missions in the years since 1999. The second part of the chapter compares some important selected aspects of Warsaw Pact and NATO forces, focusing on the nature of the changes needed to fully adopt the system of land warfare typical of modern Western states, in the context of the rapid change in the security environment in Europe. The final section considers some of the implications of the continuing evolution of combined arms tactics and operations, with a focus on the mission to deter or defeat an adversary possessing a modern combined arms land force.

Background: From Warsaw Pact to NATO

Only about 10 years separated the end of Communist rule in Poland in 1989 from its admission into NATO in March of 1999. Poland, along with the Czech Republic and Hungary, were the first new NATO members since 1982, and the land forces of these countries, particularly those of Poland and Czechoslovakia, had been key elements of the Soviet-dominated forces facing Western Europe. The fact that these armies had been trained, organized, and equipped to fight under Soviet leadership has imparted legacies that remain to the present day. This section very briefly outlines key trends over three decades that influenced the transition and development of the land forces of these former Warsaw Pact states.

During the Cold War, the land forces of Poland, Czechoslovakia, and Hungary made up three of the four main Soviet-allied forces facing the West on the inter-German border (the fourth being the National People's Army of the German Democratic Republic). Each of these countries fielded large land forces, organized typically into armies that were to operate under Soviet command and alongside Soviet groups of forces that were permanently stationed on their territory. By the end of the 1980s, as the grip of Communist Party control in these countries began to waver, there had already been indications of demobilization of the large army forces in each country, with units being deactivated or, as was the case in Hungary, resized divisions to brigade strength.[1]

---

[1] International Institute for Strategic Studies, 'The Alliances and Europe', *The Military Balance* 89, 1 (1989): 44.

Poland, Czechoslovakia, and Hungary formed the Visegrád Group in 1991 in order to promote stability and to collaborate on military, economic, cultural, and energy issues. The Group is based on their shared histories, the goals of bringing the era of totalitarian rule to an end, and developing free market economies, all on the way to pursuing their aspiration to join the European Union and eventually achieve membership in NATO.[2] In 1993, Czechoslovakia amicably divided into the Czech Republic and Slovakia, bringing the number of Visegrád countries to four. Poland, the Czech Republic, and Hungary were offered a path to NATO membership at the Madrid summit in 1997, and officially joined in 1999.[3]

More recent developments have brought about a shift in the focus of former Soviet-dominated states in Central Europe. With the 2014 invasion of Crimea and subsequent rise in tension between the Russian Federation and NATO, large-scale combined-arms ground warfare is once again a focus of the Visegrád countries, particularly Poland, which borders the Russian exclave of Kaliningrad as well as Russia's ally Belarus. Poland, the Czech Republic, and Slovakia are regular contributors to the enhanced forward presence (eFP) battlegroups in the three Baltic states and Poland.

From the perspective of the post-Warsaw Pact era, the starting point for the transition to NATO was a fully Soviet-designed and equipped, large-scale force based on Soviet-trained officers and large numbers of conscript soldiers. The task facing the military leaders of Poland, Czechoslovakia, and Hungary was monumental in scope, and it was not initially focused on tactical and operational challenges: there were far more pressing concerns at the time. The demands of transitioning ministries of defence to civilian control, dismantling military secret police and counterintelligence organizations, and screening officers who had received training in the Soviet Union prevailed.[4] With the division of Czechoslovakia into the Czech and Slovak Republics, there was also further turbulence for the newly divided armies of the two countries. All of this took place whilst a great deal of the Cold War era force structure was being cut and rebalanced.

Both before and after joining NATO, the Visegrád countries contributed units, including in some cases combat units, to multi-national missions, including to the United Nations Protection Force, and NATO-led missions in Bosnia, Herzegovina, and Kosovo; each also participated in operations in

---

[2] 'Declaration on Cooperation between the Czech and Slovak Federal Republic, the Republic of Poland and the Republic of Hungary in Striving for European Integration', The Visegrád Group, 1991, https://www.visegradgroup.eu/documents/visegrad-declarations/visegrad-declaration-110412.

[3] Slovakia joined later, in 2004, along with the Baltic states, Bulgaria, Romania, and Slovenia.

[4] See, for example, Jeffrey Simon, *NATO and the Czech and Slovak Republics: A Comparative Study in Civil–Military Relations* (Lanham: Rowan and Littlefield, 2004), 6–7.

Iraq and in Afghanistan.[5] These deployments served a number of national goals, but in particular they helped gain international expeditionary experience for their forces and domestic support for military reform. Deployments with the USA were particularly significant, as they reflected not only a political interest by these countries to become security contributors following their accession to NATO, but also the role of the USA in the transformation of their armed forces.

Poland was a particularly striking example of these trends. Poland sent a division headquarters and a brigade to Iraq, where it led Multinational Division Central-South from 2003 to 2008.[6] The Polish contingent in Afghanistan operated alongside US troops in Task Force White Eagle in Ghazni province between 2008 and 2014, with as many as 3,000 soldiers at its peak.[7]

How Land Force Capabilities Changed

This chapter's main aim is to highlight the distinct differences in the two competing systems of land combat operations experienced by the Visegrád countries over the period of interest. In addition to changes due to new approaches to land operations, the transitions took place also during a period of rapid change in the security environment. These two main influences cannot be readily deconflicted: both were important drivers of key decisions and force redesign and resizing efforts. This section considers the demands of the Soviet and of the Western systems, as well as how transition influenced the capacity and force structure of the land forces of the Visegrád countries, their command philosophy and force design; and the modernization of forces to meet the demands of high-intensity conflict.

Force Structure

The most visible trend since the end of the Cold War was the dramatic reduction in force structure and personnel. The large-scale deactivation of land units was linked in many cases to the reduction and eventual elimination

---

[5] Miroslav Tuma, 'Relics of Cold War. Defence Transformation in the Czech Republic', *SIPRI*, September 2006.

[6] Polish Ministry of National Defence, 'Missions', https://www.gov.pl/web/national-defence/missions.

[7] Polish forces in some form operated in Afghanistan from 2001 to 2021. See Kap Kim, 'Polish Army Ends Afghan Mission', U.S. Army press release, 8 May 2014, https://www.army.mil/article/125595/polish_army_ends_afghan_mission.

of conscription in all four countries.[8] Although the force reductions freed up resources for modernization and transition, the effects of the change are more complicated than a simple trade of quantity in exchange for quality.

Table 18.1 depicts the changes in the sizes of land forces of the Visegrád countries from 1990 to 2020, showing that during this 30-year period the total number of personnel in these land forces has declined by about three-quarters. Throughout this period the Polish Land Forces has remained by a fair margin the largest of the armies here. These 1990 figures also reflect some early demobilizations; for example, in 1988 the Polish People's Army was said to have 217,000 personnel and had already declined by about 8 per cent in two years.[9]

Comparing personnel figures does not fully capture the change in force structure and capability, although it does help to account for the differences in unit sizes between the armies. Although each of the Warsaw Pact armies had been designed to be Soviet compatible, in practice there were elements that made each unique. In 1987 the Hungarian People's Army had transitioned from a divisional structure to a number of corps made up with brigades, whilst the other Warsaw Pact members were generally organized

Table 18.1 Selected former Warsaw Pact states' land forces personnel, 1990–2020

	1990	2000	2010	2020
Czech Republic	(as Czechoslovakia) 87,300 personnel (69,000 conscripts)	23,800 personnel (15,500 conscripts)	7,026 personnel	13,000 personnel
Slovak Republic		19,800 personnel (10,400 conscripts)	7,322 personnel	6,250 personnel
Hungary	66,400 personnel (36,400 conscripts)	13,160 personnel (some conscripts)	10,100 personnel	10,450 personnel
Poland	199,500 personnel (127,500 conscripts)	120,300 personnel (67,200 conscripts)	47,300 personnel	58,500 personnel

Source: IISS *Military Balance*.

[8] See, for example, Cindy Williams, 'From Conscripts to Volunteers: NATO's Transitions to All-Volunteer Forces', *Naval War College Review* 58, 1 (Winter 2005): 35–62.
[9] International Institute for Strategic Studies, 'The Alliances and Europe': 49.

in groups of divisions.[10] (In the Soviet system, a manoeuvre brigade was a combined-arms formation, comparable to a smaller division in that it was considered capable of independent action as designed, in contrast with a motor rifle or tank regiment that operated as part of a division.)

Table 18.2 depicts some of the changes over time in the structure of Poland's Land Forces. The number of divisions has changed along with their contents; the Soviet-designed division of the Cold War had four manoeuvre regiments, whereas the 2020 Polish Land Force divisions have three brigades. The newer Polish brigades have larger battalions, although generally fewer than before, as late Soviet manoeuvre regiments typically had four manoeuvre battalion subunits.

The change in force structure has had several subsequent effects on land force capabilities for combined-arms operations. Forces are smaller, fewer are available as a pool for readiness, and some of the important army- or corps-level capabilities have disappeared. Tactical ballistic missiles, as well as army and national-level air defence systems, have been deactivated or reached obsolescence and were not replaced; these capabilities, not generally used in stability or peacekeeping operations, were not missed until more recently.

In parallel with the reduction in forces, the focus has shifted as well toward smaller, more deployable formations. In the armies considered here and across many NATO forces, the basic tactical combined arms formation has become the brigade; much of the division-level capability has been deactivated or reduced in size. These are also lighter formations; for example, the land forces of the Czech Republic, Hungary, and Slovakia each retained only a single tank battalion. By comparison, the Czechslovakian People's Army in 1986 had an estimated 3,400 main battle tanks available, including those assigned to the army's five tank *divisions*.[11]

Finally, of the four armies considered here, only Poland has retained the ability to deploy multi-brigade formations; it has preserved several division headquarters and associated divisional units, as well as a corps headquarters. It has deployed brigade-sized forces to Iraq and Afghanistan, contributed division headquarters to NATO and other coalition missions.[12] The greater capacity of the Polish Land Forces also enables it to rely less on attachments from allied military forces to be able to form tactically and operationally

---

[10] 'The Alliances and Europe': 44.

[11] See, for example, Office of Soviet Analysis, 'Selected Data on Soviet and Non-Soviet Warsaw Pact Ground Forces Maneuver Divisions in the Atlantic-to-Urals Zone', *Central Intelligence Agency*, Sanitized Copy Approved for Release, CIA-RDP86TO1017R0000605540001-0, 2 October 1986.

[12] Polish Ministry of National Defence, 'Missions'.

Table 18.2 Polish Land Forces structure, 1990–2020

1990	2000	2010	2020
Manoeuvre	**Manoeuvre**	**Manoeuvre**	**Manoeuvre**
9 Mechanized Divisions	5 Mechanized Divisions	2 Mechanized Divisions	1 Armoured Cavalry Division
1 Airborne Brigade	1 Armoured Cavalry Division	1 Armoured Cavalry Division	3 Mechanized Divisions
1 Coastal Defense Brigade	5 other brigades	Airborne Brigade	Airborne Brigade
		Air Cavalry Brigade	Air Cavalry Brigade
Artillery	**Artillery**	**Artillery**	**Artillery**
3 Artillery Brigades	2 Artillery Brigades	2 Artillery Brigades	3 Division Artillery Regiments
3 Tactical Missile Brigades	1 Tactical Missile Regiment	3 Division Artillery Regiments	
9 Division Artillery Regiments	6 Division Artillery Regiments		

Source: IISS *Military Balance.*

relevant battalion or brigade battlegroups. Although multi-national battalion battlegroups have a variety of political and training advantages for the improvement of interoperability, the additional frictions posed by these forces could also be to their detriment in higher-intensity conflict scenarios.

Combined-Arms and Command

One of the crucial differences between the competing systems was the role of leaders at different echelons and where battlefield decisions were made. This section discusses some of the key aspects of the training and command philosophies of Warsaw Pact forces and compares them with the approaches that they eventually adopted—or continue to attempt to adopt. The emphasis here is principally on changes in combined-arms philosophy reflected in force designs and doctrine, with attention also to the related implications of the different approaches to recruiting and training personnel, particularly leaders.

To reiterate some of the points made above, the military forces of non-Soviet Warsaw Pact members were profoundly shaped by the Soviet Army. Polish, Czechoslovakian, and Hungarian officers were trained by the Soviets. They had (approximately) common force designs, their tactics and operational doctrine were developed by the Soviets, and they planned to operate alongside Soviet formations. In each country a group of Soviet Army forces was also present: the Northern Group of Forces in Poland; the Central Group of Forces in Czechoslovakia; and the Southern Group of Forces in Hungary. Warsaw Pact militaries in wartime fell under the command of the Soviet High Command, and their primary use in practice had been to crush uprisings in Hungary and Czechoslovakia.[13]

Adjustments to doctrine did not come as quickly as reductions in force strength. Fully retraining and reorienting the officers of the former Warsaw Pact member armies, and creating a competent and experienced non-commissioned officer corps, is the work of a generation or more. A 1995 textbook on tactics from the Polish National Defense Academy contained much of the same content of Warsaw Pact doctrine, even being formatted in a similar way.[14] More interestingly, the 2008 'Regulations of the Land Forces'—which used as its starting point the NATO doctrinal publication on

[13] For more on the role of the non-Soviet Warsaw Pact states in the Soviet command structure, see chapters 1 and 5 of Jeffrey Simon, *Warsaw Pact Forces: Problems of Command and Control* (Boulder, CO: Westview Press, 1985).

[14] Zbigniew Scibiorka, et al., *Działania Taktyczne Wojsk Lądowych* [Tactical Actions of the Land Forces] (Warsaw: National Defense Academy, 1995).

land operations—still contained parts that were not far removed from prior practice, for example how forces were task organized for combat operations.[15]

Many differences existed between the approaches to warfare of the Soviets and of the NATO states that faced them, but a few important ones stand out. The Soviet approach focused on generating mass and conducting offensive operations at a very high rate of advance. Soviet doctrine emphasized offence as the main form of warfare, even in defence; their forces were organized and trained to fight mounted and attack from the march; this was most likely both out of preference and necessity given the assumption that tactical nuclear weapons would likely have been employed on the Cold War battlefield.[16] The Soviets did recognize that battles were fluid and required flexibility; however, the system they created reserved that flexibility for combined-arms commanders at a higher echelon than was the case in Western armies. It is worth laying out how they envisioned the combined-arms battle to unfold in practice.

Soviet-trained and organized units were heavily optimized for mounted combined-arms operations, with a large percentage of their forces in armoured vehicles, and with integrated air defence, anti-tank, and indirect fire units. Although they had the tools for combined-arms, they were somewhat lacking in experienced personnel, given the high percentage of conscript soldiers and lack of professional noncommissioned officers, which resulted in the creation of smaller units that trained to operate primarily with a set of battle drills and established norms. These were highly rehearsed, but key tactical decisions and planning primarily took place at the regimental level and higher.[17] Transition to more integrated combined-arms at the battalion level was beginning to emerge in Soviet forces by the 1980s but the higher-echelon focus was still the norm for large-scale conflict.

By comparison, NATO forces tended to place a higher degree of emphasis on flexibility at lower levels, being able to rely in some cases on professional soldiers, as well as on experienced noncommissioned officers. A battalion commander from a US or British Army battalion would often be a lieutenant colonel with 16 years of experience, whereas in some cases a Soviet battalion might be commanded by a captain with only 6 years in service.[18] NATO

---

[15] This is by no means a criticism; the approaches retained appear tactically sound. See, for example, Training Division, Headquarters of the Land Forces, *Regulamin działań Wojsk Lądowych* [Regulations of Activities of the Land Forces], Warsaw, DD/3.2, 2008, pp. 24–26 and subsequent sections.

[16] 'The offensive the main form of battle' is a recurring theme across Soviet and Soviet-derived doctrine. See, for example, V. G. Reznichenko, ed., 'Chapter One: Fundamentals of Combined Arms Combat,' in *Taktika* [Tactics], Moscow: Voenizdat, 1987, http://militera.lib.ru/science/tactic/index.html.

[17] David C. Isby, *Weapons and Tactics of the Soviet Army* (London: Jane's Publishing Company Ltd., 1981), 47–50.

[18] Lester W. Grau and Marcin Wiesiolek, 'Training and Mobilizing the Polish Army Reserve: A Reflection of Sweeping Change in Independent Poland,' *Journal of Slavic Military Studies* 8, 4 (1995): 767.

organizations, whilst not exactly decentralized, did attempt to leverage the ability of subordinate leaders to make decisions based on shared understanding of battlefield conditions, reducing some of the vulnerabilities of excessive centralization.

These distinctions existed because of the underlying conditions in the force designs and force generation approaches. The need for Soviet junior officers to perform tasks that would have been handled by senior enlisted personnel in a NATO force ultimately led to force designs with a smaller span of control. Transitioning former Warsaw Pact forces to NATO organizations and doctrine therefore also implied a demand for more fundamental changes, such as adopting all-volunteer forces and more dynamic training. In the case of the Army of the Czech Republic (ACR), one challenge in 1995 was addressing the imbalance of a top-heavy officer corps. The ACR had 7,000 colonels but only 2,000 lieutenants; it further had a critical shortage of warrant officers and noncommissioned officers.[19] Building experienced junior leaders, senior enlisted personnel, and warrant officers has been a central challenge for the new NATO members attempting to adopt new methods of operation.

Weapon Systems and Modernization

The new NATO members entered the Alliance with mostly old equipment. This aspect of the challenge of transition was not only that they lacked NATO-standardized systems for communication or ammunition, although this was also a challenge; it was also a near-total lack of modern weapons even by 1990s standards, a limitation that has not been fully overcome to this day, even given the dramatically reduced requirements of their smaller land forces. This section provides a brief overview of some of the developments that have taken place in key capability areas—manoeuvre systems, indirect fires, and air defences—and their implications for combined-arms operations.

In several respects, the rate of improvement in armoured vehicle designs has slowed considerably since the end of the Cold War; most of the major fighting vehicles currently in service on both sides of the former Iron Curtain are modernized versions of systems designed in or before the 1980s. However, as non-Soviet members of the Warsaw Treaty, countries like Poland, Czechoslovakia, and Hungary were primarily armed with fighting vehicles and weapons that were already obsolescent in the 1980s. The principal main

[19] Jeffrey Simon, *NATO and the Czech and Slovak Republics*: 35–36.

battle tank in use in all three countries was the T-54 or T-55, with a somewhat smaller number of older-model T-72s. The armoured fighting vehicles for infantry troops were either a mix of BMP-1 and BMP-2, as in Czechoslovakia, or BMP-1 only in Hungary and Poland.

Table 18.3 provides the IISS estimates of main battle tank and infantry fighting vehicle counts for the Visegrád countries, comparing their holdings at the end of the Cold War with those of the present day. Each of these countries has managed to shed their oldest equipment and has taken steps to replace them with modernized weapons; none have fully escaped the legacy of these now quite dated vehicles. Although Table 18.3 depicts manoeuvre platforms, similar trends are evident in other important branches, including surface-to-surface fires and air defences; in actuality, manoeuvre systems

Table 18.3 Selected former Warsaw Pact states' land forces equipment

	1990	2020
Czech Republic	(as Czechoslovakia) Tanks: 2,827 (approx. 1207 in storage) 900 T-72 1,927 T-55 Infantry Fighting Vehicles: 1,560 310 BMP-2 (53 in storage) 1,250 BMP-1 (109 in storage)	Tanks: 119 30 T-72M4CZ 89 T-72 (in storage) Infantry Fighting Vehicles: 410 185 BMP-2 (65 in storage) 127 Pandur-II 98 BMP-1 (in storage)
Slovak Republic		Tanks: 30 30 T-72 Infantry Fighting Vehicles: 256 17 BVP-M 91 BMP-2 148 BMP-1
Hungary	Tanks: 1,420 138 T-72 1139 T-55 (152 in storage) 143 T-54 Infantry Fighting Vehicles: 502 502 BMP 1 (8 in storage)	Tanks: 44 4 Leopard 2A4HU 44 T-72 Infantry Fighting Vehicles: 120 120 BTR-80A
Poland	Tanks: 2,850 757 T-72 2,093 T-54 Infantry Fighting Vehicles: 1,391 1,391 BMP-1	Tanks: 808 247 Leopard 2A4/2A5 232 PT-91 329 T-72 Infantry Fighting Vehicles 359 Rosomak 1,252 BMP-1

Source: IISS *Military Balance*.
Note: Omits over 400 WWII-era T-34 tanks still listed as in storage in Czechoslovakia and Hungary as of 1990; also accepts IISS assessment of BTR-80A as IFV (it is an 8 × 8 wheeled light armoured personnel carrier with 30 mm primary armament).

have received somewhat more attention than artillery and surface-to-air missiles that had only minor roles in peacekeeping and stability missions.

The case of the Polish Land Forces is instructive. The Poles have retained relatively more force structure than other former Warsaw Pact states and have sought for years to replace most of their legacy weapons with more modern platforms.

Poland's armoured vehicle fleet has received a number of more capable systems but much of it remains dated. The Polish Land Forces fielded a modernized T-72 variant called the PT-91 Twardy and a new wheeled infantry vehicle, the KTO Rosomak, derived from the Patria AMV. Mechanized infantry battalions are otherwise still equipped with the BWP-1, which is the Polish version of the Soviet BMP-1 that first entered service in the 1960s. Some older Leopard 2 main battle tanks—about four battalions' worth—have been procured. In July 2021, Poland announced another four battalion sets of M1A2 Abrams were being purchased, presumably in order to replace the oldest T-72s in Polish service.[20]

The distinction between legacy and new platforms is highly relevant for Poland's land combat capabilities. The export-variant Soviet vehicles that have made up the bulk of their force for decades are outmatched both by the vehicles of other Western states but also those of their prospective adversary, Russia. The older vehicles are considerably less well-protected, lack effective night fighting capability, and lack the ability to fire modern anti-armour ammunition. They also pose the additional challenge of identification, in the event that they are required to fight alongside other NATO forces against an adversary armed with T-72s and BMPs.

In the area of field artillery systems, Poland has experienced some successes. A new self-propelled howitzer, the Krab, has been developed and fielded; three squadrons each with 24 howitzers so far, with two more on the way by 2022.[21] Similarly, Poland's Ministry of National Defense in 2019 announced its intent to acquire a squadron of High Mobility Artillery Rocket System (HIMARS) launchers, capable of employing both rockets and battlefield missiles under its HOMAR ('Lobster') programme.[22] The

[20] Jaroslaw Adamowski, 'Polish Defence Ministry Confirms Plan to Buy M1 Abrams Tanks', *Defense News*, 14 July 2021, https://www.defensenews.com/land/2021/07/14/polish-defence-ministry-confirms-plan-to-buy-m1-abrams-tanks/.

[21] Jakub Palowski, 'EU Supports the European Artillery Projects: Polish Industry Involved', *Defence 24*, 10 August 2021, https://defence24.com/eu-supports-the-european-artillery-projects-polish-industry-involved.

[22] Rafał Lesiecki, '"Nowoczesna broń dla polskich żołnierzy", czyli kontrakt na HIMARS-y podpisany' ['Modern Weapons for Polish Soldiers'—a contract for HIMARS signed], *Defence 24*, 13 February 2019, https://www.defence24.pl/nowoczesna-bron-dla-polskich-zolnierzy-czyli-kontrakt-na-himars-y-podpisany.

remainder of Poland's artillery are older 2S1 and Czech-produced DANA wheeled howitzers. The increased range, rate of fire, and payload options, combined with modern digitized fire direction, mean that a few guns may do the work of many. More will be noted on this point in the final section, below.

The other Visegrád Group members' land forces do not quite parallel the Polish experience but do share a number of commonalities; in particular, each has had to make careful choices about which systems to modernize. The one potential exception to this may be Hungary. As part of its Zrinyi 2026 modernization programme, the Hungarian Defense Forces have embarked on a major modernization effort that includes the procurement of at least a brigade set of modern, mostly German-built equipment, including 44 Leopard 2A7+ tanks, 218 Lynx infantry fighting vehicles, and 24 PzH 2000 self-propelled howitzers.[23] More recently, Hungary has also placed an order for a hard-kill active protection system (APS) to equip its new Lynx fighting vehicles; when completed it will be one of the first armies in Europe so equipped.[24]

Summary

Armies balance the requirements of force structure, readiness, and modernization in different ways based on factors that are generally aligned with their national requirements. If resources are held constant, then choices must be made among these elements. The land forces of the Visegrád countries have evolved both toward desired capabilities as NATO members and away from mass conscript-based armies with the end of the Warsaw Pact and the resulting change in the security environment in Europe. The scale of the challenge, modernizing and changing virtually their entire armies, was simplified somewhat by the dramatic reduction in size, but even so the evolution of force designs, doctrine, personnel recruitment, and training as well as equipment modernization are in each case still in some respects very much ongoing.

The final section, below, considers the future, and how the reemergence of demand for large-scale land warfare capabilities has evolved since the Cold War.

[23] Andras C, 'Modernization and Rearmament—Hungary's Zrinyi 2026 Program', *Overt Defense*, 3 April 2020, https://www.overtdefense.com/2020/04/03/modernization-and-rearmament-hungarys-zrinyi-2026-program/.
[24] 'Active protection system for Lynx IFV: Market breakthrough for Rheinmetall's new StrikeShield—€140 million order from Hungary', Press release, Rheinmetall Protection Systems GmbH, 18 May 2021, https://www.rheinmetall.com/en/rheinmetall_ag/press/news/latest_news/index_25216.php.

Looking to the Future

At the July 2016 NATO Warsaw Summit, the Alliance agreed to move forward with a number of measures to reinforce defence and deterrence of Russian aggression against its member states. The resulting communique outlined the formation of the Enhanced Forward Presence (eFP) battlegroups and expanded on the Alliance's renewed focus on collective defence since Russia's invasion of Crimea and support of separatist republics in the Donbas in 2014.[25] This section outlines some of the implications of current and emerging trends in land warfare capabilities focusing on the challenge for states like Poland posed by a capable modern adversary like the Russian Armed Forces.

The Visegrád states have each changed considerably since the end of the Cold War, but the Russian military has also evolved a great deal, and in some ways has also reversed some of the practices of its Soviet predecessor. Most analysts of the Russian Federation do not consider an opportunistic large-scale military attack against a NATO member state to be at all likely.[26] That said, it is generally accepted in defence ministries in countries like Poland and the Baltic states that preparing for a serious defence is one path toward making a conflict less likely. It is, moreover, also noted by analysts of the Russian Armed Forces that, in parallel with a range of unconventional and non-military influence efforts, Russia is building capability and capacity for large-scale warfare.[27] It is useful to consider how technology, the threat, and changing views of the security environment combine to affect how an army like the Polish Land Forces thinks about land combat in the near to mid-term future.

One of the important factors that profoundly shapes how a future land war might unfold is the much lower force density of land forces in Europe. As described above, the common post-Cold War experience of armies has been dramatic reductions in end strength; countries that once fielded armies of multiple divisions now field brigades. Even those countries—Poland and Russia, for example—that have retained more substantial forces now often have brigades where once there were divisions. There is simply no comparison with the 1980s balance of forces.

---

[25] North Atlantic Treaty Organization, 'Warsaw Summit Communiqué', Press Release, 9 July 2016, https://www.nato.int/cps/en/natohq/official_texts_133169.htm.

[26] See, for example, Michael Kofman, 'Getting the Fait Accompli Problem Right in U.S. Strategy', *War on the Rocks*, 3 November 2020, https://warontherocks.com/2020/11/getting-the-fait-accompli-problem-right-in-u-s-strategy/.

[27] The Swedish Defense Research Institute FOI has been making this clear in its *Russian Military Capability in a Ten Year Perspective* series; for more on this issue in specific, see Johan Norberg, *Training for War: Russia's Strategic-level Military Exercises 2009–2017*, FOI-R—4627—SE, October 2018.

This change in density has several implications for land operations. Units will likely be required to operate in a more dispersed manner, aggregating or disaggregating based on conditions on the ground. Some countries have already transitioned to a battalion-centric approach; this is probably the lowest level at which the full range of combined-arms forces are likely to be integrated.[28] Contiguous lines of operation would likely be a rarity, at least prior to the period of mobilization. Forces on the move may be required to manoeuvre with open flanks, and defenders may place greater reliance on strongpoint defences and natural obstacles. Russian forces already place particular importance on anti-tank guided missiles; most of their long-range direct fire anti-armour capability, including from tanks as well as from motor rifle troops, comes from missiles, implying an incentive for more armies to join the USA as well as soon Hungary in fielding active protection.

Compounding the lower density of land forces is the technology trend toward improved reconnaissance and precision strike systems. Ground forces, increasingly down to the lower tactical levels, have greater ability to perceive and attack adversaries at increasing distances. The relevance of longer-range fire systems has increased by a large margin, and the importance of range in particular has grown. Long-range howitzer and rocket systems permit units to provide mutual support; greater effective ranges also allow artillery units more options for concealment (by allowing a particular location to be attacked from a larger number of firing positions). The Polish Land Forces' acquisition of new, much-longer-ranged howitzers like the Krab and rockets like HIMARS mirrors Russian fielding of longer-ranged rockets for its 122 mm and 300 mm Tornado multiple rocket launchers, as well as newer howitzers like 2S19M2 and 2S35 Koalitsiya.[29]

A further key area that has gained in importance since the Cold War is the role of intelligence, surveillance, and reconnaissance systems. A dazzling array of sensors have reached technological maturity in the intervening years, including new generations of thermal imagers, and smaller and more portable radars.[30] These, along with satellite navigation, have in turn been combined with battlefield management systems, which are

---

[28] Wlademar Skrzypczak, 'Proposed Changes to the Land Forces Battalions', Casimir Pulaski Foundation, Policy Paper No 14 2017, 31 July 2017, https://pulaski.pl/en/analysis-proposed-changes-the-land-forces-battalions/.

[29] Rafał Lipka, *The Future of the Missile Force and Artillery—Poland's 'Homar' Program*, Pulaski Policy Casimir Pulaski Foundation, Policy Paper number 5, 2018, 29 March 2018, https://pulaski.pl/en/pulaski-policy-paper-the-future-of-the-missile-force-and-artillery-polands-homar-program/.

[30] For an example of just the Russian capabilities for tactical reconnaissance along these lines, see Lester W. Grau and Charles K. Bartles, 'The Russian Reconnaissance Fire Complex Comes of Age', Changing Character of War Centre, May 2018. http://www.ccw.ox.ac.uk/blog/2018/5/30/the-russian-reconnaissance-fire-complex-comes-of-age.

powerful tools to provide increased awareness and coordination in land operations. Particularly from the perspective of Poland attempting to deter Russia, there are compelling incentives to gain these advantages and to deny them to the adversary, through the use of camouflage, deception, and electronic warfare. The electronic spectrum will be a crucial domain for high-intensity conflict and the side that develops an asymmetrical advantage in information and visibility could translate that into concrete gains on the battlefield.

Another key factor is air power. Since fewer land forces are available and those that are available but not at high readiness may be slow to activate and move to the battlefield in a crisis, control or denial of the air domain is of central importance for land forces. This is particularly true when considering the relative strengths of current NATO members and the Russian military; Russian forces have substantial land-based fires available but are disadvantaged in an aerial contest. This disparity is likely to grow even further as more NATO members field greater numbers of fifth-generation fighters.[31] The combination of much more capable aircraft delivering a new generation of more capable precision munitions like the Brimstone missile or Small Diameter Bomb-II may be the principal factor enabling NATO asymmetrical advantages in information and mobility. The role of ground forces—focusing in particular on difficult terrain where sensors will be less effective, for example—will inevitably evolve along with this trend.

Finally, the years to come will see a step change increase in the fielding of robotic capabilities on the battlefield. Military drones of various sizes are already widespread, but greater numbers of more capable systems are on the way and will be combined with loitering munitions.[32] The interaction between drones and counter-drone systems as part of the broader fight for information is emerging as a major and possibly decisive factor in land combat. Ground robotic platforms may eventually also become pervasive, but in the near- and mid-term future the remotely-piloted—or someday autonomous—aerial vehicle will be a crucial capability. Based on the experiences of its forces and allies in Eastern Ukraine, Syria, and Nagorno-Karabakh, Russia's military is clearly moving forward with both drones and improved defences against them.[33]

---

[31] The current Supreme Allied Commander Europe, General Tod Wolters, recently predicted that NATO members will have 450 fifth-generation F-35 fighters by 2030. U.S. European Command, 'Transcript: Gen. Wolters remarks at the Atlantic Council Competition and Deterrence in Europe event', 9 June 2021, https://www.eucom.mil/document/41348/transcript-of-gen-wolters-at-atlantic-council-on-june-9-2021.

[32] See, for example, the Warmate loitering munition in use by the Polish military. 'Warmate Loitering Munition', *WB Group*, https://www.wbgroup.pl/en/produkt/warmate-loitering-munnitions/.

[33] See, for example, Timothy Thomas, *Russian Lessons Learned in Syria: An Assessment*, The MITRE Corporation Center for Technology and National Security, June 2020; as well as Ruslan Pukhov, ed., *Burya*

Taken together, these conditions will result in ground operations where manoeuvre will be highly dynamic; land forces will be required to operate with greater uncertainty, potentially intermingled with hostile forces and empowered by massed fires from long-range aerial and land-based systems. The problems of sustainment, reconnaissance, and command and control of forces in these environments will force land units to adapt, and that adaptation will require experienced and well-trained personnel.

Conclusion

From the end of the Cold War to the beginning of the 2020s, the military forces of the Czech Republic, Hungary, Poland, and Slovakia have followed a winding and occasionally abrupt path from a mass conscript force subject to the control of a foreign power to a smaller but more modern and flexible land force capable of making contributions to international missions and to collective defence. A military can be cut in size relatively quickly, and these were, but growing new senior enlisted soldiers and officers who are fully able to operate in a different style of warfare is the work of decades. As this work continues, it will be instructive to see how these armies make their own way toward developing the forces and capabilities they need to meet their nations' aims in the future.

na Kavkaze [*Storm in the Caucasus*] (Moscow: Centre for the Analysis of Strategies and Technologies, 2021), 80–83.

PART III
CONCLUSIONS

19

Towards a Versatile Edge

Developing Land Forces for Future Conflict

Mikael Weissmann and Niklas Nilsson

Introduction

This chapter draws together and builds on the thematic and empirical chapters to develop a synthesized vision of the contemporary opportunities and challenges of land warfare research, as well as the conceptual challenge of developing and utilizing land forces in future conflict. The chapter's aim is to provide a comprehensive picture of current theory and practice, identify fruitful pathways for future research, and develop an integrated understanding of the dynamic evolution of truly versatile land forces.

The chapter seeks to navigate and clarify the multiple and often contradictory forces at play as land forces develop to meet the future challenges outlined in the introduction to this volume. These forces include inherited and established understandings of combat principles as well as new and purportedly revolutionary concepts enabled by diverse developments in, for example, technology, digitalization, and C2 methods. Potential future operational environments are also diverse, and inextricably connected to the character of future wars, the opponents whom land forces must develop capabilities to fight, and the circumstances of tactical engagements.

In addition to several thematic and generic competing influences on land forces, the localized contexts of specific national land forces will be a fundamental determinant of change and continuity. Diverse military cultures, financial constraints, history, and tradition will affect the development of land warfare, as will national perceptions of future war, threats, and interpretations of security interests.

These forces are set to work in tandem at times, but also collide and compete for the attention of national decision makers and defence spending. The contribution of this chapter is therefore to establish a modicum of analytical

Mikael Weissmann and Niklas Nilsson, *Towards a Versatile Edge*. In: *Advanced Land Warfare*. Edited by Mikael Weissmann and Niklas Nilsson, Oxford University Press. © Mikael Weissmann and Niklas Nilsson (2023).
DOI: 10.1093/oso/9780192857422.003.0019

order in the complex field of land warfare. It seeks to establish where current academic research stands on the subject, and to outline a forward-looking agenda for describing and understanding the perspectives and problems facing academics as well as practitioners interested in the evolution of land forces in the years to come.

The chapter is outlined as follows. The first section sums up and discusses the volume's findings on the dynamics of twenty-first-century land warfare. We then describe a continuum of land operations, an expression of the heterogeneity of potential conflict environments in which land forces must be capable of operating. This operational complexity is visualized in a schematic model that takes account of how different levels of conflict intensity translate into demands and constraints on the utilization of land forces. The chapter then moves on to locating land operations in the broader operational environment before synthesizing the findings in what we label *the integrated versatility model*, outlining the preconditions for securing land warfare capability and the requirements for achieving a versatile edge in the future operational environment. Finally, we discuss ways forward for land warfare and outline an agenda for future research.

Dynamics of Twenty-first-century Land Warfare

The individual chapters of this volume have contributed a set of distinct perspectives on the past, present, and future of land warfare. These perspectives put to the fore both general observations regarding the enduring relevance of warfighting concepts and specific challenges relating to the evolution of tactics, technology, battlefield intelligence, and information flows as well as weapon range and lethality. Christopher Tuck makes a compelling argument for the continued relevance of manoeuvre warfare in the contemporary operational environment. Originating as a concept aiming to maximize the efficiency of mechanized, mobile warfare, the manoeuvrist approach can be extrapolated to the conduct and synchronization of warfare across domains, denoting the significance of tempo, surprise, and exploitation of own strengths and the opponent's vulnerabilities. However, as Tuck points out, whilst manoeuvre warfare as a concept or intellectual construct may remain well attuned to contemporary and future war, its actual conduct might become more circumscribed and problematic, particularly in the contexts of limited and irregular wars and urban operations.

A similar discrepancy between a warfighting concept's utility in the abstract and its applicability as a concrete military method can be observed in the utilization of mission command. The command philosophy was widely adopted by Western militaries and is indeed a trope of military leadership and command. It denotes a distinct ethos and desired qualities in military officers, as well as an approach to warfighting that acknowledges the human nature of warfare and the necessity of dealing with uncertainty as a constant characteristic of war. Yet, as Nilsson points out, it is simultaneously unclear to what extent mission command can be applied consistently in the context of contemporary and future land operations. The increasing demands for informational superiority and synchronization, taken to new heights with the introduction of multi-domain operations and related concepts, seemingly create conflicting demands on command for vertical as well as horizontal coordination between services and domain capabilities. In turn, this creates incentives for redefining or delimiting the use of mission command, which is nevertheless set to retain a distinct utility in military practice.

The extent to which received truths will remain applicable to future land warfare is in question. As Friedman and Paulsson argue, there is a need for tactical theory, here presented as a set of tactical 'tenets' to provide a common language and understanding of what tactics may imply (mass, manoeuvre, firepower, and tempo, deception, surprise, confusion, shock, and moral cohesion). Yet how these tenets are to be translated and exercised in the future operational environment is a different question, and several chapters bring to our attention prospective trends in modern warfare that extensively change the preconditions under which land forces will operate, as well as the tactics they will have to exercise.

Disruptive technologies are doubtless among the most important factors affecting tomorrow's battlefield. As argued by Watling, paramount technological shifts, including but not limited to autonomous systems, layered precision fires, pervasive sensors, and AI, are expected to change many of the preconditions for land operations. The timeframes for developing and introducing these systems may be overly optimistic. However, their potential consequences include extremely quick reaction times between discovery and highly precise kinetic effect, potential information supremacy, and extreme information processing capacities. Should these developments be realized, land forces, particularly in their manoeuvre elements, will need to develop much more comprehensive means to avoid destruction, even to enter the battlefield. The challenge posed by competitors to US military hegemony, particularly China and Russia, also underlines the increasingly significant collective element of land warfare, in the form of increasing interdependence between allies and partners. In turn, concepts developed by the USA and

NATO to address this challenge will, as Curtis points out, put an increasing premium on the interoperability of Western forces.

Besides technology and interoperability, significant shifts in the human terrain create new challenges, given the likelihood that cities will be major battlefields in future wars. As Weissmann argues, urban terrain should be considered an increasingly salient part of the future operational environment due to the rapid ongoing urbanization process, as a majority of the world's population resides in large cities. This assertion is valid regardless of whether we are considering high-intensity conflict against peer-opponents, or localized irregular warfare. Indeed, urban warfare poses a different set of challenges compared to open manoeuvre-based warfare, raising questions regarding current applications of technology and tactics, weapons systems, and considerations of casualties and destruction.

A more comprehensive consideration of the role of land forces in future conflicts, beyond just their conduct on the battlefield, should also include a reconsideration of what motivates land forces to fight. As Sandman argues, our understanding of will and cohesion among soldiers tends to be limited to their immediate context, but should be considered in much broader terms, as a product of the societies to which they belong. This broader and more complex view of morale is indeed more in tune with a more integrated and complex view of future conflict.

In a similar vein, further specific questions regarding the organization and sustainment of Western land forces need to be addressed. For example, Storr argues that the current exercise of military command is ill-equipped to function in an environment that requires tempo and initiative on the battlefield. In this regard, headquarters are overstaffed and immobile, extensive planning processes tend to produce unnecessarily long and complex orders, and the entire command structure tends to generate and promote officers based on preconditions other than competence to lead in battle. Combat logistics is another area in need of attention, being highly vulnerable to Long Range Precision Fires (LRPF). Logistics will require increased mobility and protection in a future high-intensity conflict—a problem that transcends borders, governments, and military organizations across the European continent, according to Kinsey and Ti. Moreover, Bricknell presents the provision of military health services as a frequently overlooked component of land power, which will become increasingly important given the potential rate of casualties in future wars, but also increasingly vulnerable due to the increasing potential of deep battle. Simultaneously, the functions of military health systems (MHS) are expanding, as demonstrated by their utility in providing broader societal assistance during the COVID-19 pandemic.

The changing realities of land operations, as well as the anticipation of the future preconditions for land forces, have driven continuous adaptation, which has taken on different features depending on national context. As shown in the case studies presented in this volume, variations in strategic outlook, previous experiences of war, and military cultures shape adaptation in land forces, resulting in differences and similarities. Gudmundsson outlines how the diverging operational cultures of the US Army and US Marine Corps have played out in the US conduct of land operations in the twentieth century. As the US Army now undertakes yet another adaptation to new realities, in the form of multi-domain operations, this reflects a reaction to the prospective need to conduct high-intensity warfare against near-peer competitors in the form of China and Russia. Brad Marvel demonstrates in his chapter how China's warfighting concepts to a large extent mirror those of the USA, by introducing an equivalent of multi-domain thinking. China nevertheless places an even higher emphasis on dominating the cognitive and information dimensions, whilst pursuing the capability to rapidly mass capabilities for employing combat power that aims to exploit enemy weaknesses in all domains. Whether China's ambitious concept will prove feasible in practice is another question.

Göransson's chapter on Russia highlights that a narrow focus on high-intensity warfare in multiple domains provides a far too limited view of the prospective utilization of land forces. In Syria, the land component of Russia's operations has been marginal, deployed in support of an operation that has otherwise been dominated by an extensive air campaign. However, these forces have not been unimportant, as the use of special forces, military police, and military contractors, as well as auxiliary forces, served particular roles, indicating the flexibility of land forces in limited wars. The engagement seen in Syria is a likely model for future endeavours, especially for larger military powers and should therefore not be discarded as irrelevant. This underlines that the utility of land forces is not limited to regular fighting in high-intensity conflict.

Hecht and Shamir demonstrate how Israel has gone through several stages in its thinking and practice regarding the exercise of land power. Of particular importance in recent decades was Israel's inability to defeat Hezbollah in Lebanon during the 2006 war, the effect of prioritizing LRPF capabilities at the expense of manoeuvre units. As the authors point out, a debate continues in Israel regarding whether land forces should provide sensor capability for LRPF or whether land manoeuvre capability is a key capability that cannot be replaced. The UK case demonstrates the dilemmas involved in developing a multi-purpose army and how different visions of future conflict give rise to competing priorities and demands, depending on what type of units are deemed to be needed. According to Galbreath and Neads, the UK faces the key questions of whether to prepare for large-scale modern conflict with significant adversaries and/or sub-threshold competition and hybrid warfare, which require different forces, units, and equipment. These approaches also differ in terms of the scale and costs of military procurement and preparation and drive the development and envisioned future use of land forces in different directions. As shown by Schmitt and Tenebaum, the French army is facing a similar dilemma, as its forces undergo transition from a paradigm of expeditionary deployment to preparation for strategic competition and high-intensity warfare.

Finally, Boston's exposé of the massive transformation processes of Poland, Czech Republic, Hungary, and Slovakia, simultaneously indicates that substantial change and adaptation are possible in military forces, and that these processes indeed take time. The journey from being part of the Warsaw Pact military infrastructure to NATO membership, with smaller but more professional forces, has been an arduous undertaking. Yet the fact that tactics, techniques, and procedures (TTPs) and other aspects of military practice have proven resilient to new circumstances speak to the slow process of military adaptation and reform. Also, the current process of readapting to new missions and roles as NATO allies underlines that downsizing is easy but growing and especially training cadres of new officers and soldiers take generations.

In sum, the chapters of this volume highlight the growing complexity of land operations in the twenty-first century, whereby strategic campaigns, operational contexts, and tactical preconditions for fighting have become highly heterogeneous. In turn, the contemporary and future operational environments will present many converging and competing demands on land forces. The conclusions speak to the need for developing truly multi-purpose capabilities for warfare on land, with extensive implications for the organization of land forces, as well as equipment procurement and training for the entire spectrum of possible land operations in the future, including the capability to operate across the entire conflict spectrum, from peacetime tasks to high-intensity combat, and across domains.

The Continuum of Land Operations

As shown in this book, comprehending the challenges of future land warfare is a highly complex, but very important endeavour. Views on how land forces should be organized, equipped, and trained have historically varied with the perceived character of future wars and the missions that land forces are expected to accomplish. Several paradigmatic shifts in perspective regarding the utilization of land forces can be identified in the last century, akin to pendulum swings, which have had substantial effects on defence planning and on armies across the Western world. The massive conventional confrontation with large peer adversaries that formed the key dimensioning challenge of the Second World War and the Cold War was replaced by the US experience in Vietnam, which made clear the need for different forces and tactics than those possessed by the USA at the time. The US defeat in Vietnam nevertheless disqualified the need for capabilities to fight insurgencies, in US thinking—and the conclusion was instead that US land forces should not be utilized in this type of conflict. The focus then shifted to developing agile conventional capabilities to confront the numerically superior USSR in Europe, via the AirLand battle and Manoeuver Warfare Doctrines, only to face seeming irrelevance as the Warsaw Pact disbanded in 1991. The wars of the 1990s, including the Gulf War, wars in the Balkans, and the Kosovo campaign, and of the 2000s in Afghanistan and Iraq, provided a different set of conclusions. Although conventional land power played a significant role in the two wars in Iraq, the experience of post-Cold war conflict implied a less significant role for large, heavily equipped, and manoeuvrable land forces than that envisioned in the 1980s, and underlined the importance of land warfare against unconventional opponents, as well as lighter and more modular expeditionary forces. In the 2010s, the pendulum swung again. Russia's invasion of Ukraine and its aggressive posturing against eastern NATO members, as well as China's emergence as a determined competitor in the global arena, underlined once more the significance of modernized conventional land forces dimensioned to fight high-technological peer or near-peer adversaries, and an unprecedented need to integrate their capacity with capabilities in other warfare domains.

However, these historical shifts in focus also underline that the future of land warfare cannot fully be anticipated and that land forces may be deployed in different types of conflict environments, requiring different sets of capabilities. Land forces face multidimensional challenges and demands and therefore need to be organized, structured, and trained in a manner that highlights versatility as a key property of land forces. This includes an integrative approach in relation to both domains and allies and partners, as well as awareness and readiness to evaluate and adapt to new developments in technology, integrating these technologies where appropriate or discarding them as needed. It also includes a dynamic and proactive approach to TTPs, which must evolve in sync with realities in the operational environment as well as capabilities, and must not stagnate into dated and static checklists or set rules.

This will enable land forces to adapt in a changing and evolving security environment, where several external forces impact the preconditions for warfare. This will also enable land forces to take a proactive and flexible approach to existing challenges and devise new approaches to problems that are yet to emerge.

However, before outlining the framework for a versatile approach to land warfare, we must establish a modicum of structure for the myriad elements and factors that influence land forces. It is useful to envision these questions/challenges in terms of two distinct spectra, namely the intensity of conflict and the role and purpose of land operations. These two dimensions are sketched out in Figure 19.1, where the horizontal axis denotes the level of conflict intensity, the vertical axis indicates the land operations continuum, and the diagonal line—with intentionally blurred edges to illustrate the approximation involved—denotes the resulting utilization of land forces.

Conflict intensity. Naturally, a conflict's level of intensity will be a key determinant of the role of land forces therein. This applies to (1) the tasks and objectives to be accomplished, (2) the scale of deployment, and (3) the capabilities required to deliver the desired effects. It is common (and useful) to envision the possible range of conflict intensity in an ideal-typical spectrum

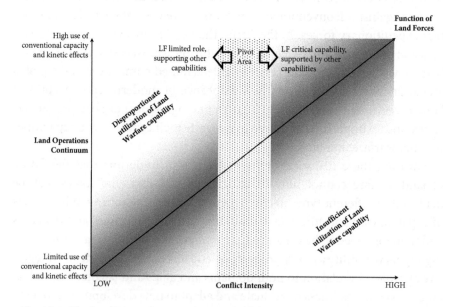

Fig. 19.1 The land operations continuum model

of low- to high-intensity conflict.[1] Low-intensity conflict here denotes conflict with no or very limited utilization of kinetic force. This applies to, for example, post-conflict environments where forces mainly serve peacekeeping or peace-supporting missions. Low-intensity counterinsurgency missions also belong to this side of the spectrum, as do several thinkable activities relating to competition below the threshold of armed conflict or in a grey zone context, where land forces mainly function as a deterrent and no or few significant combat actions take place. More intense expeditionary missions, such as those in Afghanistan and Iraq after the conclusion of major fighting, fall in the middle of the spectrum but remain located on the 'left' side. This since combat involving land forces, although occasionally intense, for the most part took place on the lower tactical levels and predominantly aimed to support the objectives of agencies other than the military.

On the 'right' side of the spectrum is limited war, that is, a war which in large part takes on regular/conventional features, but where the antagonists (or the dominant one) takes precautions to keep the war limited and avoid escalation beyond a specific theatre in order to accomplish objectives without risking a confrontation with major adversaries. Of course, a war's limited nature is a matter of perspective; a stronger party may fight a limited war, whilst the weaker party fights an existential one. Post-Cold War examples include the 1991 Gulf War, the 2003 invasion of Iraq, and Russia's wars in Georgia and in Ukraine during its first invasion in 2014-15. Whilst a limited war can indeed be of high intensity, it is delimited to a localized geography and/or delimited in the means utilized.

In contrast, high-intensity conflict in this case denotes all-out war between peer or near-peer competitors, without any clear geographical delimitation and utilizing all means at their disposal, across all warfighting domains and throughout the ladder of escalation.

In theory, any conflict can be located at any point on the spectrum at a specific time.

Land operations continuum. To visualize approximately what these different levels of conflict intensity might entail, in terms of the demands on land forces and their expected utility, the land operations continuum denotes the emphasis on conventional capability for land warfare, in the respective categories of conflict. The logic implied is that the emphasis on conventional

---

[1] See, e.g. Frank G. Hoffman, 'The Contemporary Spectrum of Conflict: Protracted, Gray Zone, Ambiguous, and Hybrid Modes of War', Heritage Foundation, 2015, accessed January 26, 2022, https://www.heritage.org/military-strength-topical-essays/2016-essays/the-contemporary-spectrum-conflict-protracted-gray; *Allied Joint Doctrine for Land Operations (AJP-3.2 Edition A)* (Brussels: NATO, 2016); Linton Wells II, 'Cognitive-Emotional Conflict—Adversary Will and Social Resilience', *PRISM* 7, 2 (2017): 5–17.

land warfare capabilities, or the demand on land forces to deliver coordinated kinetic effects, will grow exponentially depending on the intensity of conflict. In a low-intensity environment, land forces, if they are deployed at all, will mainly serve functions other than combat, for example peacekeeping or peace support missions or other supporting activities. Counterinsurgency missions require land forces to deliver kinetic force, but often limited to lower tactical levels. Limited (conventional) wars may indeed involve a substantial land component, but in a delimited theatre. Conversely, high-intensity conflict will employ land forces to their full conventional capacity and require that these as well as other capabilities are coordinated across domains.

The continuum also envisions a pivot area, in which the role of land forces shifts from being limited and mainly supportive of other activities, to constituting a critical capability that is instead supported by other capabilities and activities. This area is intentionally wide, symbolizing the problem for the military as well as political leadership in determining when a conflict has transformed, or should escalate, into a higher level of intensity demanding a more substantial involvement of land forces. The point, however, is that on the left-hand side of the continuum, the centre of gravity in the conflict focuses on activities other than conventional military ones. For example, these might include diplomacy, narrative promotion, economic sanctions, or various unconventional means of warfare. On the right-hand side of the spectrum, however, conventional military force instead comprises the main activity and focus of the actors involved, and other available means deployed in the conflict are geared toward supporting the military effort.

Indeed, the continuum attaches great importance to perceptions regarding the limits of legitimate or appropriate utilization of land forces, which are expected to change in pace with the intensity of conflict. The areas above and below the continuum respectively symbolize the disproportionate and insufficient utilization of land warfare capability. The area above the continuum denotes disproportionate use of force, perceptions of which are envisioned to shrink in pace with conflict intensity. Conversely, the potential for insufficient utilization of land forces is projected to grow as conflicts intensify, symbolized by the area below the continuum. Questions pertaining to these two areas relate to the specific implications of deploying land forces in a conflict and the tasks they are expected to fulfil. The deployment of land forces is a sign of commitment with acceptance of risks far more substantial than those of a more limited utilization of air or maritime power alone. Deploying 'boots on the ground' in a conflict drastically increases the likelihood of human casualties, which, in turn, becomes more acceptable as conflicts become more intense. This risk also relates to the political ability to motivate engagement in the conflict, where engagement in a high-intensity conflict will likely justify far greater human sacrifice than one of lower intensity. Examples of this dynamic include the tolerance of US casualties in the Second World War, compared to the Vietnam War.

The deployment of land forces in a conflict also implies a risk of conflict escalation, since their presence introduces a unique dynamic in a conflict. Warfare on land is more difficult to control in terms of resources and capabilities that ultimately become involved in fighting. This also relates to conflict intensity, since the perception of disproportionate use of force will have consequences for the ability to engage in a conflict. Consider, for example, a minor tactical engagement in a low-intensity conflict environment, which nevertheless escalates and ultimately produces substantial casualties as well as collateral damage in terms of civilian lives and property. This could prove detrimental, politically and in the eyes of the public, to continued engagement in that conflict. In a high-intensity conflict, on the other hand, this would likely appear legitimate given the mission of land forces and the stakes involved.

The question of perceptions and legitimacy also has relevance for the utilization of irregular or proxy forces in different types of conflict. Low-intensity conflict can be assumed to incentivize disassociation of kinetic violence from governments and regular militaries. Thus, assigning tasks that potentially involve disproportionate use of force, for example controversial combat missions to proxy forces or military contractors is seen as particularly attractive in low-intensity conflict environments, in order to establish a distance and deniability in relation to kinetic violence and the risks involved. Examples include Russia's warfare in Ukraine before 2022 and in Syria, which has significantly relied on local proxy forces as well as military contractors like the Wagner Group. Another example is the US strategy to decrease its own military commitment in Iraq and Afghanistan by training and equipping national land forces in these countries. Of course, similar forces may very well be deployed in high-intensity conflicts but are expected to be less important as a military-political tool. That is, conflict intensity relates to the relative significance of unconventional and conventional forces, strategies, and tactics in the conflict.

Locating Land Operations

The land operations continuum visualized the heterogeneity of possible conflict environments in which land forces might be deployed, and the wide spectrum of possible tasks they might be required to carry out. In sum, land operations could take place virtually anywhere and everywhere across the spectrum of conflict intensity, yet with substantial variations in their expected effects. This section seeks to locate land forces in the broader operational environment—an exercise intended to provide a basis for discussing both the particular issues pertaining to land forces, and their role as an integrated part of military strategy and operations, inseparable from other components. Indeed, the role of land forces cuts across the strategic, operational, and tactical levels of a military operation.

As visualized in Figure 19.2, the operational environment can be schematically outlined as five concentric circles that symbolize, respectively, the overarching strategic context of the operational environment; the possible range of conflict intensity; interoperability; multi-domain operations; and the land forces themselves. Together, the four outer circles can be considered constitutive for land operations, since they decisively enable and/or constrain activities in the land domain.

The model is land forces-centric, placing these in the middle. However, land forces do not operate in a vacuum; they never act alone in any form of operation or conflict, since military operations always include other warfighting domains. Whilst this does not necessarily imply fully integrated multidomain operations, the land domain can never be completely separated from the air, land, cyber, and space dimensions of the contemporary battlefield; nor does a land operation include only land capabilities. The third circle denotes the interoperability dimension, which is the ability to work together with allies and partners—an increasingly important aspect in military operations. NATO is a case in point, where interoperability is defined as 'the ability for Allies to act together coherently, effectively and efficiently to achieve tactical, operational and strategic objectives [enabling] forces, units and/or systems to operate together and allows them to share common doctrine and

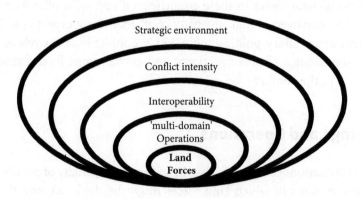

Fig. 19.2 The operating environment model

procedures, each other's infrastructure and bases, and to be able to communicate.[2] Whilst interoperability may not always be a requirement in specific
scenarios, the ability to coordinate land forces of different national origins is
nevertheless an important dimension of the operational environment.

The outer circles symbolize, respectively, the strategic environment and
conflict intensity, and thus frame the environment in which land forces
operate. The strategic environment is the overarching context of the operational environment, consisting of the interests and perceived needs that
inform motives, possibilities, and constraints and affect all other levels of
the operating environment. Conflict intensity, in turn, narrows the operational environment, as varying levels of conflict intensity give rise to different
operational needs.

Having located land forces in the operational environment, we now zoom
in on some of the factors of immediate concern to land forces, arising from
the peculiarities of the land domain. Indeed, whereas the most basic issue
pertaining to land forces relates to their own capabilities, that is, what they
can and cannot do, this question cannot be understood or discussed in isolation but needs to be contextualized. Since warfare and combat are always
an interaction between intelligent opponents, land capabilities and military
power cannot be measured with a one-sided scale but is always relative and
relational. In other words, one's own capabilities have no intrinsic or inherent value—they must be assessed in relation to an adversary. The German
spring offensive through the Ardennes is a case in point; although the Mag-
inot Line provided solid defensive positions, these were of little value against
an adversary that could utilize its superior capability for movement to simply circumvent these positions. Likewise, in the 1967 and 1973 wars, Israeli
forces proved capable of defeating several numerically stronger armies chiefly
through more competent force employment—a far more efficient utilization
of own capabilities relative to the opponent.

Moreover, different specific dimensions of the land domain remain crucial
to land warfare. The physical terrain, and utilizing it to one's advantage, has
always been a hallmark aspect of land warfare. Yet aspects of the human terrain and the information environment are also features that must be seriously
considered in the contemporary land environment. Whilst these dimensions
have not been unimportant historically, they are arguably becoming increasingly crucial in the contemporary land domain. Indeed, a battle cannot be

[2] 'Interoperability: Connecting Forces', NATO, accessed 14 December 2021, https://www.nato.int/cps/
en/natolive/topics_84112.htm. See also Backgrounder: Interoperability for Joint Operations, NATO, July
2006, https://www.nato.int/nato_static_fl2014/assets/pdf/pdf_publications/20120116_interoperability-
en.pdf.

won exclusively in the physical terrain; defeat and victory are increasingly defined in the human terrain and the information environment.

Nor can these three dimensions be analytically separated, since actions and development in one will unavoidably affect the others. Actions in the physical and human terrain will affect narratives in the information environment, which are today disseminated at lightning speed via a plethora of information outlets. In turn, consideration of the information environment delimits the possible range of actions in the physical and human terrain, how military activities can be carried out, and what type of forces and capabilities are required.

One prominent example is the trajectory of the Vietnam war, where a one-sided US focus on defeating the enemy in the physical terrain proved decidedly counterproductive and eventually led the superpower to defeat. Measures of capability and success including force ratios, body counts, carpet bombings, and capacity for herbicidal de-leafing of jungles only strengthened the resolve and military recruitment of the North Vietnamese Army and South Vietnamese guerrillas. Simultaneously, US and international media provided graphic documentation of the indiscriminate and disproportionate use of force, eventually making the war not only unwinnable, but also utterly illegitimate in the eyes of the US public.

The illustration in Figure 19.3 shows how, in the land domain, the capabilities of land forces should be understood as a function of the interaction between one's own capabilities and those of the adversary, the physical and human terrains, and the information environment.

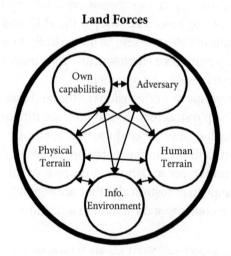

Fig. 19.3 Land forces capabilities in the land domain model

Towards a Versatile Edge: Securing Land Warfare Capability

Against the backdrop of the multidimensional demands placed on land forces in contemporary and future operational environments, the development of land warfare capabilities will require a conscious multi-pronged approach toward gaining a versatile edge on tomorrow's battlefields. In turn, this concerns both the build-up or construction of capabilities and how they are deployed and utilized in future conflict.

We argue that the achievement of *versatility* should be a crucial aim of contemporary land forces. To retain a competitive edge, the land forces of tomorrow must be capable of resolving a wide spectrum of tasks, and of providing utility across a complex operational landscape consisting of innumerable thinkable situations and circumstances. This is not to say that certain components or parts cannot be considered in narrow focus—in fact, this might even be a requirement—but versatility must be the main point of departure for land forces at the organizational level.

Versatility builds on two *interrelated and mutually reinforcing* qualities of a military organization (Figure 19.4). These are *adaptability* and *flexibility*, which together compose the underlying preconditions for truly versatile land forces. In simplest terms, adaptability concerns the ability of organizational change to efficiently deal with a new situation. Adaptability is a property

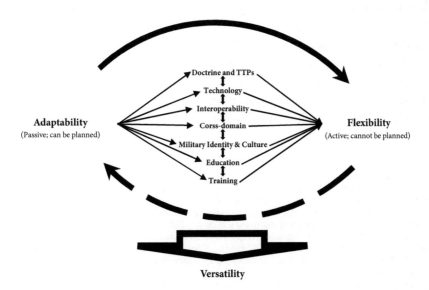

Fig. 19.4 The integrated versatility model

of the organization, a capacity that can be planned and constructed, and concerns the development of land force capabilities. It can also be termed a passive property, in that it enables a spectrum of possible actions, rather than constituting action *per se*. Flexibility, on the other hand, denotes the active utilization of the options granted by adaptability.[3] It constitutes the capacity for active adaptation in the face of unforeseen circumstances and therefore cannot be planned (although capacity for flexibility can be developed in preparation). Flexibility is therefore an acquired quality that is ready to be executed when needed, as new conditions or situations require change in actions and behaviours. Whilst the two concepts may be independent in theory, we argue that adaptability is a *de facto* precondition for flexibility. If land forces lack adaptability, this will at best delimit and at worst deny them the ability to be flexible. A highly adaptable force will be more flexible, and in turn, a flexible force will be more adaptable.

We envision adaptability and flexibility to develop through seven interrelated factors: (1) doctrine and tactics, techniques and procedures (TTPs), (2) technology, (3) interoperability, (4) cross-domain ability, (5) military identity and culture, (6) education, and (7) training. Doctrine and tactics, techniques and procedures (TTPs), refers to conceptual constructs, including instructions and guidelines in different doctrines and handbooks, along with tactics, techniques, and procedures outlined in these, as well as their execution in military practice. Technology denotes how different technologies are adopted, implemented, and utilized in the force. Interoperability and cross-domain ability refer to how well-prepared and experienced land forces are to operate jointly with others across borders, functions, and domains. Military identity and culture refers to the traditions and culture that foster and create innate beliefs and identities among military staff, their self-perceptions, and the sets of values, conventions, or social practices associated with being an officer or soldier.[4] Finally, education concerns the process through which individual members of the force are taught, trained, and learn, whilst training

---

[3] The view on flexibility presented here draws on Finkel's argument that various historical attempts to prevent surprise on the battlefield, through research and intelligence gathering, have consistently failed. Thus uncertainly about future attacks and the course of action an opponent will follow are factors which defence forces have to live with and manage, by building sufficient conceptual/doctrinal, organizational and technological, and cognitive and command and control flexibility into military organizations so as to be able to deal with surprises during an ongoing war (Meir Finkel, *On Flexibility: Recovery from Technological and Doctrinal Surprise on the Battlefield* (Stanford, CA: Stanford University Press, 2011). See also Meir Finkel, *Military Agility: Ensuring Rapid and Effective Transition from Peace to War* (Lexington: University Press of Kentucky, 2020).

[4] Mikael Weissmann and Peter Ahlström, 'Mirror, Mirror on the Wall, Who Is the Most Offensive of Them All?—Explaining the Offensive Bias in Military Tactical Thinking', *Defence Studies* 19, no. 2 (2019), https://doi.org/10.1080/14702436.2019.1599287.

is focused on the force as a collective and the skills it learns to be able to operate as such.

Adaptability in land forces results from the combination of the seven listed factors. The ability of a force to change or be changed is an aggregate function of the seven factors, where some might be positive, and others negative in each unique case. For perfect adaptability, there would in theory be a perfect match between the doctrine and TPPs that regulate and guide the force, the way that technology is utilized, the capacity for interoperability and cross-domain operations, and the way the force is trained, its officers and soldiers educated and the identity and culture, that is, the *esprit de corps*. This is, of course, never the case in reality, and the level of adaptability will by definition be a suboptimal function of the seven factors. In addition to these, the level of flexibility that is achieved will in turn influence the level of adaptability, either in a positive or negative sense.

Since flexibility is an acquired capability building on the adaptability of the forces, the capacity to be flexible can increase (or decrease) depending on how the seven factors are utilized and coordinated in the development of land forces. The ability to successfully shift or transform preparedness to action, or to turn an adaptable force into a flexible one, relies on doctrine and TTPs that allow for, and preferably encourage, flexibility on the battlefield.

Technology can hinder or empower flexibility and needs to be adopted in a way that is non-constraining. For example, the implementation of command-and-control systems can both encourage battlefield flexibility, through shared situational awareness, and establish rigid decision systems that counteract this purpose. Interoperability and cross-domain capability are also key features of flexibility on an increasingly integrated battlefield. The final three factors provide an interconnected set of preconditions for a force's ability to be flexible. A military culture and identity that allows and preferably promotes independent and creative thinking is a key cognitive precept for flexibility, which will always be hampered in the absence of such. In turn, education and training of the force and its members is important to build the capacity and ability to successfully handle new conditions or situations that emerge at short notice. And, of course, education and training are also crucial components in creating the *esprit de corps* referred to above.

In sum, adaptability and flexibility are mutually reinforcing, and both depend on the integrated effects of the seven factors. In turn, it is the integration of adaptability and flexibility that awards land forces versatility. Hence, the integrated versatility model.

It should be noted that the seven factors do not indicate discrete processes but rather an interrelated system of functions. Together, the functions make up the capability of a land force to wage war. They cannot be separated since training builds on doctrine and TTPs, whilst technology is a component in all training as well as doctrine and TPPs. Technology is also important for interoperability, as equipment and systems need to be compatible. Likewise, similarities in identity and culture facilitate interoperability and are affected by education and training. In short, whilst possible to delimit and analyse individually, only together can the significance of these factors be fully understood and their combined impact seen. Moreover, the combined result can build an adaptable force capable of flexibility. In combination, these interconnected concepts form the basis for the versatile edge, the precondition for success on tomorrow's battlefield.

In sum, securing land warfare capability requires a land force with a high level of adaptability and flexibility. Adaptability and flexibility will together create the versatility needed on tomorrow's battlefield. To succeed in developing the required qualities, all seven forces/factors must be included in the development of tomorrow's land forces. Of course, victory may be possible without them, but such victories will be harder to reach, less decisive, less likely, and perhaps far more costly.

Ways Forward: Practical Implications and Agenda for Future Research

The previous sections presented a conceptual understanding of land forces in relation to their roles in different types of conflicts, their location in the operational environment, and the inherent properties of a military organization fostering versatility. We now sketch out some practical implications of the findings presented in the volume for the land domain, and suggest a set of focus areas for future research on land operations and land warfare.

Practical implications. One key challenge for military forces seeking to address the complexity of contemporary and future warfare is finding ways to translate knowledge and experience epitomized in conceptual understandings of future war and ways to fight into practice. Succeeding in this area requires considerable work to develop actionable conceptual approaches to warfare, and to ensure their comprehensive representation in doctrine and handbooks. However, doctrinal development must also be thoroughly implemented in the fabric of the organization, which is frequently a slow and arduous process, particularly if this involves rethinking of established practices and norms. One of the most important areas of doctrine implementation is the education and training of soldiers and officers, at all levels.

In an operational context that increasingly requires international collaboration and collective action, practical implementation presents specific challenges, as national militaries must develop approaches that permit interoperability, whilst simultaneously ensuring that they are optimized for national security needs. These requirements might be largely compatible, but can at times be contradictory. One example is decisions on material procurements, where narrow short-term national defence needs may compete with broader strategic and political considerations as well as interoperability.

Besides practical implications, there is a need for future research in several areas. Whilst we do not claim to present an exhaustive list of themes in need of more research, we have identified the following six areas as topics that stand out as particularly important in relation to the development of versatile land forces.

First, we need to develop a more dynamic understanding of the opponent. This includes better anticipation of the agency of a thinking opponent. It is also paramount to acknowledge the diversity in the types of opponents that land forces are likely to face in future conflicts, which may range from high-technological peer or near-peer adversaries in the form of states, to irregular forces in the form of low-technological but resilient insurgents and terrorists. Considering that the combined Western military forces also include states that are NATO partners but not allies, these states must also develop the means to resist a numerically as well as technologically superior opponent.

Second, urban areas will be an increasingly important arena for future land warfare. Urban operations and warfare should therefore acquire a greater significance in our understanding of the operational environment. Large cities are the centre of gravity for political and economic interaction and although urban warfare is a nightmare that one reasonably hopes to avoid, it is not always possible to choose the battlefield and it is therefore better to prepare thoroughly for this eventuality.

Third, developing functional and efficient command of land forces will be key to achieving a versatile edge. This challenge relates to a number of ongoing transformation processes affecting military forces, including the evolution of existing command concepts, technological shifts most importantly connected with information management, and the considerable challenge of cross-domain synchronization and force integration.

Fourth, emerging and breakthrough technologies are set to have a major impact on, and will possibly revolutionize, warfare. The relationship between technological development and warfare has been and must continue to be a significant area of research. Of particular significance are the proliferation of sensors and unmanned aerial systems, the advent of artificial intelligence, robotics, and automation, and the anticipation of quantum computing, which have the potential to challenge the very nature of land warfare as it is currently conceived.

Fifth, intelligence will be of crucial importance in the future operating environment and is in need of further attention. Closely related to areas two and three, the collection, analysis, and dissemination of intelligence faces new challenges in the contemporary information environment, which places significant demands on capabilities to identify crucial information in the proliferation of available data. Nevertheless, the demand for reliant and actionable intelligence is greater than ever before, and informational advantage is a decisive factor to victory. This is further underscored by the prevalence of grey zone problems; as contemporary antagonistic competition to a very large degree takes place in a spectrum between peace and war, where an interconnected world makes the battlespace difficult to confine geographically.[5]

Sixth, this leads us to the question of what roles, functions, and actions land forces can perform in the grey zone. Land forces are indeed an important component in comprehensive and total defence and the engagement of hybrid threats, yet their potential and utilization in the grey zone remains underexplored.[6] This is equally a question of what land forces could and should be doing, and what they should not do.

Finally, whilst not necessarily a research agenda per se, it is essential to ensure the continual integration of research findings in these areas, together with experience and best practices, in professional military education (PME). Together, these components constitute a formula for developing land warfare capabilities with a versatile edge, ready for tomorrow's battlefields.

---

[5] Niklas Nilsson, Mikael Weissmann, Björn Palmertz, Per Thunholm, and Henrik Häggström, 'Security Challenges in the Grey Zone: Hybrid Threats and Hybrid Warfare', in *Hybrid Warfare: Security and Asymmetric Conflict in International Relations*, edited by Mikael Weissmann, Niklas Nilsson, Björn Palmertz, and Per Thunholm (London: I.B. Tauris, 2021).

[6] Mikael Weissmann, 'Conceptualizing and Countering Hybrid Threats and Hybrid Warfare: The Role of the Military in the Grey Zone', in *Hybrid Warfare: Security and Asymmetric Conflict in International Relations*, edited by Mikael Weissmann, Niklas Nilsson, Björn Palmertz, and Per Thunholm (London: I.B. Tauris, 2021).

Index

Note: The following abbreviations have been used:- 'f' refers to figures/illustrations; 'n' refers to footnotes; 't' refers to tables

A

A2/AD *see* Anti-Access/Area Denial systems
Abrams, General 120
acceleration 30, 32, 33, 160, 170, 327
ACE *see* Allied Command Europe (ACE) Mobile Force
ACR *see* Army of the Czech Republic
Action for Peacekeeping initiative (United Nations) 218
Active Defence 26, 260–1
Active Defense (United States Army) 323
Active Electronically Scanned Arrays (AESA) 163–4
active protection system (APS) 385
Adamsky, Dima 282, 291
adaptability 21, 407, 408, 409, 410
additive manufacturing (AM) 16, 64, 77–9, 80–3, 84, 229
advance C2ISR systems 282, 294
advance groups 264
advantage, positions of 27–8, 33, 38, 40, 91
adversaries 69, 94–5, 128t, 139, 149, 411
  manoeuvre warfare 26, 27, 28, 39, 41
  mission command 43, 48, 50, 56
adversary precision strikes 159
aerial warfighting domains 18, 49, 138, 188, 302, 402, 404
  China and 262, 263
  France 358, 362, 364
  manoeuvre warfare 30, 43
  military health services (MHS) 216, 227
  Russia 280, 281, 284, 294–5
  Visegrád States 378, 381, 382, 388
Aerospace Forces (Russia) 281, 291
AESA *see* Active Electronically Scanned Arrays
Afghanistan 1, 5, 16, 49, 53, 135, 190, 222
  British Army 321, 322, 327, 330, 331, 332, 333, 334, 347

combat logistics 75, 77
command forces 91, 96, 99
France 355, 357, 358, 359, 365
future of land warfare 399, 401, 403
manoeuver warfare 28, 35, 36, 38, 42
military health services (MHS) 215, 217, 223
morality of fighting 210, 211
Russia 283, 288, 298
tactical theories 119, 120, 121, 127, 130
Visegrád States 376, 379
Afghan National Army battalions (*kandak*) 357
Africa 354, 355, 361
African Union 223
aggression 28, 37, 51, 64n2, 121, 139, 360, 385
agility 28, 34, 268, 278, 367
Agincourt, Battle of (1415) 108
Aimpoint 2035 (multi-domain army) (United States) 180, 184
AirLand Battle doctrine (United States) (1982) 4, 28, 181, 323, 324, 399
AI-RMA *see* Revolutions in Military Affairs
'Air-Sea Battle' (United States) 266–7, 340, 349
Ajax tanks 336–7, 343, 344, 346, 350
Aleppo 288, 289, 296
Alexander the Great 30, 109, 115
Alford, Brig. Gen. Julian 131
algorithms 59, 60, 141, 168, 169
al-Hassan, Brigadier-General Suheil Salman 296
Allied Command Europe (ACE) Mobile Force 185n65
Allied Command Europe Rapid Reaction Corps (NATO) 87
Allied Joint Doctrine for Land Operations (NATO) 6

Allied Joint Doctrine for Medical Support (AJP 4.10(C)6) 217
Allied Rapid Reaction Corps (ARRC–ACE/RRC) 325, 336
All-Volunteer Force model (United States Army) 323
ally coordination 18, 296
ALMRS *see* Autonomous Last Mile Resupply System (United Kingdom)
alpine regiments 363
Al-Qaeda 8, 127, 358, 359
AM *see* additive manufacturing
ambulance companies 222
ambushes 109, 118, 147, 166, 358
American Civil War 122
ammunition 63, 72, 157, 343, 364, 384
amphibious operations 121
Antenne Chirurgicale (French) 222
Anti-Access/Area Denial (A2/AD) systems 30, 32, 49, 50, 74, 180, 345
anti-armour projectiles 344
Antill, Peter D. 66
anti-tank mines/weapons 155–6, 158, 263, 308, 309, 314, 317, 329, 344, 345, 381
anti-terrorist operations 281
Apache helicopters 345
APCs *see* armoured personnel carriers
'applicatory method' 17, 234, 236, 237, 240, 242, 243, 245, 248
APS *see* active protection system
AR *see* Augmented Reality
Arab-Israeli Wars 30, 115
Ardant du Picq, Charles 112
Ardis, John A.S. 33
armed quadcopters 154
Armistice Agreements (Israel) 302, 303n2–3
'Armored Army Corps' (France) 353
armoured personnel carriers (APCs)
    Boxers 337, 339, 343, 344, 345, 346, 350
    British Army 328, 337, 338, 339, 342, 350
    Israel 302, 305–6, 307, 308, 309, 310, 313
    Visegrád States 381, 382, 383, 384
armoured vanguard vehicles (VABs) 363, 368
Army 2020 (British Army) 334–41, 347
Army 2020 Refine (British Army) 336
Army Command Standing Order (British Army) 189
Army of the Czech Republic (ACR) 381
Army Reserve 2020 (British Army) 335

ARRC-ACE/RRC *see* Allied Rapid Reaction Corps
Arthashastra (Kautilya) 108, 109
Article IV (Outer Space Treaty) 154
Article V (North Atlantic Treaty) 64n2–3, 65, 66, 67, 68, 69, 70, 71, 77, 80, 228
artificial intelligence (AI) 184, 336, 395, 411
    China and 262, 276
    combat logistics 79, 80–1, 84
    land warfare 2, 3, 7, 8, 9, 13, 15
    manoeuver warfare 30, 32, 33, 36, 37, 42
    mission command 43, 44, 59, 60, 61
    technology 153, 154, 165, 166–9, 170, 171
    urban warfare 126, 134, 141, 142, 143, 144, 149
artillery warfare 63, 110
    British Army 321, 333, 345, 346
    China 262, 263, 269, 270
    France 358, 363, 364
    Israel 302, 303, 305, 307, 309, 310n18, 312–13
    Russia 280, 282, 285, 286–7, 290
    Visegrád States 378t, 382, 384, 387
Art of War, The (Sun Tzu) 108, 109, 111, 113, 117, 128, 262, 271
Art of War in the Western World, The (Jones) 106
asymmetrical warfare 11, 139, 150, 367
Asymmetric approach (China) 267
asymmetric strategies 38
Atlantic Alliance 353
Attacks on Healthcare Initiative (World Health Organisation) 226
attrition warfare 4, 314, 326, 342
    manoeuvre warfare 26, 28, 29, 30, 31, 36, 41
'Au Contact' model (France) 362
audacity 28
Auftragstakti (Germany) 45
Augmented Reality (AR) 142
Australia 162n37, 217, 221, 227
authoritarianism 101n34, 102, 103
automation 31, 32, 388, 395, 412
    combat logistics 78, 79–80, 84
    land warfare 3, 13, 15, 16
    mission command 43, 44, 49, 59, 60–2
    technology 154–61, 170, 171
    urban warfare 142, 143
autonomous aerial vehicles 159

Autonomous Last Mile Resupply System (ALMRS) project (United Kingdom) 73n28, 80n51
autonomous navigation systems 160
aviation medicine 220
Azerbaijan 342

B
Bagnall, General Sir Nigel 323, 324
ballistic missiles 74, 163, 264, 269, 378
Baltic and Balkan States 175, 347, 355, 360, 399
BAOR *see* British Army on the Rhine
Barkawi, Tarak 208
Bartles, Charles 281, 282, 283, 285–6, 288
Bartov, Omer 197
base-of-fire teams 157
base security 295
battalions 222, 262, 264, 335, 342
    command forces 91, 94, 97, 100, 102
    France 357, 359
    Israel 307, 308, 309, 310, 313, 314
    Russia 284, 288, 295, 296
    Visegrád States 378, 379, 381, 383, 386
battle confrontational tension/fear 198
battle-planning and management 1, 5, 6, 13, 73, 110, 182, 387
    command forces 90, 93, 94–5, 99, 102, 103
    manoeuvre warfare 31, 32, 33, 41
    mission command 43, 52, 58
    transformation of 9–11
Battle Studies (Ardant du Picq) 112
Belarus 7
Belgium 221
Bellavia, David 144
Berenhorst, Georg Heinrich von 110
Berkovich, Ilya 195
Berlin Wall (1989) 185, 325
Bhuta, Nehal 155
Biddle, Stephen 40
big data analytics 31, 52, 134, 143, 150, 228
bio-engineered biological agents 226
biometrics 142, 143
biotechnology 229
'black boxes' 80
'blitzkrieg' operations (Germany) 115
blood transfusions 222, 228
'Blue Brigade' (China) 270
Blue Force Tracker 57–8, 329

body counts 406
boredom: operational 204–5
Bosnia 326, 375
Bosser, General Jean-Pierre 362
Boulding, Kenneth E. 74
Bowman digital radios and communications systems 330, 339
Boxers (armoured personnel carriers) 337, 339, 343, 344, 345, 346, 350
Boyd, Colonel John 26, 45, 112, 122
Bradford, Jeffrey P. 66
'brain control' (制脑权) 272, 273
brain–nerve–machine interfaces 229
Brazil 8
breaching vehicles 160
brigades 144, 189, 222
    British Army 325, 328, 329, 333, 334, 335, 342, 343, 345–6, 349
    China and 262, 264, 269, 270, 276
    command forces 91, 93, 94, 96, 97
    France 362, 363
    Israel 308, 309, 310, 311, 313n23, 315, 318, 319
    technology 162, 163, 164, 170
    Visegrád States 374, 377, 378t, 379, 386
Brimstone missiles 388
British Army 12, 18, 87, 93, 101, 117, 321–2, 397–8
    capability development/counterinsurgency 327–34, 340, 350–1
    interoperability 175, 185, 186n75, 189, 321, 335
    manoeuvre warfare 321, 322–7, 329, 333, 339, 340, 341–9, 350, 351
    organizational change (Army 2020) 334–41
    see also United Kingdom (UK)
British Army Medical Services 217
British Army on the Rhine (BAOR) 185
British Defence Doctrine 324, 327
British Special Forces 347
Buddy First Aid Course (United Nations) 223
budgets *see* defence spending
Building Stability Overseas Strategy (British Army) 348
Bundeswehr 134
Burkhard, General 369
Byzantine Empire 108, 109–10

C

C2ISR systems 282, 294

C4ISR technology 296, 298

Calibrated Force Posture 180

camaraderie 196, 197, 205

camouflage 229, 387

campaigns *see* operational environments

Canada 182, 217, 221, 227

capability development 19, 33, 36, 67, 327–34

capacity-building 18, 20, 224, 296, 348

captains 99, 100, 101, 381

cardiac signatures 143

career structure/progression 100–2

Caribbean 361

Carleton-Smith, General Sir Mark 189–90, 347

Carmel Armoured Combat Vehicle 142

carpet bombings 406

Carson, Austin 297

Carter, General Sir Nicholas 347

casualties 16, 26, 222, 294, 295

cavalry charges 117

CBRN *see* chemical biological, nuclear, or radiological medicine

Central African Republic 223

Central Group of Forces (Soviet Union) 380

centralized command 83, 381

Chad 358

Challenger 2/3 main battle tanks 325, 342, 346

Challenger Life Extension Programme (LEP) 343

character of war 106

Chasing Multinational Interoperability (RAND) (2020) 176, 177

Chechnya 210, 298

chemical biological, nuclear, or radiological (CBRN) medicine 220, 224, 225

China 109, 136, 137, 164, 180, 221

  future of land warfare 395, 397, 399

  intelligentized warfare 259, 260, 261–4, 266, 267, 268–73, 274, 276, 277, 278

  land warfare 3, 7, 8, 10, 12, 17

  manoeuvre warfare 30, 35, 36, 39

  mission command 43, 48, 49

  Multi-Domain Operations (MDO) 257, 261, 265–8, 273, 278

  see also People's Liberation Army (PLA)

Chinese Civil War 260

Chinook helicopters 364

Chirac, Jacques 369

chivalry 108

Chodoff, Elliot P. 204

Christine de Pisan 119

Churchill, Winston 41

CIFS *see* Common Indirect Fire System

city-avoidance doctrine 129

civilians 10, 150–1, 323, 370

  Israel 304, 305, 314

  military health services (MHS) 216, 218, 221, 224, 227, 229

  technology 157, 160, 164

Clark, Brigadier John 343

Clausewitz, Carl Philipp Gottfried von 5, 29, 45, 81

  tactical theories 106, 107, 109, 111–12, 113, 117, 119, 122

climate change *v*, 9

clinical registries 223

close-air support 284

closed data systems 167

close fight 170

Close Quarter Battle (CQB) 125n2

cloth model exercises 95

cloud storage 32–3, 52

coastal cities 132

coded communication 146

Codner, Michael 178, 184

'cognitive domain operations' (认知域作战) 269, 271–3, 278

cognitive scaffolding 114

'cognitive security' (国家认知空间安全) 272

'COIN slide' (ISAF Joint Command briefing) (2009) 210

Coker, Christopher 36, 205

Cold War/post-Cold War 66, 97, 129, 230, 354, 373, 399, 401

  British Army 322, 324, 325, 332, 335, 350

  combat logistics and 68–9, 72, 84

  interoperability 175, 184, 185n62–6, 185n68, 186, 190, 191

  land warfare 1, 3, 4, 12, 16

  Visegrád States 376, 378, 389

Cold War Reforger (Return of Forces to Germany) 183

collaborative combat 368

collaborative working 182

collective defence 64, 65, 181, 360, 369, 385, 389
Collins, Randall 196, 202, 205
colonels 100
colonial troops ('colo') (France) 354n2, 355, 356, 363
combat-engineer vehicles 302
combat experience 274, 277–8
combat logistics 12, 13, 14, 56, 63n1, 76, 77, 85, 106
   AM/Robotics/AI 80–2, 83n56–7
   applications of 68–80
   character/scope of 65–8
   North Atlantic Treaty Organization (NATO) 64–72, 73n26, 74, 75, 80, 81–3, 84
combat motivation 16
combat training centres (CTCs) 268, 277
combined arms operations 4, 258, 358, 359, 362
   Visegrád States 374, 375, 378, 379, 380–1, 382, 386
Combined Joint Expeditionary Force (Europe) 189
COMINT see Communications Intelligence
command see mission command
command and control systems (C2) 9, 13, 14, 70–1, 367
   British Army 324, 349
   China and 274, 276
   manoeuvre warfare 31, 32, 40
   Russia 279, 282, 291
   Visegrád States 376, 388, 393
'commander's intent' 92
command forces 87–9, 102–3, 410
   people within 99–102
   processes of 93–5
   products of 91–3
   purpose of 89–91
   structures of 96–7
   systems of 97–9
command orders 91–3, 102
command post exercises (CPXs) 95
Command Post of the Grouping of Forces (Russia) 291
commercial contracted logistic firms 72
commercial health services 224
Common Indirect Fire System (CIFS) 369
communication and information systems (CIS) 97–8, 103

Communications Intelligence (COMINT) 134
communications technology 31, 43, 56–8, 62
Communist Party of China (CPC) 259, 260, 261, 272, 274, 277
compatible interoperability 188
complexity of war 90
'Comprehensive Approach' (British Army) 330, 340
'Comprehensive Operational Planning Directive' (COPD) (NATO) 93
'Comprehensive Preparation of the Operational Environment' (NATO) 93
computed tomography (CT) 220, 223
computer technology 5, 30, 141, 167, 270
comradeship 203, 205
concealment 40, 286
concept drills 93
concentration 28
conflict intensity 20, 400f, 404, 405
conformity 51–2, 59
confusion 114, 117, 118, 395
Connor, James et al. 206–7
conscious objectors 201
conscription systems 199, 200, 369, 370
   Visegrád States 375, 377, 381, 385, 389
control 70–1, 136
convergence 35, 62, 145, 180
conviction 272
coordination 50, 70–1
COPD see 'Comprehensive Operational Planning Directive' (NATO)
Cordesman, Anthony 38
Cornish, Paul 334
corruption 274, 276–7
counter-fire 168
counterinsurgency (COIN) 1, 38, 53, 120, 210, 284, 290
   British Army 327–34, 340, 350
   France 358, 359
   future of land warfare 401, 402
cover 4, 40, 265
cover-ups 197
COVID-19 pandemic 16, 183, 215, 218, 225–6, 227, 230, 396
CPC see Communist Party of China
CPXs see command post exercises
CQB see Close Quarter Battle
Crécy, Battle of (1346) 108

Creveld, Martin van 45, 65
crewed/remotely-crewed systems 156
Crimea: annexation of (2014) 336, 375, 385
critical war studies literature 208
cross-domain/cross-conflict-spectrum
    fighting 7, 10, 11, 126, 139
  China 267, 268
  future of land warfare 408, 409, 411
cross-European border movement 69, 72
CSP *see* Warrior Capability Sustainment
    Programme
CTCs *see* combat training centres
Cuban missile crisis (1962) 51, 279
Cuche, General Bruno 365
Cunaxa, Battle of (401 BC) 109
Curtis, Andrew 396
cyberattacks 10, 31, 137
cyber capabilities 2, 31, 145, 341
cyberspace warfighting domains 31, 33, 85,
    112, 169, 188, 404
  China 266, 270, 271
  land warfare 6, 7, 8, 9
  military health services (MHS) 225, 227,
    229
  mission command 43, 49, 50, 55, 57
  urban warfare 126, 128, 130, 134, 138,
    146, 147
Cyrus the Younger 109
Czechoslovakia 374, 375, 380, 382
Czech Republic 20, 374, 375, 377t, 379,
    383t, 384, 389, 398
Czechoslovakian People's Army 379

D

Dangerous Cargo 72
data networking and management 31, 165,
    229, 262, 276
Dayan, Moshe 117
'death of distance' 7, 126
decentralization 27, 29, 33, 77, 95, 184
  mission command 44, 46, 47, 51, 54, 55,
    56, 57, 62
deception 29, 114, 116, 159, 166, 286, 338,
    387, 395
decision-making 82, 169, 228, 272, 280, 333,
    347, 381, 394, 409
  command forces 88, 94, 98n27–28, 99,
    103
  interoperability 173, 182
  land warfare 27, 29, 31

  mission command 53, 58, 59, 60
decision trees 262
deconflicted interoperability 188
Defence in a Competitive Age, CP 411 (United
    Kingdom) 187, 343, 345
defence in depth 4
Defence Forces (Israel) 12
Defence Health Agency (United States) 224
Defence Lines of Development
    (DLODs) 188n87
Defence Logistics Organisation (British
    Army) 186n75
Defence, Ministry of (France) 353, 354
Defence, Ministry of (United Kingdom) 19,
    76, 80n50, 98, 130
  British Army 321, 328, 330, 332, 335, 341,
    344
defence planning and policies 5, 38, 173,
    186, 276, 321, 325, 326
Defence Science and Technology Laboratory
    (DSTL) (United Kingdom) 73n28
defence spending 69, 182, 185, 312, 331,
    333, 341, 393
'Defence Studies' 101
Defence White Paper (United Kingdom)
    (2003) 174
Defense, Department of (United States) 120,
    176, 327, 328
defensive combat 4, 127, 128t
delegation 83
*Delivering Security in a Changing World
    Defence White Paper, Cm 6041-I*
    (United Kingdom) 174, 328
demobilization 374, 377
democratic warrior 199
Deng Xiaoping 258, 259
deniability 31, 292, 293, 296–7
Denmark 189
dentistry 221
Deployed Hospital Care (DHC) 218, 221,
    222
deployment and redeployment 6, 93, 195,
    337, 348, 367, 376
  combat logistics 66, 69, 77
  future of land warfare 398, 400, 402
  Israel 303, 308, 316
  military health services (MHS) 221, 225,
    230
  Russia 280, 283, 284, 285, 289, 294, 296,
    297, 298

depth attack/defence groups 265
DePuy, General William 323
De Rei Militari (Vegetius) 108, 109
deterrence 296–7
DHC *see* Deployed Hospital Care
digitized command systems 5, 393
'Dinner Out' operation (France) 358
diplomacy 402
directed energy weapons 169
direct fire zone 157, 269, 306
dispersion 4, 40, 115, 338
'distancing' 198
diversity 51–2
divisions and divisional orders 336, 346, 362
    command forces 92, 97, 102
    Israel 308–9, 319
    technology 163, 170–1
    Visegrád States 374, 377, 378t, 379
Djibouti 361
DLODs *see* Defence Lines of Development
DNA technologies 143
Donbass 341, 385–6
Dorman, Andrew 334
drones 9, 57, 75, 153, 198, 229, 270, 388
    manoeuvre warfare 31, 37
    Russia 286, 295
    urban warfare 140–1, 146
DSTL *see* Defence Science and Technology
    Laboratory
Dubai 8
Dvornikov, Colonel General Aleksandr 287,
    290

E
'Eastern' warfare 81–3, 109, 110, 113
EBAO *see* 'effects-based approach to
    operations' (United Kingdom)
EBO *see* Effects-Based Operations (EBO)
    (United States)
Ebola breakout (2014) 16, 215, 224
'echeloned maneuver' 34
economic sanctions 402
education *see* military education and
    training
'effects-based approach to operations'
    (EBAO) (United Kingdom) 330, 337
Effects-Based Operations (EBO) (United
    States) 328, 330
eFP *see* enhanced forward presence
Egypt 117

Eichler, Maya 210
'Eight Rules of Urban Warfare and Why
    We Must Work to Change Them, The'
    (Spencer) 133–4
Eisenhower, Dwight D. 93
election monitoring 137, 355
electrical medical equipment 229
electromagnetic spectrum (EMS) 6, 30, 31,
    137, 229, 338, 387
    China 266, 270, 271
    mission command 43, 49, 50, 57
Electronic and Network Warfare Group
    (China) 271
electronic warfare 31, 33, 49, 112, 134, 227,
    387
    China 264, 269, 270–1, 272, 273, 274
    technology 159, 162, 169
electro-optical sensors 157
emissions analysis 168
EMS *see* electromagnetic spectrum
endurance 64, 65, 66, 68, 69, 84
enemies *see* adversaries
enhanced forward presence (eFP) 360, 375,
    385
Enlightenment, The 110
environmental medicine 220
equalization 30
escalation management 18, 294, 296–7
espionage 113
Estonia 136
European Union (EU) 72n25, 136, 186, 223,
    227, 326, 331, 375
Exercise Defender-Europe 20 (military
    exercise) 183
Exercise Warfighter 19-4 (Command Post
    Exercise) 189
exoskeletons 79, 229
expeditionary operations 2, 12, 16, 19, 279,
    288, 398, 401
    British Army 325, 328, 330, 333, 337
    China 261, 267, 277
    France 354–5, 356, 357, 358, 362, 363,
        364, 371
    interoperability 175, 186–7
exploitation 28
explosive ordnance disposal 79
'expose/attack' company 317

F
facial/ocular measurements 143

Falklands conflict 167n58, 185n68
Fallujah (Iraq) 38, 132, 133, 142, 144
Fallujah, Second Battle of (2004) 133
Farrell, Theo 329, 333
Fennell, Jonathan 205
feral cities *see* urban warfare
FHP *see* Force Health Protection
fifth-generation warfare 2
fighting in built-up areas (FIBUA) 125n2
'fighting smart' 29
'fight to move'/'move to fight' 325
fingerprints 143
Finkel, Meir 56, 408n3
Finland 364
Finnish Army 100
firepower 30, 31, 134, 269, 282, 349, 395
    tactical theories 114, 115, 118, 120, 121,
    122
    technology 158, 165, 168, 169–70
firepower groups 265
Fire Weaver ('networked sensor-to-shooter
    system') 141
first aid 222
first-generation warfare 3
First World War 4, 73, 111, 121, 217
    morality of fighting 195, 196, 200
'Five Incapables' (Xi Jinping) 274, 275
flagship exercises 269
Fleet Marine Force Manual 1 (US Marine
    Corps) (1989) 28
FLET *see* forward line of enemy troops
flexibility 21, 55, 56, 407, 408n3, 409, 410
FLOT *see* Forward Line of Own Troops
FMFM-1 *Warfighting* 122
Foch, Marshal 366
'fog of information' 135
'fog of war' 5, 9, 36, 45, 48, 53, 166
Follow-On Forces Attack (NATO) 324, 326,
    345
foot patrols 366
force employment 2, 127, 128t
Force Health Protection (FHP) 220
force interoperability 176
force posture 67, 180
force protection 170
force ratios 406
force-to-space ratios 41
'force transformation' 322, 328, 331, 354,
    355, 376
Foreign Legion 355, 361, 363

fortification tactics 110
fortified trench lines 4
'forward edge of battle' 73
forward ground operations 292
forward line of enemy troops (FLET) 165
Forward Line of Own Troops (FLOT) 161
forward resuscitative surgical system (US
    Marine Corps) 222
forward surgical teams (US Army) 222
4CISR technology 280
'Four Modernizations' (China) 258
fourth-generation warfare 3
Foxhound (protected mobility patrol
    vehicle) 346
fragging 197
France 117, 138, 306, 337, 353–4, 355n3
    force structure and capabilities 361–2,
    363f, 364
    French Army doctrine 364, 365t, 366,
    367t
    interoperability 188, 189
    land warfare 3, 12, 19
    military health services (MHS) 221, 222
    military interventions 355–6
    network-centric/high intensity
    warfare 354, 368–9
    operational experiences 356–61
    provision of soldiers 369–71
Frederick the Great 110–11, 115
'free-market' economies 69
French Army 364, 365t, 366, 367t, 398
French Guyana 361
FRES *see* Future Rapid Effects System (FRES)
    (United Kingdom)
Friedman, B.A. 193, 209
FROG rockets 99
frontier defence groups 265
frontline 73, 74
frontline attack groups 265
Fry, Royal Marine Brigadier Robert 326
FT-01-05 (doctrinal documents)
    (France) 365t, 366, 367t
fuel supplies 63, 72
Fuller, J.F.C. 29
'full-spectrum' warfighting 332
Fundamental Threat (Israel) 303, 305, 312,
    319
Funk, General Paul 189
futility of war 89
Future Force 2020 programme 334

Future Rapid Effects System (FRES) (United Kingdom) 328, 331, 332, 350
'futuristic fad' phenomenon 17

G
Gabon 361
Gaius Julius Caesar 109
Galvin, General John R. 42
garrison healthcare 216, 221, 224
'General Review of Public Policies' (*Révision Générale des Politiques Publiques* (RGPP) (France) 354
generals 100
Geneva Convention 216, 226
GEOINT *see* Geospatial Intelligence
geometric tactics 110
Georgia 283, 298, 401
Geospatial Intelligence (GEOINT) 134
Gerasimov, General Valeriy 279, 284, 290–1
German Army ('Bundeswehr') 65, 96
Germany 4, 17, 30, 44, 221, 307, 374, 405
    France and 357, 364
    interoperability 183, 185
    tactical theories 115, 117, 118
Ghost Unit (experimental unit) (Israel) 318
global health engagement 220
globalization 69
Global Trends 2040 (National Intelligence Council) 132
Global War on Terror 327, 358, 359, 360
Glonass navigation system 282
Google Maps 146
GPS-tracking 57, 141, 143
Graham, Stephen 138
Grau, Lester 281, 282, 283, 285–6, 288
Gray, Colin S. 119
Gray, J. Glenn 203
great-power competition 43, 62
Greece 364
green men (China) 10, 136, 137
grey zones 15, 49, 51, 351, 361, 401, 412
    urban warfare 126, 127, 130, 131, 136–9, 148, 149
Griffon (APC) 368
ground manoeuvre surgical groups (UK) 222
Ground Moving Target Indication 163
group anonymity 198
Groupements Tactiques Interarmes (GTIA) 359, 362

groups 264, 265, 269–70, 271
groupthink 59
GTIA *see* Groupements Tactiques Interarmes
guerilla forces 115, 301, 312, 313, 319, 359, 406
Guide to Tactics (Clausewitz) 112
Gulf War (1990–1991) 5, 30, 115, 260, 326, 399, 401
gunpowder 110
Guthrie, George James 217

H
'hacking' 81
Hadès Force (France) 355
Halutz, Dan 318n38
Hamas 315
Hart, Basil Liddell 29
Headquarters, Allied Forces Southern Europe 88
Healthcare in Danger project (International Committee of the Red Cross) 226
Healthcare Management and Occupational Safety and Health, Division of (United Nations) 218
health promotion 216
health services *see* military health services
Health Services Support (HSS) 215, 216, 218, 226–30
Heavy Equipment Transporters 345
helicopters 329, 345, 364
HEMEX-Orion project 369
Henriksen, Rune 198, 203, 205
Herzegovina 375
Hezbollah (2006) 146, 296, 313, 314, 319, 357, 397
High command 45
high fidelity layered sensors 154, 163–6
high fidelity radar 163
high-/low-intensity fighting 302
    British Army 339, 341
    France 354, 357, 363, 366, 367, 368–9, 370, 371
    future of land warfare 396, 397, 398, 401, 402–3
    Visegrád States 376, 387
High Mobility Artillery Rocket System (HIMARS) launchers 384, 387
high-mobility tactical transport 263
high-risk targets 360
high-value tasks 295

HIMARS *see* High Mobility Artillery Rocket
System launchers
HNS *see* Host Nation Support
HOMAR ('Lobster') programme 384
home front 194, 205, 210, 212, 213n137
Homer 107
Hong Kong 133
hospitals and hospital ships 216, 220, 221,
224
Host Nation Support (HNS) 72–3, 84
House to House: A Soldier's Memoir
(Bellavia) 144
house-to-house fighting 130, 147
Houthi rebels (Yemen) 146
howitzers 384, 385, 387
HSS *see* Health Services Support
human agency 107, 110
human augmentation 30, 79–80
Human Intelligence (HUMINT) 134
humanitarian and disaster relief operations 16, 220, 223, 227, 288,
297
humanization: military power 366
human–machine teams 160
HUMINT *see* Human Intelligence
Hungarian Defense Forces 384
Hungarian People's Army 377
Hungary 20, 398
Visegárd States and land warfare 374,
375, 377t, 379, 380, 382, 383t, 385, 387,
389
Hunter, John 217
hybrid warfare 10
future of land warfare 398, 412
manoeuvre warfare 30, 31, 35
urban warfare 136, 137, 138, 149
'hyperactive battlefield' 33
hypersonic glide vehicles 163

I
ideology 196, 197, 202, 205, 208, 212
IDF *see* Israel Defence Forces
Idlib Provinces 289
IEDs *see* improvised explosive devices
IFVs *see* infantry fighting vehicles
Imagery Intelligence (IMINT) 134
Imperial Japanese Army 260
improvised explosive devices (IEDs) 142,
331, 359
India 8, 109, 217, 221

Indian Ocean 361
indirect fire 29, 73, 74, 224, 226, 286, 339,
381, 382
Indonesia 8, 132–3
infantry fighting vehicles (IFVs) 263
infectious disease 220
infiltration techniques 4
influence operations 112
information environment 97–8, 103, 229,
281, 324, 330
China and 262, 270, 271, 272, 273, 276,
278
France 354, 367, 368
future of land warfare 405, 406
land warfare 6, 7, 8, 10, 11, 13, 20, 394,
395
manoeuver warfare 30, 31, 32, 33
mission command 43, 44, 47, 49, 50,
52–6, 57, 58
urban warfare 130, 137, 138, 147, 149
Informationized Warfare 261, 263–73, 275,
276, 278
'Information Manoeuvre and Unconventional Warfare' (British
Army) 341
'ink spot' deployment 77
Instruments of Military Medical Care 218,
219t
insurgency 11, 15, 38, 42, 53, 120, 411
urban warfare 127, 131, 145–8, 151
Integrated Action (British Army) 340, 341,
348, 350
integrated interoperability 188
Integrated Operating Concept (IOpC)
(Ministry of Defence) (UK) 19, 187,
188, 321, 340, 348, 350
Integrated Review (United Kingdom)
(2021) 16, 174, 175, 187
British Army 340, 341, 342, 343, 345, 347,
348, 351
integrated versatility model 12, 21, 394,
407f, 408, 409
intelligence *see* military intelligence
Intelligence, Surveillance and Reconnaissance Brigade (British
Army) 341
intelligence, surveillance and reconnaissance
(ISR) 140, 141, 148, 170, 270, 282, 387
intelligence, surveillance, target acquisition
and reconnaissance (ISTAR) 32

British Army 329, 333, 335, 341, 347
technology 163, 165, 166
Intelligentized Warfare (智能化作战) 261–2, 276, 277
intelligentized warfare 261–2, 263–73, 276, 294, 394
'intent-based mission command' 33
Interim Force in Lebanon, United Nations (UNIFIL) 356–7
International Committee of the Red Cross 226
International Humanitarian Law 226
International Security Assistance Force (ISAF) (NATO) 357
interoperability 51, 61, 173–6, 190–1, 217, 227, 327, 379
  British Army 175, 185, 186n75, 189, 321, 335
  combat logistics 67n13, 69, 70–1, 72, 81, 83
  France 188, 189
  future of land warfare 396, 404, 405, 408, 409, 410, 411
  land warfare and 7, 12, 15–16, 20
  Multi-Domain Operations (MDO) 175, 179, 180, 181, 182, 183, 184, 190
  systemic competition 179–84
  understanding of 176–7, 178t, 179
  United Kingdom's approach to 184, 185n62–6, 185n68, 186n73–7, 187–90
interpersonal solidarity 196
intra-service interoperability 188
IOpC *see* Integrated Operating Concept
Iran 7, 49, 287, 291, 294, 296
Iraq 1, 5, 16, 259
  British Army 321, 322, 327, 329, 330, 331, 332
  command forces 91, 96, 99
  France and 355, 359, 365
  future of land warfare 399, 401, 403
  manoeuvre warfare 28, 30, 35, 36, 38
  military health services (MHS) 215, 217, 222, 223
  mission command 49, 53, 56
  tactical theories 119, 120, 127, 130
  urban warfare 132, 133, 142, 144, 146
  Visegrád States 376, 379
Iraqi Counter Terrorism Service (TF Narvik) 359
Iraq, Invasion of (2003) 115, 328

Iraqi Regular Army 201, 359
Irastorza, Elrick 361
irregular warfare 37, 39, 42, 145, 394, 396
ISAF *see* International Security Assistance Force
Islamic State (ISIL/ISIS) 8, 38, 41, 127, 146, 359
ISR *see* intelligence, surveillance and reconnaissance
Israel 3, 12, 26, 30, 142, 162n37, 319, 397
  evolution of ground forces 305–9, 310n18, 311–18, 319
  historical role of ground forces 301–2
  National Security Strategy 303, 304–5
  perception of threat 301, 303, 304n6
  tactical theories 115, 117
  urban warfare 134, 136
Israel Defence Forces (IDF) 18, 142, 201
  role of ground forces 305, 307–9, 313n23, 314, 315, 317, 318n38–9, 319
Israeli Air Force 316
ISTAR *see* intelligence, surveillance, target acquisition and reconnaissance
Italy 221, 364
Ivory Coast 357, 358, 361

J

J3 ('operations' directorate) (France) 356
JADO *see* Joint All Domain Operations
Jaguar (wheeled reconnaissance tank) 368
Jakarta (Indonesia) 132–3
Janowitz, Morris 195, 198
Japan 7
Jewish defence organizations 302
JFC-NF *see* Joint Forces Command, Norfolk
JOA *see* Joint Operational Area
Joint All Domain Operations (JADO) (NATO) 181, 190
joint approach: defence 186n73–6
Joint Doctrine and Concepts Centre (United Kingdom) 327
Joint Doctrine Publication 0-01 (UK Defence Doctrine) 187
joint exercises 156n13, 188, 190
Joint Force (NATO) 70
Joint Forces Command, Norfolk (JFC-NF) 70, 84
joint land–air–sea operations 6
Joint Medium Weight Capability concept (British Army) 346

Joint Operational Area (JOA) (NATO) 68,
    71
Joint Rapid Reaction Force (United
    Kingdom) 327
Joint Support and Enabling Command
    (JSEC) (NATO) 70, 84
Jomini, Antoine-Henri 111, 112, 121–2
Jones, Archer 106
Jordan 221
JP 3-06 (Joint Urban Operations) (2013) 133
JSEC *see* Joint Support and Enabling
    Command (JSEC)
'just cause' 194, 209

K
Käihkö, Ilmari 196
kandak see Afghan National Army battalions
Kaspersen, Iselin Silja 207
Kautilya 108, 109
Keegan, John 200, 201
Keene, Shima D. 33
Kellett, Anthony 204
key leader engagement 189
Khe Sanh, Battle of (1968) 115–16
Khrushchev, Nikita 279
Kilcullen, David 120, 132
'kill chain' 33, 75
kinetic/non-kinetic warfare 261, 281, 330,
    339, 360
  land warfare 395, 401, 402, 403
  urban warfare 129, 130, 136, 149
King, Anthony 39, 47, 198, 199
King, Professor Anthony 130
King, Admiral Ernest 173
Kiszely, Major General John 324, 325
KMW 368
Kochavi, General Aviv 317, 318
Kofman, Michael 282, 283
Konaev, Margarita 143
Korean War 30, 129, 262
Kosovo War (1999) 5, 87, 88, 190, 224, 280,
    312, 326, 357, 375, 399
Krab (self-propelled howitzer) 384, 387
Kramer, Eric-Hans 54
Kurdish militias (Iraqi Peshmerga/Syrian
    YPG) 359
Kurdish People's Defense Units (YPG) 289
Kurds 206, 289
Kuwait War (1991) 87, 312

L
'Land Action Force' (France) 353
Land Environment Tactical Communica-
    tion and Information Systems (LE
    TacCIS) 339
'Land Logistic Force' (France) 353
Land Operations (British Army) 321, 339,
    350
Land Power doctrine (United Kingdom) 189
land warfare 43, 49, 138, 189
  C21st dynamics 394–8
  continuum of 398–9, 400f, 401–3
  current/future challenges 7–9
  future research 410–12
  generational development of 3–6
  location of operations 403, 404f, 405, 406f
  securing capability 407f, 408n3, 409–10
  Visegrád States 377t
Land Warfare Research Group (LWRG)
    (Swedish Defence University) 12
'large-scale contingent operations'
    (LSCO) 226, 230
Larrey, Baron Dominique Jean 217
lasers 226
'last tactical mile' 14, 69, 73–5, 76, 77, 80, 84
layered air defence systems 345
layered precision fires 15, 154, 161–3, 171,
    395
'Leader-follower' systems 79
leadership 88
Lebanon 314, 315
Leclerc Main Battle Tanks 364, 368
legacy systems 81–3
legitimacy 39, 194, 209, 210, 212, 402, 403
Leishman, William 217
Leo IV, Emperor of Byzantine Empire 109
Leonhard, Lt. Col. 126
LE TacCIS *see* Land Environment Tactical
    Communication and Information
    Systems
Letterman, Jonathan 217
Light Equipment Transporters 345
Li Jang 110
limited war 37, 42, 401, 402
Lind, William 26
linear tactics 107, 110–11
local forces/militias 39, 296
'local wars' 261
'logistic disaggregation' 75, 77
logistics 33, 173, 309, 362, 367, 396

'logistic-strategic nexus' 82
loitering munitions 162n38, 163, 338
long-distance weapons systems 1, 2, 5, 50, 303
longer range precision fire 162, 165
Longmore, Sir Thomas 217
Long Range Precision Fires (LRPF) 396, 397
long-range surface-to-air/surface-to-surface fire 266, 270
long-range weapons systems 273, 338, 387
Lonsdale, David J. 66
'Loss of Strength Gradient' (LSG) theory 74
Lower Echelon Leadership 274–5
lower echelon planning 167
low-intensity fighting see high-/low-intensity fighting
LRPF see Long Range Precision Fires
LSCO see 'large-scale contingent operations'
LSG see 'Loss of Strength Gradient' theory
LWRG see Land Warfare Research Group
LXXIV Infantry Corps (Germany) 197
Lykke Model 119
Lynn, John A. 66, 110, 211

M
McChrystal, Stanley 48
Machiavelli, Niccolo 109, 118, 119
machine learning (ML) 7, 8, 9, 32, 42, 59, 79, 134, 336
McNair, General Leslie 91
Mahan, Dennis Hart 121
Main Battle Tank fleet (British Army) 343
Main Events List (MEL) 95
Main Ground Combat System (MGCS) 268
mainline groups 265
'Major Engagement' contract (France) 369
Major Joint Operation (NATO) 369
Malešević, Sinisa 199, 203, 206
Mali, Republic of (2013) 1, 223, 224, 337, 358, 361, 364
Maneuver in Multi-Domain Operations (United States Army) 34
Manoeuver Warfare Doctrines 399
manoeuvre warfare 25–9, 41, 43, 133, 270, 396, 397
    alternative perspectives 35–42
    artificial intelligence (AI) 30, 31, 32, 33, 35, 36, 37, 38, 42
    British Army 321, 322–7, 329, 333, 339, 340, 341–9, 350, 351

future importance of 29–35
Israel 310–11, 315–16, 317, 319
land warfare 2, 4, 6, 12–13, 18, 394, 395
mission command 33, 43–4, 324, 330
Multi-Domain Operations (MDO) 30, 34–5, 36, 37, 41, 42, 321, 350
Russia 30, 35, 36, 336, 337
tactical theories 113, 114, 115, 116, 118, 120, 121, 122, 134
technology 30, 31, 166, 167, 170, 326, 331
Visegrád States 378, 382, 387
Mao Zedong 260, 261, 267
Marcoussis accord (2003) 357
Marine Corps see United States Marine Corps
Marine Corps Planning Process (Marine Corps) (United States) 93
maritime warfighting domains 30, 43, 49, 138, 216, 227, 302, 402
    interoperability 185, 189
    Russia 280, 281, 285–6, 295
Marshall, Andrew 5
Marshall, S.L.A. 195, 203
Marten, Kimberley 292, 293
'martinet' 110
MASCAL see mass casualty events
masculinity 198, 208
MASINT see Measurement and Signatures Intelligence
masked soldiers 10, 136
mass 196, 338, 380, 395, 397
    tactical theories 114, 115, 116, 118, 120
mass casualty events (MASCAL) 228
Mastiff (mine resistant protected patrol vehicles) 335
Mattis, Jim 48
Maurice, Emperor of Byzantine Empire 108, 109
MCDP-1 *Warfighting* 122
MDI see multi-domain integration
MDO see Multi-Domain Operations
Measurement and Signatures Intelligence (MASINT) 134
mechanical tactics 107
Mechanised Infantry Vehicle (MIV) 337
MedC4I see Medical command, control, communication, computers, and information
MEDEVAC see medical evacuation

Medical command, control, communication, computers, and information (MedC4I) 220
medical confidentiality 229
medical education programmes 228
medical evacuation (MEDEVAC) 220–1, 222, 223, 224, 226, 229
medical fitness evaluation 221
medical information technology 16
'medical planning timelines' 220–1
medical standardization 227
Medical treatment facilities (MTFs) 218, 220, 223, 226, 228
megacities *see* urban environments
MEL *see* Main Events List
mental health 220, 221
mercenaries 109, 291
metropolitan ('metro') army (France) 354
MGCS *see* Main Ground Combat System
MHS *see* military health services
micro-/nano-drones 9, 141
middle ranking officers 100
militarism and anti-militarism 205, 370
military advisors 280, 282, 285, 290–1
military biomedical manufacturing and research 225
military contractors 291, 292, 294, 403
Military Decision Making Process (NATO) 93
military diplomacy 230
military education and training 14, 47, 135, 198–9, 230
  China 268–9, 270, 277
  command forces 100, 101, 103
  France 355, 360, 369
  future of land warfare 403, 408–9, 410, 412
  interoperability 183, 189
  Russia 290, 296
  Visegrád States 379, 380, 385
military engineers 363
military exercises 183, 189
military headquarters 96, 102, 396
military health services (MHS) 12, 16, 215–18, 396
  COVID-19 pandemic 225–6
  definition of 218, 219f, 220–2
  future of Health Services Support (HSS) 215, 216, 218, 226–30
  twenty-first century context 222–4

military history 114
military identity/culture 408
military intelligence and counterintelligence 9, 57, 89, 94, 116, 316, 375
  urban warfare 133, 134, 135, 137, 138, 141, 150
military interoperability 176
military interventions 355–6, 364–5, 369, 370
Military Medical Centre of Excellence (NATO) 227
'military mobility' project (European Union) 72n25
military operations in urban terrain (MOUT) 126n2
military police 167, 280, 284, 285, 288–90, 295, 299, 397
Military Programming Law (France) 353, 354, 361, 362
military secret police 375
Milley, General Mark 126
MILREM UGV (vehicle) 80n50
'mind superiority' 272
mine warfare 142, 288
missiles 224, 317, 364, 387, 388
mission command 27–9, 62, 83, 184, 340, 395
  automation/robotics 43, 44, 48, 49, 59, 60–2
  contemporary/future operational environment 48–50
  culture/method 44–8
  diversity/conformity 51–2
  information environment 43, 44, 50, 52–6
  land warfare 2, 5, 6, 12, 13
  manoeuver warfare 33, 43–4, 324, 330
  Multi-Domain Operations (MDO) 43, 44, 49, 50–1, 52, 55, 56, 57
  technology 47, 48, 52, 54, 56–8, 59–60, 62, 153–4, 163
mission creep 298–9
mission statements 92
Mistral missiles 364
MIV *see* Mechanised Infantry Vehicle (MIV)
Mixed Human–Robot teams 79
MK60 Captor Mine 156n12
MMCC/EMC *see* Multinational Medical Coordination Centre/European Medical Command

mobility 14, 65, 68, 84
'model 2015' (France) 353, 354
modern tactics 107, 111–13
momentum 28
'Momentum' (*Tnufa*) (Israel) 318, 319
Montgomery, Field Marshall 215
Moore, David M. 66
moral cohesion 16, 40, 273, 395, 396
    morality of fighting 193, 194, 195, 199,
        202, 203, 206, 207, 208, 212
    tactical theories 114–15, 117–18
morale 16, 224, 272, 273, 370
    morality of fighting 193, 194, 195, 199,
        202, 203, 207, 208, 212
morality of fighting 193–5, 212–13, 216
    reasons why soldiers fight 195–201
    role of society and 201–11
Moskos, Charles C. 197, 209–10
Mosul (Iraq) 133
motivation to fight 193, 194
    reasons for soldiers fighting 195, 196,
        197, 198, 199, 201
    society and 202, 203, 207, 208, 209, 212
MOUT *see* military operations in urban
        terrain
movement (combat) 64, 66, 69, 70, 84
MRAV programme 328, 329, 337
MRLS *see* Multiple Rocket Launcher Systems
MTFs *see* Medical treatment facilities
Multi-Domain Battle in Megacities
        Conference (2018) 133
multi-domain capability set 17
Multi-Domain Formations 180
multi-domain integration (MDI) 180, 181
Multi-Domain Operations (MDO) 155,
        395, 397, 404
    China and 257, 261, 265–8, 273, 278
    France 367, 369
    interoperability 175, 179, 180, 181, 182,
        183, 184, 190
    land warfare 6, 7, 13, 14, 16, 19, 20
    manoeuver warfare 30, 34–5, 36, 37, 41,
        42, 321, 350
    mission command 43, 44, 49, 50–1, 52,
        55, 56, 57
    urban warfare 126, 127, 130, 131, 136–9,
        140, 145, 147–8
multi-national cooperation 188
Multinational Division Central-South 376

Multinational Logistic Coordination Centre
        (Prague) 67
Multinational Medical Coordination
        Centre/European Medical Command
        (MMCC/EMC) 228
multi-national operations 366
Multiple Rocket Launcher Systems
        (MRLS) 74, 342, 346
multi-territorial battlefields 10
mutiny 196

N
Nagorno-Karabakh 1, 35, 224, 341, 388
Napoleon Bonaparte 30, 115
narrative promotion 402
National Defense, Ministry of (Poland) 384
National Defense Strategy (United States)
        (2018) 174
National Intelligence Council 132
nationalism 205, 208, 212
national mobilization 216
National People's Army (German
        Democratic Republic) 374
National Security Advisor (United
        Kingdom) 174
National Security Capability Review
        (NSCR) 174
National Security Strategy (Israel) 303,
        304–5
National Security Strategy (United
        Kingdom) (2010) 187
national service *see* conscription
NATO *see* North Atlantic Treaty
        Organization
NATO Supply and Procurement Agency
        (NSPA) 73n26
natural resources *see* resource competition
nature of war 106
naval domains *see* maritime warfighting
        domains
navigation 31
NCW *see* Network-Centric Warfare (United
        States)
near-peer adversaries 69, 369, 397
    interoperability 180, 182
    mission command 43, 48, 49, 51
NEC *see* Network Enabled Capability
        (United Kingdom)
nested orders 91, 92
Net Assessment, Office of 5

Netherlands 221
network attack/protection 271
Network Centric Capability (United Kingdom) 186
Network-Centric Warfare (NCW) (United States) 5, 31, 32, 33, 36, 328, 368–9
Network Centric Warfare (United States) 186
networked centric combat force ('combat info-valorisé') (France) 354, 368–9
networked missile systems 60, 316
'networked sensor-to-shooter systems' 141
Network Enabled Capability (NEC) (United Kingdom) 330
New Labour 325–6, 331
New Public Management 354
'New Wars' 37–8, 332
New Zealand 227
Nexter 368
Nightingale, Florence 217
night vision googles 143
Nilsson, Marco 206, 395
9/11 terrorist attacks 186, 327, 328
19th Brigade (British Army) 87
node (节点) 263, 264, 265, 339
non-commissioned officers 274–5, 381, 382
non-linearity 27
'non-linear maneuver battles' 28
non-military activities 10
non-peer adversaries 139, 145, 150
non-state actors 8, 146
'normalization': armed forces 369
North Atlantic Alliance 173, 191
North Atlantic Treaty (1949) 64n2–3, 65
North Atlantic Treaty Organization (NATO) 48, 51, 93, 155, 215, 268, 280, 357, 369
    combat logistics 64–72, 73n26, 74, 75, 80, 81–3, 84
    command forces 87, 97
    development of land warfare 5, 6, 9, 13, 14, 15–16, 20, 26
    future of land warfare 396, 398, 399, 404, 411
    interoperability 174–5, 176, 179n32, 180–4, 186, 187, 188, 190, 191
    manoeuvre warfare 322, 323, 324, 325, 326, 327, 331, 345, 347, 349
    military health services (MHS) 221, 223, 227, 230

urban warfare 126, 136, 139, 141
Visegrád States 373, 375, 379, 381, 382, 385, 386, 388
Northern Army Group (NORTHAG) (NATO) 323, 324, 325, 326
Northern Group of Forces (Soviet Union) 380
North Korea 7, 49
Norway 221
NSCR see National Security Capability Review
NSPA see NATO Supply and Procurement Agency
nuclear armoury and deterrence 37, 354, 355, 381
NWCC see Warfighting Capstone Concept (NWCC)

O
object recognition 167
OBUA see operations in built-up areas
OEM see Original Equipment Manufacturer
offence–defence balance 171
offensive combat 4, 127, 128t
offensive electronic warfare 169, 380
Ogarko, Nikolai 5
Ogarkov, Nikolay 282
On War (Clausewitz) 111–12, 113
OODA-loop 45, 135
Open-Source Intelligence (OSINT) 134
operational assessments 11
operational environments 119, 204, 367, 394
    China and 263, 264, 268–73
    command forces 89, 92, 98
    France 355, 356–61
    future of land warfare 404f, 405, 406, 407, 411
    land warfare 2, 4, 5, 7
    mission command 43, 48–50, 52
Operational Groups of Advisers (Russia) 291
Operation Allied Force (Kosovo) (1999) 87, 280
Operational Manoeuvre Groups (Soviet Union) 323
'operational military partnership' 361
Operation Anady (Cuba) (1962) 279
Operation Barkhane 359, 361, 364
Operation Corporate (1982) 185n68

Operation Desert Storm (Persian Gulf War) 259
Operation *Eagle* (France) 361
Operation *Enduring Freedom* 186n77
Operation Harpie (2008) 362
Operation *Iraqi Freedom* 186n77
Operation *Licorne* (France) 357
Operation Overlord (1944) 91, 92
operations in built-up areas (OBUA) 125n2
Operation Serval (France) 359
operations extérieures ('OPEXs') 355
opposing force (OPFOR) 269
optimal firing positions 167n58
ORBAT (NATO) 182
organizational inertia 274, 276–7
Original Equipment Manufacturer (OEM) 80
Ortal, Brigadier General Eran 315n29
OSINT *see* Open-Source Intelligence
Outer Space Treaty 154
oxygen production 220

P
Pacific 361
Palazzo, Albert 31
Palestinian Terror Offensive 313
Pantazopopulos, Stavros-Evdokimos 155
Parachute Marine Infantry (France) 358
paratroopers 307, 309, 310, 358, 363
Pare, Ambrose 217
Parliamentary Joint Committee on the National Security Strategy (United Kingdom) 174
passive EW collection 164
patriotism 208
'peace disease' (和平病) 277
peacekeeping missions 16, 36, 215, 223, 230, 288
    France 355, 356, 357, 358
    future of land warfare 398, 401, 402
    Visegrád States 379, 382
Peace Support Operations (PSOs) (United Kingdom) 325, 332
peacetime activities 6, 47
peer bonding 196
peer/near-peer adversaries 14, 182, 224, 226
    combat logistics 56, 69, 76, 77
    future of land warfare 396, 399, 401, 411
    urban warfare 128t, 135, 137, 139, 140, 145, 150

penetrating attacks 115
People's Liberation Army (PLA) (China) 12, 17, 257
    building of 258–63
    concept of *Stratagem* 273–4
    informationized/intelligentized warfare 263–73
    ongoing challenges/lessons 274–8
    see also China
People's War: active defence (人民战争) (Mao) 260, 261, 267
People's War in Conditions of Informationization (China) 261, 267
People's War in Modern Conditions (现代条件下的人民战争) 259, 260, 261
'peri-urban'/'rurban' areas 132
'Permanent Structured Cooperation' program (PESCO) 72n25
Pernin, Christopher G. et al. 176
Persian Gulf War (Operation Desert Storm) 259
personal exchange 189
PESCO *see* 'Permanent Structured Cooperation' program
Phalanx close-in defence system 61
PHEC *see* Pre-Hospital Emergency Care
Philippines 127
photo sharing 146
PLA *see* People's Liberation Army
PLA Daily 273
Plato 107
platoons 94, 100, 101, 157
PMCs *see* private military companies
PME *see* professional military education
Point of Injury (PoI) 218
Poland 3, 12, 20, 175, 182, 337, 360, 398
    Visegrád States 376, 377t, 380, 382, 383t, 384, 386, 389
    Warsaw Pact/NATO 373, 374, 375
Polish Land Forces 373, 378t, 383, 386, 387
Polish National Defense Academy 380
Polish People's Army 377
positional warfare 41
post-Soviet succession wars 5
power symmetry 127, 128t, 135
power systems 30
precision bombing 5
precision guided missiles 280

precision targeting 140, 162, 335, 337, 387
preemption 28
Pre-Hospital Emergency Care (PHEC) 218, 221
preventive medicine 216
primary group theory 197, 198, 199, 202, 206
primary health care (PHC) 218
'principles of war' 114–18
'Principles of War' (Foch) 366
private military companies (PMCs) 8, 281, 285, 287, 291–3, 298, 299
privatization 69, 72
problem-solving 46
procedures: interoperability 179, 180n38, 181
procurement 331, 347, 350, 398, 411
professional military education (PME) 353, 354, 369–71, 381
Project Morpheus (British Army) 339
Project Theseus (Ministry of Defence) (United Kingdom) 80n50
prolonged field/hospital care 16
propaganda 139, 147, 208
prosthetics 220, 223
proxy warfare 8, 10, 31, 39, 137, 403
Prussia 44, 110–11
PSOs see Peace Support Operations
psychoactive drugs 229
psychological issues 2, 28, 101
psychological warfare (心理战) 112, 272, 273
public-private partnerships 80
punishment 200, 201
Putin, Vladimir 288
pyrotechnics 72

Q
Qatar 8
Quadrennial Defense Review (QDR) (US Department of Defense) 327, 328
quantum computing 412

R
Race to the Swift (Simpkin) 324
radar sensors 157
radar systems 263, 345, 387
radio signallers/intelligence 285
RAND Corporation 176, 177
Ranger Regiment (British Army) 347

rank representation 96
Rapaport, Amir 313n23
'Rapid Action Force' (France) 353
Rapid Reaction Corps (France) 362
RAS see robotics and autonomous systems
Raska, Michael 59
Reactive Forces (British Army) 334–5
Readiness Action Plan (North Atlantic Treaty Organization) 175
'rearward' combat logistic personnel and systems 74, 76
Reception, Staging, and Onward Movement (RSOM) 69, 71–3
reconnaissance 5, 9, 157, 287, 294, 309, 362
  British Army 328, 335
  China 262, 264, 270, 271
  fire and strike complexes 282
  urban warfare 133, 140, 141, 146, 147
  Visegrád States 387, 388
'reconnaissance–target/acquisition–targeting–battle damage assessment' loop 76
reconstructive surgery 220
Red Army (China) 260
Reed, Walter 217
regiments 362
Regional Command South (Afghanistan) 91
'Regulations of the Land Forces' (2008) (Poland) 380
rehabilitation 220, 221
remote teleconsultation 225
research and development (R&D) 9, 141
reserve forces 265, 304, 308, 310, 318, 335, 353
'reserve' medical capacity 229, 230
resource competition 5, 9
Response Force (NRF) (NATO) 362
resuscitation 222
Revolutions in Military Affairs (AI-RMA) 5, 59, 326, 331
Rheinmetall 368
Rigden, Ian 129
righteousness 194, 209, 212, 213
riot control 357
Ripley, Tim 287
risk-taking 46
Ritchie, Marnie 211
RMA see Revolution in Military Affairs
Robertson, George 186

robotics and autonomous systems (RAS) 3, 30, 60–2, 157n13, 184, 388, 412
  China and 262, 276
  combat logistics 78, 79, 80n51, 81, 83, 84
  urban warfare 142, 144
rocket technology 303, 342, 364, 387
Rojansky, Matthew 282, 283
Roles of Medical Care from Point of Injury (PoI) 218
Romania 364
Romanticism 107
Rommel, Erwin 47–8
Roper, Daniel 181, 182, 184
Rossiskaya Gazeta 290
Routine Threat (Israel) 303, 305, 312
Royal Air Force (United Kingdom) 185n63, 186, 324, 332, 345
Royal Navy (United Kingdom) 185, 186, 324, 332
Rozhdestvenskiy, Iliya 292
RSOM *see* Reception, Staging, and Onward Movement
Rush, Robert S. 197, 200
Russia 164, 210, 221, 279–81, 343
  artillery warfare 184, 280, 282, 285, 286–7, 290
  combat logistics 64, 68, 73, 74, 76–7, 84, 85
  future of land warfare 395, 397, 401, 403
  ground-based forces in Syria 281–93
  interoperability 179, 187
  land warfare 3, 7, 8, 12, 17
  manoeuvre warfare 30, 35, 36, 336, 337
  mission command 43, 48, 49
  strategic functions of ground-based contingent 293–7
  Ukraine 280, 291–2, 294–5, 296, 298–9, 336, 342, 348–9, 360
  urban warfare 137, 139, 142, 146
  Visegrád States 375, 384, 385, 387, 388
Russian Armed Forces 279, 386
Russian Ground-Based Contingent in Syria, The (Bartles and Grau) 282, 283
Russian Military Police 295
Russian Navy 281

S
sabotage 8, 10, 137
SACEUR *see* Supreme Allied Commander Europe

Sahel (Africa) 355, 361
SAM *see* Surface-to-Air Missile
sanctions 10, 136
sand table exercises 95
sappers 285
Sarkozy, Nicolas 354, 357–8
satellites 32, 43, 57, 387
'saturated battlefield' 311
Saudi Arabia 87
Scattered Islands (Indian Ocean) 361
SCORPION Information System (SIC) (France) 368
Scout Specialist Vehicles 333, 335, 337
SDR *see* Strategic Defence Review (United Kingdom) (1998)
SDSR *see* Strategic Defence and Security Review (2010) (United Kingdom)
Search and Destroy operations 120
Second French Armoured Division (Deuxième Division Blindée (2DB) 92, 93
second-generation warfare 3
Second Lebanon War (2006) 18, 316, 318n38, 319
Second World War 4, 5, 173, 217, 226, 399, 403
  command forces 94, 96
  morality of fighting 195, 196, 198, 200, 203, 213
  tactical theories 111, 115, 120, 121
  urban warfare 129, 130
Secretary General (United Nations) 154
sectarian wars 38
Security Assistance Group (British Army) 341
Security Force Assistance Brigade (British Army) 347–8
security sector reform 220
seizing the initiative 28
self-driving vehicles 60
Senegal 361
senior officers 96, 98, 100, 101, 103
sensor technology 2, 5, 8, 9, 18, 30, 33, 224, 267, 387
  British Army 328, 337, 338, 341
  emerging technology 157, 158–9, 160, 161, 162n36, 162n40, 164–5, 166, 169, 171
  future of land warfare 395, 397, 411
  high fidelity layered sensors 154, 163–6

sensor technology (*Continued*)
  urban warfare 129, 140, 141, 143, 144, 145, 146
Sentinelle (anti-terrorist internal security mission) 360
September 11 attacks 127
Serdyukov, Anatoliy 288
Shamalan Canal (Helmand, Afghanistan) 87
Shamir, Eitan 44–5 '
Sherman, Nancy 209
Shils, Edward A. 195, 198
shock 114, 117, 118, 395
short-range air defence systems 270
short-range ballistic missiles 74
short troop rotation times 285
Siebold, Guy L. 196
siege tactics 110, 130
Sierra Leone 327
SIGINT *see* Signals Intelligence
signal companies 89
Signals Intelligence (SIGINT) 134
Simpkin, Brigadier Richard 324
simultaneity 28
Sino-Japanese War 260
Sino-Vietnamese War (1979) 258
situational awareness/understanding 5, 99, 134, 200, 272, 329, 409
  interoperability 182, 184
  mission command 50, 57, 58
  technology 158, 162, 165
Six-Day War (1967) 115, 405
Six Secret Teachings (T'ai Kung) 108
16 Air Assault Brigade (British Army) 334
Skynet 5 satellite communications system 330
Slavonic Corps 291
Slovakia 20, 375, 379, 389, 398
Slovak Republic 377t, 383t
Small Diameter Bomb-II 388
small-group loyalty 202, 205
small-unit independent manoeuvre 4, 40
'small war' 332
smart-sea mines 156n12
Smith, General Sir Rupert 332, 365
sniper fire 142, 161
social media 130, 143
software engineers 229
soldiers 10, 12, 156, 369–71
solidarity 197, 202, 205
Somalia 223, 224

Somme, Battle of (1916) 200
Southern Group of Forces (Soviet Union) 380
South Sudan 223
Soviet-Afghan War (1979–1989) 279, 294
Soviet Army 115
Soviet Military Economic Science 82
Soviet Union 4, 129, 259, 288, 312, 323, 354, 399
  interoperability 181, 182, 184
  Visegrád States 373, 374, 375, 376, 378, 380, 381
space warfighting domains 30, 31, 188, 266, 338, 404
  land warfare and 6, 7, 8, 10
  mission command 43, 49, 55
  technology 155, 164, 169
  urban warfare 126, 138
'spaghetti' 210–11
Spain 221
Special Operations Forces (SOF) (Task Force *Arès*) (France) 357
special operations forces (*spetsnaz*) 10, 136, 138, 228, 313, 362, 397
  Russia 280, 282, 284, 285, 287, 289, 290, 292, 294, 295, 297
'speed' *see* tempo
Spencer, John 133–4
spetsnaz see special operations forces
stability missions *see* peacekeeping
staff officers 14, 45, 96, 97, 100, 102, 168, 396
staff selection and training 97, 100, 103, 114, 139
STANAGS (standardization agreements) (NATO) 182, 185n62, 190
standing militia 109, 303
Standing Naval Force Atlantic (STANAVFORLANT) (NATO) 185n64
standoff 180n38, 282
standoff enemy surface vessels 264
standoff fire systems/weapons 18, 36, 50, 281, 314, 315, 316, 318
standoff ISR platforms 159
stand-off/stand-in sensors 170
state-on-state conflict 373
Stavridis, James 211
stereotypes 101
stop-and-start systems 259, 262
Storr, Jim 117

Stouffer, Samuel A. et al. 195
Strachan, Hew 200
stratagem (计策) (People's Liberation Army)
    (China) 273–4, 278
strategic communication 8, 10
strategic competition 12, 20, 174, 370, 398
Strategic Defence Review (SDR) (United
    Kingdom) (1998) 186, 325, 327, 328
Strategic Defence and Security Review
    (SDSR) (2010) (United Kingdom) 332,
    333–4, 335, 336, 348
strategic MEDEVAC (STRATEVAC) 220
Strategic Review (2017) (France) 369
Strategikon (Maurice) 108, 109
strategy 55, 65, 107, 118–19, 122
STRATEVAC *see* strategic MEDEVAC
'Strike Brigades' (British Army) 336–7, 339,
    343, 345, 346, 350, 351
Stryker Brigade (United States) 77
subordinate command 45, 88
subterranean warfare *see* underground
    warfare
subversion 8, 10, 136
Suez, Battle of (1973) 117
Suez War (1956) 306, 307
Suez, withdrawal from (1968) 174, 185
Sun Tzu 29, 63, 128–9, 262, 271
    tactical theories 105, 108, 109, 111, 113,
    117
supply chains 78, 225, 229
Supplying War (Creveld) 65
suppressive fire 4, 40
Supreme Allied Commander Europe
    (SACEUR) (NATO) 69, 70
Surface-to-Air Missile (SAM) systems 169,
    364, 382
surface to surface missile systems 74
surprise 28, 114, 116–17, 118, 394, 395
surveillance 9, 31, 43, 75
    urban warfare 133, 136, 140, 141, 146,
    147
survivability 33, 142, 168, 345
sustainment (combat) 14, 120, 163, 170,
    183, 266, 367, 388
    combat logistics 64, 65, 66, 69, 70, 84
'swarms' *see* drones
Sweden 221
Swedish Army 14, 100
Swedish Defence University 12
synchronization 28, 50, 62, 95, 395, 411

synthesis 270, 271
Synthetic Aperture Radar 163
Syria 1, 12, 17, 224, 359, 388, 397, 403
    manoeuvre warfare 35, 38, 41
    Russian ground-based forces 281–93
    strategic functions of ground-based
        contingent 293–7
    urban warfare 127, 130, 146
Syrian Armed Forces 290
Syrian Democratic Forces 359
system-based thinking 27, 32
systemic competition 15–16, 174, 175,
    179–84, 187, 188, 191
'system of systems' 5
'system warfare' 263, 264, 268, 270, 273

T
table-top exercises 95
TACEVAC *see* tactical MEDEVAC
Tactical Air Force (Central Region) 185
tactical augmented reality (TAR) 142
tactical decision games 95, 114
tactical MEDEVAC (TACEVAC) 220, 221
tactical system warfare 263–5
tactical theories 40, 105–6, 122–3, 338, 349,
    366
    command forces 89, 90, 95
    emerging technologies 162, 163n39
    future of land warfare 393, 394, 395, 396,
        404
    history of 106–13
    land warfare 12, 14, 19, 20
    mission command 52, 54, 55
    practical application 120–2
    principles/tenets 114–18
    purpose of 113–14
    relationship with strategy 118–19
    urban warfare 127, 130, 134
tactics, techniques, and procedures
    (TTPs) 179, 181, 398, 399, 408, 409,
    410
T'ai Kung 108
Taktika (Leo IV) 109
Taliban 87, 127, 358
tank warfare 63, 117, 364
    British Army 321, 325, 329, 336–7, 342,
        343, 344, 345, 346, 350
    Israel 302, 305–6, 307, 308, 309, 310, 311,
        312–13, 314, 317, 318

tank warfare (*Continued*)
  Visegrád States 379, 381, 382, 383t, 384,
    387
TAR *see* tactical augmented reality
targeting 2, 5, 31, 39, 140, 338, 349
  technology 161, 168, 169
Task Force *Arés see* Special Operations
  Forces (SOF)
Task Force White Eagle (Ghazni) 376
task-oriented groups 270
technology 153–4, 182–3, 262, 286, 328,
  336, 363, 367
  artificial intelligence (AI) 153, 154, 165,
    166–9, 170, 171
  autonomous systems 154–61, 170, 171
  combat logistics 77–80
  combination of emerging
    capabilities 169–71
  future of land warfare 394, 396, 399, 408,
    409, 410, 411
  high fidelity layered sensors 154, 163–6
  Israel 302, 305, 311, 315, 316, 319
  land warfare 1, 2, 5, 8, 12, 13, 15, 18, 19
  layered precision fires 154–63, 170, 171
  manoeuvre warfare 30, 31, 166, 167, 170,
    326, 331
  military health services (MHS) 224, 228,
    229, 230
  mission command 47, 48, 52, 54, 56–8,
    59–60, 62, 153–4, 163
  urban warfare 127, 128t, 131, 135–6, 137,
    138, 139, 140–5, 147, 148, 149
  Visegrád States 387, 393
tempo 66, 67, 95, 188
  land warfare 394, 395, 396
  manoeuvre warfare 27, 28, 32, 39
  tactical theories 114, 115, 118, 120, 121,
    122
  technology 158, 161, 162, 168, 170
'10-1-2(2)+2' timelines 221
Tenufa (Momentum) (Israeli Defence
  Forces) 18
TEPIDOILS *see* Defence Lines of
  Development (DLODs)
territorial competition 5, 26
terrorism 8, 10, 137, 312, 313, 314, 358, 360,
  411
Tetlock, Philip 35
Teutoburg Forest, Battle of (9 AD) 118
TF Wagram (French artillery) 359

Theater Command (TC) 262
thermal imagers 387
thermodynamics 111
'soldier's dilemma' 200
third-generation warfare 3
Thomas, Timothy 281, 292
3D printing *see* additive printing
'360 degree threat' 74
Three Strategies of Huang Shihkung 108
thrust manoeuvre groups 265
Thucydides 107
Tiananmen Square (1989) 259
Tiger Forces (25th Special Missions Forces)
  (Syria) 296
TITAN project (France) 369
TOS-1A Solntsepyok system 286
TRACER programme 328
tracking systems 345
TRADOC *see* Training and Doctrine
  Command
training *see* military education and training
Training and Doctrine Command
  (TRADOC) (US Army) 323
'Transforming the British Army' *see* Army
  2020
transport systems 79, 80
trench warfare 73
Troupes de Marines (Marine Units)
  (France) 355n3
Tsyganok, Anatoliy 281
TTPs *see* tactics, techniques, and procedures
Tuck, Christopher 181, 394
Turkey 8, 289n51, 297, 341

U
UAVs *see* unmanned aerial vehicles
UGV *see* Unmanned Ground Vehicles
UK Air Defence Ground Environment
  (UKADGE) 185n66
Ukraine 1, 48, 77, 127, 146, 224, 360, 370,
  388
  land warfare 399, 401, 403
  manoeuvre warfare 35, 38, 41
  Russia 280, 291–2, 295, 298–9, 336, 342,
    348–9, 360
underground warfare 18, 129, 134, 144, 314
underwater medicine 220
unfamiliarity 90
UNFIL *see* Interim Force in Lebanon, United
  Nations

Unitary Rocket Launchers 364
United Arab Emirates 361
United Kingdom (UK) 3, 12, 19, 34, 98,
    156n13, 162n37
  combat logistics 73n28, 76, 80n50
  defence doctrine 187, 343, 345
  interoperability 172–3, 174, 180, 184,
    185n62–3, 185n68, 186n73–7, 187–90,
    189
  military health services (MHS) 217, 221,
    227
  urban warfare 130, 138
  see also British Army
United Nations 16, 154, 215, 218, 223, 227,
    356
United Nations Protection Force 375
United States Army 4, 6, 13, 17, 137, 144,
    189, 222
  command forces 96, 100, 101
  future of land warfare 397, 399
  interoperability 177, 180, 181, 182, 183,
    184, 186
  manoeuver warfare 26, 28, 32, 34, 41, 322,
    323, 324
  mission command 43, 44, 49
  tactical theories 105, 115, 120, 121, 127
United States Marine Corps (USMC) 14, 17,
    26, 28, 93, 100, 222, 397
  tactical theories 105, 120, 121, 122, 127
United States (US) 77, 80, 120, 210, 293,
    336, 354
  'Air-Sea Battle' 266–7, 340, 349
  China 259, 266, 268, 273
  development of land warfare 3, 5, 7, 9, 12,
    13, 14, 16, 17, 18
  future of land warfare 395, 399, 403, 406
  interoperability 176, 179, 180, 184, 186,
    187, 188, 190
  manoeuvre warfare 26, 38, 321, 324, 332
  military health services (MHS) 217, 222,
    224, 227
  technology 153, 155, 156n13, 162n37,
    164
  urban warfare 129, 134, 139, 141
unmanned aerial and ground systems 2, 49,
    60, 63, 75–6, 136, 271
  British Army 336, 338
  military health services (MHS) 224, 229
  technology 157, 158

unmanned aerial vehicles (UAVs) 313n23,
    364, 411
  British Army 327, 338, 339, 341, 342
  technology 153n1–2, 159n27, 161, 162,
    164, 168
  urban warfare 139, 140, 141, 144, 146
Unmanned Ground Vehicles (UGV) 78, 80,
    142
UORs see urgent operational requirements
Urban Reconnaissance through Supervised
    Autonomy (URSA) 140
urban warfare 125n2, 126–7, 128t, 284, 342
  approaches to 128t, 129–30
  future challenges for 131–6
  future of land warfare 394, 396, 411
  insurgency and 127, 131, 145–8, 151
  land warfare and 2, 7, 9, 10–11, 12, 15
  lessons learnt about 148–51
  manoeuvre warfare 37, 38, 39, 42
  multi-domain operations/grey zones 126,
    127, 130, 131, 136–9, 140, 145, 147–8
  technology 127, 128t, 131, 135–6, 137,
    138, 139, 140–5, 147, 148, 149, 150
urgent operational requirements
    (UORs) 331, 335
URSA see Urban Reconnaissance through
    Supervised Autonomy
US Defence Advanced Research Projects
    Agency 140
USMC see United States Marine Corps
Utility of Force, The (Smith) 365
Utility Vehicles (FRES) 333, 335, 337
Utkin, Dmitry 292

V
VABs see armoured vanguard vehicles
vaccination centres 225
van Bezooijen, Bart 54
Vauban, Sébastien Le Prestre de 110
Vegetius 108, 109
Very High Readiness Joint Task Force (VJTF)
    (NATO) 360
veterans 195, 227
Vietnam 323, 399, 403, 406
  morality of fighting 197, 198, 210, 212
  tactical theories 115–16, 119, 120, 121,
    122
virtue tactics 107–10

Visegrád Group (Czech Republic, Hungary, Poland and Slovakia) 3, 12, 20, 373–4, 385–9, 393
   land force capabilities 376, 377t, 378t, 379–82, 383t, 384–5
   Warsaw Pact/NATO 374–6
VJTF *see* Very High Readiness Joint Task Force
voice/gait patterns 143

W
Wagner Group 283, 291, 292, 293, 295, 296, 297, 403
war, character of 9–11
war failure 210
Warfighting Capstone Concept (NWCC) 181
wargaming 93, 95, 114
warrant officers 381
Warrior Capability Sustainment Programme (CSP) 343, 344, 346
Warsaw Pact 20, 26, 29, 68, 323, 398, 399
   combat logistics 81–3, 84
   interoperability 184, 186
   Visegrád States 373, 374, 377, 379, 380, 381, 382, 383t, 385
'wars of choice'/'wars of necessity' 210
Washington Treaty *see* North Atlantic Treaty (1949)
Watling, Jack 181, 182, 184

wearable sensor technology 9
Wehrmacht (Germany) 92, 100, 197, 200
Welland, Julia 210
Wessely, Simon 201
'Western' warfare 81–2, 83n57, 109, 110, 113, 147, 210
Westmoreland, General 121
West Point 122
Wildcat Lynx helicopters 345
Wilhelm, Brigadier Robert 73
will to fight 16, 325, 396
   morality of fighting 193, 194, 195, 199, 202, 203, 204, 207, 209, 213
winning and losing 90
Wong, Leonard et al. 196, 201, 202
'world-class military' 261, 278
World Health Organisation (WHO) 226

X
Xi Jinping 274, 275, 276–7

Y
Yemen 1, 146, 224
Yom Kippur War (1973) 26, 309, 310, 405
YPG *see* Kurdish People's Defense Units
Yugoslavia 5

Z
Zrinyi modernization programme (2026) (Hungarian Defense Forces) 384